ROYAL BC MUSEUM HANDBOOK

RODENTS
& LAGOMORPHS
OF BRITISH COLUMBIA

DAVID W. NAGORSEN

Illustrations by Michael Hames,
Donald Gunn and Bill Adams

Volume 4:
The Mammals of British Columbia

ROYAL **BC** MUSEUM
Victoria, Canada

Published by the Royal British Columbia Museum,
675 Belleville Street, Victoria, British Columbia, V8W 9W2, Canada.

Edited, designed and typeset by Gerry Truscott, RBCM.
Typeset in Plantin 10/12 (body) and Optima (captions).

Cover design by Chris Tyrrell, RBCM.

Printed in Canada by Hignell Printing.

Front cover photograph of a Vancouver Island Marmot by Andrew Bryant.
Back cover photograph of an American Pika by Robert Cannings.
See page 403 for photograph and illustration copyrights and credits.

The Royal BC Museum gratefully acknowledges funding received from the Forest Innovation Investment program of British Columbia.

National Library of Canada Cataloguing in Publication Data

Nagorsen, David W.
    Rodents and lagomorphs of British Columbia

    (The mammals of British Columbia; v. 4)

    At head of title: Royal BC Museum handbook.
    Includes bibliographical references: p.
    ISBN 0-7726-5232-5

    1. Lagomorpha – British Columbia. 2. Rodents – British Columbia.
    3. Lagomorpha – Classification. 4. Rodents – Classification.
    I. Royal BC Museum. II. Title. III. Series: Royal BC Museum handbook.

QL737.L3N34 2004   599.32   C2004-960093-1

CONTENTS

To
Hamilton Mack Laing,
Charles J. Guiguet
and
Ian McTaggart Cowan
for their pioneering work
on the small mammals of British Columbia.

PREFACE

This is the fourth in a series of six handbooks revising the Royal BC Museum's Handbook 11, *The Mammals of British Columbia* by Ian McTaggart Cowan and Charles J. Guiguet, which is long out of print. Similar to the previous three volumes (*Bats of British Columbia*; *Opossums, Shrews and Moles of British Columbia*; and *Hoofed Mammals of British Columbia*), this handbook emphasizes natural history, distribution and identification. It has been nearly 40 years since *The Mammals of British Columbia* was revised and, since then, a large body of literature has been published on the biology of rodents and lagomorphs. I tried to incorporate these recent findings into the general biology sections and species accounts.

Readers will note several changes in taxonomy and nomenclature from the original handbook. Distributional data are strongly emphasized and a detailed provincial range map is provided for each species based on a comprehensive review of museum specimens and observational records. The identification keys will assist in identifying animals in the hand and museum specimens with skulls.

This book covers two rather unfamiliar groups of mammals: the lagomorphs (hares, rabbits, pikas) and rodents (squirrels, mice, voles, rats). Although some of the larger diurnal species, such as the ground squirrels, marmots and chipmunks, have popular appeal, many of the mice and voles are often regarded as pests. Nevertheless, the lagomorphs and rodents are major components of the ecosystems of the province. Because of their abundance, they are important prey species for birds of prey and various mammalian carnivores. Burrowing species play an important role in aerating soils. Rodents such as chipmunks, voles and flying squirrels disperse seeds and spores in forest ecosystems. Hopefully, this book will promote an awareness of these often misunderstood mammals and stimulate more research on their biology.

GENERAL BIOLOGY

Introduction

This book covers two superficially similar but evolutionarily distant mammalian orders: the lagomorphs (Lagomorpha) and rodents (Rodentia). Both orders share the common traits of having gnawing incisor teeth and grinding cheek teeth, but they demonstrate a number of morphological differences. The exact evolutionary relationship of the lagomorphs to the rodents and other mammalian orders is unclear. Early taxonomists considered the lagomorphs to be closely related to the rodents, classifying them as a suborder of the Rodentia. Another early scheme was to group the lagomorphs and rodents in a superorder known as Glires. This old classification has recently gained favour with some biologists, but molecular studies comparing DNA sequences among mammalian orders fail to show a close affinity between the lagomorphs and rodents.

The lagomorphs are a relatively small order of about 80 species with fossils that that can be traced back to the late Paleocene Epoch, some 40 million years ago. They are grouped into two families: the pikas (Ochotonidae) and the hares and rabbits (Leporidae). About 26 species of pikas are recognized, most occurring in the Old World; 2 species are found in North America, both in British Columbia. There are 54 species of hares and rabbits worldwide, and 3 are native to B.C. Some of the diagnostic traits of the lagomorphs include: short indistinct tails; five digits on the front feet and four or five on the hind feet; the soles of the feet covered in fur (except for a few toe pads on the feet of pikas); a fenestrated skull (figure 1); two pairs of incisors on the upper jaw, the smaller, second pair behind the first; high-crowned, rootless molar teeth; and males lack a genital bone (baculum). All lagomorphs are adapted to a running or hopping mode of locomotion. It is curious that, although they are an ancient group of mammals, the lagomorphs have failed to undergo the spectacular adaptive radiations shown by rodents. One idea is that the lagomorphs are ecological equivalents of miniature ungulates, and that competition with ungulates and rodents has limited the evolutionary potential of the lagomorphs to a narrow range of specializations.

In contrast, the rodents comprise the most successful mammalian group – the number of rodent species exceeds all other mammalian orders. Of the 4629 species of mammals recognized in *Mammal Species of the World*, about 44% are rodents; in British Columbia, 30% of the native mammal species (40 of 133) species are rodents.

Fossils of primitive rodents can be traced back to a few isolated teeth from the late Paleocene Epoch, but the origins of this group are obscure. Some of the distinctive traits of rodents include: five digits on the front feet (although some species may lack a thumb) and three to five digits on the hind feet; a conspicuous tail (on most); a single pair of incisors on the upper and lower jaws (figure 2); no canine teeth; cheek teeth with complex grinding surfaces; and males usually have a baculum.

Rodents demonstrate a great range of adaptations for locomotion and feeding. This diversity is reflected in the taxonomy of the order, which has 28 families. The rodent order is a difficult group for biologists to classify; its higher level classification and relationships remain contentious. A traditional arrangement is to group rodents

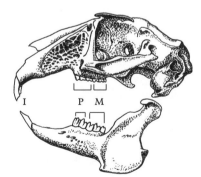

Figure 1. Lagomorph skull with three types of teeth: incisors (I), premolars (P) and molars (M).

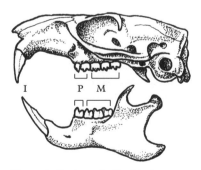

Figure 2. Rodent skull with three types of teeth: incisors (I), premolars (P) and molars (M).

into three suborders based on their jaw musculature and associated skull features: the Sciuromorpha (squirrel-like rodents), the Myomorpha (rat-like rodents) and the Hystricomorpha (porcupine-like rodents). It's far from clear that these groups realistically reflect evolutionary relationships in the rodents. But, it is a useful classification for demonstrating the major rodent groups in British Columbia. The Sciuromorpha is represented by five families in B.C.: Aplodontiidae (mountain beavers), Castoridae (beavers), Geomyidae (pocket gophers), Heteromyidae (pocket mice) and Sciuridae (tree squirrels, ground squirrels, chipmunks, marmots). The Myomorpha is represented by two families: Muridae (rats, mice, voles, lemmings) and Dipodidae (jumping mice). The Hystricomorpha is a group found mostly in the New World tropics and represented here by only one family: Erethizontidae (New World porcupines).

In the following sections, I give a brief overview of the general biology of the lagomorphs and rodents, focusing on western North America. For more detailed information, consult the General Books listed on page 387.

Selected References: Anderson and Jones 1984, McKenna et al. 1997.

Adaptations

Feeding and Nutrition

The lagomorphs and rodents are generally herbivorous, feeding on green plant material, bark, seeds, fruits and fungi, although some rodents also eat invertebrates and other animal material. Consuming plant material presents some challenges. Because plants are generally low in energy, rodents and lagomorphs have to eat a great amount to meet their daily energy and nutritional requirements. Tough fibrous leaves and stems have to be well macerated before they can be digested. Bark has to be scraped from woody stems or branches before it can be chewed and digested. Once swallowed, plant material presents another problem for rodents and lagomorphs, because they do not possess an enzyme that will break down cellulose, the main carbohydrate in plants. To overcome these challenges, rodents and lagomorphs have adapted a number of features in their skulls, teeth and digestive tracts.

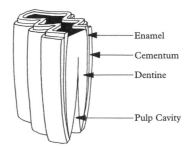

Figure 3. Douglas-fir branches found in the food cache of a Bushy-tailed Woodrat: incisor gnaw marks tell how this rodent removed the bark.

Figure 4. The structure of a vole's molar.

The skulls (figures 1 and 2) are equipped with a pair of chisel-like incisor teeth in the upper and lower jaws. Lagomorphs have a second smaller pair of upper incisors, but they may not be functional. The incisors grow continuously throughout the animal's life. The outer surface of the incisor is harder than the inner surface, because its enamel is strengthened by iron, which produces a yellow or orange colour. Unequal wear on the outer and inner surfaces of the incisors creates a sharp, chisel-like edge-ideal for gnawing bark, cutting woody stems of shrubs or clipping the stems of grass and forbs (figure 3). A large gap (the diastema) separates the incisor teeth from the cheek teeth.

The jaw muscles are designed for front-to-back or side-to-side movements of the upper and lower jaws. The lagomorphs have up to three pairs of premolars and molars in the upper jaw, and as many as two pairs of premolars and three pairs of molars in the lower jaws. Rodents have fewer premolars, and some species have completely lost them, their cheek teeth consisting of only three upper and lower pairs of molars. Because they function to grind and crush plant material, the height of the cheek teeth and the structure of their chewing surfaces varies with diet. For rodents such as the squirrels and mice that eat seeds, fruits or invertebrates, the cheek teeth have prominent cusps adapted for crushing. In other species the chewing surfaces of the cheek teeth are flattened with complex folds of enamel and dentine.

Voles have some of the most specialized cheek teeth. Their high-crowned molars (figure 4) have complex patterns on the chewing surface consisting of inner dentine basins and outer enamel triangles

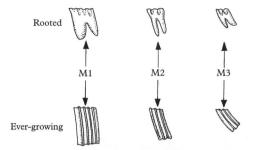

Figure 5. Lateral view of the two types of vole molars from the upper jaw: M1, first molar; M2, second molar; M3, third molar.

Rooted

M1   M2   M3

Ever-growing

that form a prism pattern. The number and shape of the triangles on the chewing surface of the molars are an important taxonomic character used for identifying vole species. The Heather Vole, Southern Red-backed Vole and Muskrat have rooted molars (figure 5) that gradually wear down with age. Other voles (all species of *Microtus*) have "rootless" molars that grow continuously throughout the life of the animal. Generally considered to be the more primitive condition in voles, rooted molars lend themselves to a more omnivorous diet that includes seeds and invertebrates in addition to plant material.

Some rodents possess cheek pouches that they use to carry food. The external, fur-lined cheek pouches of the Northern Pocket Gopher open near the mouth (figure 6) and enable it to transport pieces of roots, rhizomes, bulbs or cuttings of green plant material in its underground tunnels. The Great Basin Pocket Mouse also has external, fur-lined cheek pouches. Ground squirrels and chipmunks have cheek pouches inside their mouths that enable them to carry large quantities of seeds and plant material to their burrow systems.

Figure 6. External cheek pouches and long claws of the Northern Pocket Gopher.

Seeds, fruits and fungi are relatively easy to digest. But green plant material is difficult to digest because it contains large amounts of cellulose. All rodents and lagomorphs are "hind gut" fermenters. Although some rodents have multi-chambered stomachs, they digest cellulose in their large intestine and caecum, which houses a rich flora of microbes that ferment and break down cellulose to digestible sugars.

Many rodents and all lagomorphs eat their feces to extract additional nutrients. This behaviour, called coprophagy, is most developed in hares and rabbits. These mammals produce both soft and hard fecal pellets. They routinely eat the soft pellets, which are rich in vitamins and proteins, but they only ingest the low-nutrient hard pellets when they face food shortages.

Selected References: Chivers and Langer 1994, Hirakawa 2001.

Locomotion and Digging

British Columbian rodents and lagomorphs demonstrate various modes of locomotion: hopping on hind limbs, walking on all four limbs, climbing, gliding. Adaptations to different modes of locomotion are reflected in the relative size of their limbs and the structure of their feet (figure 7).

Most rodents and the pikas move by walking and running on all four limbs. Because many species also dig burrows or climb in trees, they tend to have five digits on the hind feet and four or five digits on the front feet. Tree squirrels have long tails for balance, because they spend part of their time climbing trees and leaping from branch

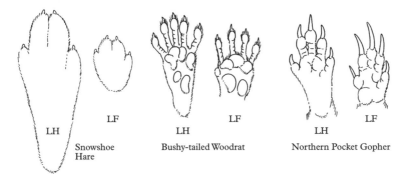

Figure 7. The soles of the feet (LH, left hind; LF, left front) can tell much about how an animal moves: the Snowshoe Hare pads over deep snow; the Bushy-tailed Woodrat climbs well; and the Northern Pocket Gopher spends most of its life digging tunnels.

to branch. One of our most arboreal rodents is the Porcupine. It is equipped with sharp claws and rough pads on its feet that enable it to grip tree trunks and branches as it climbs. Marmots and ground squirrels, being partly fossorial, have powerful front limbs and relatively shorter hind limbs and tails than other species of the Sciuridae family.

A specialized mode for moving on the ground is hopping. Our best hoppers are the tiny jumping mice (*Zapus*) that use their long hind limbs to leap and their long tail for balance. Hares and rabbits also move by hopping, propelling themselves forward with their powerful hind limbs and landing on their front limbs. Their hind legs are much longer than the front, their tail is short and the soles of their feet tend to be heavily furred.

A highly specialized mode of locomotion is gliding. Our only glider, the Northern Flying Squirrel, has a membrane of skin that extends between the front and hind limbs; a smaller membrane is stretched between the cheek and wrists of the forelimbs. To glide, the Northern Flying Squirrel leaps from a branch, fully extending its four limbs to stretch out the gliding membrane. Its flattened tail acts like a rudder during the glide to the ground.

Most rodents dig burrows with their front feet. Our most fossorial rodent, the Northern Pocket Gopher, lives in subterranean tunnels. It is a powerful digger with strong muscles in the shoulders and front limbs. The feet, particularly the front feet, are equipped with long claws (figure 6) for digging. The pocket gopher also uses its sharp incisors (figure 6) to loosen hard soil and cut through roots. Heavy jaw muscles and forward projecting incisor teeth are adaptations for tooth digging. The Northern Pocket Gopher can close its lips behind the incisors to seal off the back of the mouth and avoid ingesting soil while digging. Other adaptations for living in subterranean tunnels include tiny eyes and external ears.

All of our rodents and lagomorphs can swim. Even the Northern Pocket Gopher has been documented swimming 90 metres across a river. But only two species are specialized for an aquatic existence: the Beaver and the Muskrat. Both swim with their webbed hind feet while holding their front feet against the side of the body, and both have a prominent tail – flattened dorsally in the Beaver and laterally in the Muskrat – that acts as a rudder while swimming. Several physiological adaptations enable these animals to remain submerged for lengthy periods.

Selected References: Hildebrand et al. 1985.

Burrows and Shelters

All rodents and lagomorphs rely on some type of shelter for concealment, and for protection from extreme environments or predators. A
nest or den also provides a home site for rearing young and storing
food. The Snowshoe Hare, White-tailed Jackrabbit and Nuttall's
Cottontail make simple shelters that are little more than depressions
in the ground. Most other species construct more elaborate
dwellings. Many rodents, from small voles and mice to marmots and
ground squirrels, excavate burrows. The Northern Pocket Gopher
carries out much of its day-to-day activity in subterranean tunnels.
Other rodents use a burrow for shelter but emerge above ground to
feed or interact with others.

A typical burrow has one or more entrances (figure 8) to underground tunnels. Some of the tunnels lead to chambers containing a
nest, food caches or toilet areas. Burrow entrances and tunnels are
often located under rocks or the roots of trees to provide extra support and to make them difficult for a predator to dig out. Burrows
also provide a more moderate and stable environment than above
ground. Diurnal rodents such as marmots, ground squirrels and
chipmunks often retreat to their burrows during the heat of the day
to escape temperature stress. The Great Basin Pocket Mouse, a nocturnal rodent that inhabits hot, arid environments in British
Columbia's interior, spends the entire day underground, emerging
only at night when the temperature above ground is cooler. Burrows
also serve as hibernation sites (figure 9) for marmots, ground squirrels, chipmunks and jumping mice.

Figure 8. Burrow entrance of a
Mountain Beaver.

Figure 9. Emergence opening of a
Vancouver Island Marmot's
hibernaculum.

Figure 10. Stick den of a Bushy-tailed Woodrat.

Figure 11. Beaver lodge.

A few species, such as the pikas, construct nests above ground under protective rock debris. The Bushy-tailed Woodrat carries this a step further by constructing an elaborate stick den over a nest that is wedged in a cliff or rock crevice (figure 10). The aquatic Beaver and Muskrat often build lodges (figure 11) made from sticks or plant stems. These complex structures have a dry nest chamber situated above water level and one or more openings in the floor that allow the animal to access the lodge from underwater. Both of these species also dig burrows in banks in areas where the water runs too swiftly to construct a lodge.

Figure 12. A Red Squirrel's tree nest.

Arboreal rodents build dens or nests in trees. Tree squirrels often construct a summer nest made from leaves and twigs in the branch of a tree (figure 12), and for winter move to an underground burrow that will be insulated by deep snow. Although they excavate burrows in some habitats, the arboreal Red-tailed Chipmunk, Deer Mouse and Keen's Mouse use tree cavities for dens.

Selected References: Reichman and Smith 1990.

Winter Survival

It is a common misperception that most small mammals hibernate over winter, but most rodents and all of our lagomorphs remain active throughout winter. Only the marmots, ground squirrels and jumping mice are true hibernators. Hibernation is a form of prolonged torpor where the animal lowers its body temperature and slows down its metabolism for days or weeks. During summer, hibernators accumulate large amounts of fat, their vital energy store for winter. In late summer or autumn, they go below ground to their hibernation burrow where they remain until spring.

Two critical requirements for these hibernators are a sufficient food supply in summer to gain weight and a hibernation burrow that insulates the animal from extreme temperatures. Although male Arctic Ground Squirrels seem to store food in their burrows, most true hibernators do not store food but rely solely on their fat reserves. Hibernation or deep torpor can last for several weeks, with periodic arousals. Most hibernators lower their body temperature to 2 to 5°C, but the Arctic Ground Squirrel can reduce its temperature to -2.9°C.

Generally there is a predictable sequence for the different age and sex groups to enter and emerge from hibernation. In autumn, adult males usually enter hibernation before the adult females and younger animals. In spring, adult males typically emerge several weeks before adult females. Most rodents appear to hibernate singly, though some alpine marmots hibernate together in family groups, sharing the same burrow, a strategy that may conserve heat and reduce energy demands over winter.

A Western Jumping Mouse living in harsh subalpine or alpine conditions may hibernate as long as 10 months. Like most ground squirrels and marmots, it does not store food in its winter burrow. Remarkably, this small rodent, weighing only about 25 grams, is able to accumulate sufficient fat reserves to survive this lengthy hibernation period.

Chipmunks and the Great Basin Pocket Mouse appear to be shallow hibernators. Although they spend the winter underground in their burrows, their periods of torpor last only a few days. In contrast to the true hibernators, these mammals do not accumulate fat deposits – instead, they rely on food stored in their burrows for their winter energy demands. Between periods of torpor they arise for 12 to 24 hours and eat from their food stores. One of the most critical requirements for shallow hibernators is collecting and storing an adequate food supply in their burrow to survive the winter.

Other rodents (mice, voles and rats) and the lagomorphs do not hibernate, although some use shallow daily torpor to deal with cold temperatures or food shortages. The Deer Mouse and Western Harvest Mouse, for example, can enter short periods of torpor (less than 12 hours) to lower their energy demands. They typically enter torpor in the early morning hours and arouse late in the day, before their nocturnal activity period. While torpid, the animal sits on its haunches with the head tucked forward against its abdomen. Although the animal's body temperature drops, it remains higher than the low body temperatures achieved by deep hibernators. Although these mammals can enter shallow torpor at any time of the year in response to food shortages and cool temperatures, torpor has its greatest survival value in winter. Short bouts of torpor are an adaptation for coping with brief periods of extreme conditions. Most of the small rodents that remain active through winter adjust their physiology for winter. They often lose weight, and a smaller body mass requires less energy. They generate heat by metabolizing brown fat, a special type of fat associated with heat production.

Other strategies used by rodents and lagomorphs to cope with the harsh and variable climate of winter are acquiring a thick winter pelage, constructing well insulated nests or shelters, nesting together in groups, confining activity to areas under the snow and storing food near winter nests. The winter activity of small rodents is generally confined to the subnivean space, between the snow and ground surface (figure 13). When snow depths reach 15 to 20 cm, this space provides an environment with nearly constant temperature and

Figure 13. Burrow casts of Northern Pocket Gopher made under the snow.

Figure 14. Fresh plant cuttings from an American Pika's hay pile.

Figure 15. Cone midden of a Red Squirrel.

humidity. Temperature differences between the subnivean area and the exposed surface of the snow may be as great as 25°C. The most stressful times for small rodents are periods when the snow cover is sparse or during spring snowmelt. Burrows buried in deep snow are also buffered from extreme elements, and winter nests constructed from dried plant material provide additional insulation from the cold. Many voles and mice, even species that are solitary in summer, will nest communally in winter with small groups huddling together to conserve heat.

Another survival tactic is storing food in or near the burrow or nest. Although many rodents and the pikas store food throughout the spring and summer, winter food stores are essential because they allow the animal to avoid costly foraging in the winter cold outside its insulated nest. Among the lagomorphs, only pikas are known to store food. During spring and summer they collect grasses, sedges, forbs and shrubs, and store them in hay piles (figure 14), which are essential for winter survival. Pikas are highly territorial and defend their hay piles aggressively. The Red Squirrel and Douglas' Squirrel also defend their stores of winter food – middens of conifer cones (figure 15). Shallow hibernators, such as chipmunks and the Great Basin Pocket Mouse, that spend the winter underground, require seed stores in their burrows to meet their energy requirements. The aquatic Beaver and Muskrat construct food caches near their winter lodges, accessible underwater to avoid foraging outside in extreme temperatures.

Selected References: Lyman et al. 1982, Merritt 1984, Vander Wall 1990.

Behaviour and Social Structure

Communication is essential to warn other individuals of predators, attract or identify a mate and maintain territories or spacing among neighbours. Lagomorph and rodent species communicate through their senses of smell, sight, touch and hearing. Diurnal species such as the pikas, marmots, ground squirrels and tree squirrels rely heavily on vocalizations and visual signals. They produce loud alarm or warning calls when predators approach. Beavers sound the alarm with a loud slap of the tail on the water's surface. Colonial species, such as ground squirrels and marmots, use anti-predator calls to signal danger to an entire colony or family group. The Red Squirrel, Douglas' Squirrel and the pikas use distinct calls to delimit and defend their territories from neighbours or intruders. Nocturnal rodents also make an assortment of sounds for communication. They warn others or threaten adversaries by squeaking, chattering their teeth or stamping their feet. Evidently, some rats can hear sounds above the range of human hearing; their young emit ultrasonic calls that may be outside the hearing range of predators.

Visual signals such as tail flicking tend to be most developed in the diurnal rodents and are often made in association with territorial or alarm calls. The conspicuous white fur patches on the underside of the tails of hares and rabbits briefly distract a predator when these animals are flushed from cover. Many of the small nocturnal rodents have an assortment of postures and movements as part of their sexual or combative behaviours.

Many rodents and lagomorphs live in a world dominated by smell, obtaining much of their information from chemical signals (pheromones). Although chemical cues are less immediate than an alarm call or visual signal, they persist for hours or days and are effective cues for species that are active in the dark. For voles and mice, small mammals that have a long list of predators, communicating with smells is also less risky than making conspicuous sounds. Urine is one of the most important chemical signals – animals use it to mark their territory or signal their sexual status. Male House Mouse urine contains chemicals that stimulate the reproductive condition of females. Perhaps the most unusual use of urine is by the male Porcupine, who sprays the female during courtship. Rodents and lagomorphs also possess special scent glands that produce pheromones. These glands can be located on the cheeks, chin, flanks, abdomen or hips. Anal glands and other specialized glands, such as the castor glands of the Beaver, also produce odours for marking territory.

The basic social unit of lagomorphs and rodents is a mother and her young. Males play no role is raising the young. Biologists recognize two basic types of social structure in mammals: solitary and communal. Solitary species typically live separately. The adults usually avoid each other, except during the breeding season when a male and female come together briefly to mate. After mating, the male maintains no association with the female and her young. The young remain with their mother until they are weaned, then disperse from their birth site to establish their own territories. Many rodents, the pikas and most of the hares and rabbits have this simple, solitary social structure, although there are some minor variations on the theme. Young American Pikas attempt to establish territories near their birth site; as a result, many pikas living in an area may be related. Some rodents that are solitary in the breeding season will nest communally in small groups during winter to conserve heat.

Communal rodents and lagomorphs usually have a monogamous or polygamous social structure. In the monogamous system, an adult male and female mate only with each other and form a unit with their young. The Beaver is a good example of a monogamous rodent that maintains a long term pair bond. A Beaver lodge houses a single family group composed of an adult male, adult female and their young. When the young reach their second year they disperse from the lodge.

There are several types of polygamous social structures, where a male is associated with two or more females and their young. Among the rodents and lagomorphs, the best examples are shown by the European Rabbit and some of the marmots and ground squirrels. Polygamous species may form a loose unit where a male defends a territory that encompasses the home range of several females and their young, or it can be more structured where the male defends a harem of several females dominating other members of the harem. Some voles appear to have a flexible social structure. In spring, male and female Townsend's Voles form monogamous pairs, excluding other voles of the same sex from their home range. Young-of-the-year males disperse from their birth site after they are weaned, but young females typically remain near their birth site, increasing the number of females in a local area. By summer, the social system becomes polygamous, with the home ranges of several related females overlapping the home range of a male.

Selected References: Eisenberg and Kleiman 1977, 1983.

Biogeography

The Landscape

Encompassing some 950,000 km² and spanning 11 degrees of latitude and 25 degrees of longitude, British Columbia has the most diverse physiography and climate of any Canadian province. A series of north-south oriented mountain ranges dominate the landscape (figure 16). They play a major role in the climate of the province by intercepting Pacific weather systems as they move eastward and

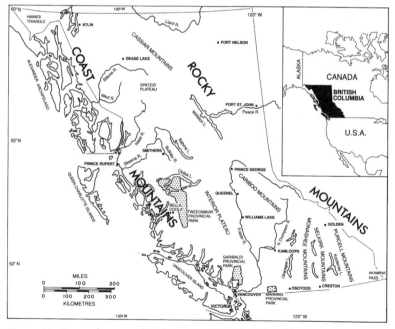

Figure 16. General geographic features of British Columbia.

creating alternating wet and dry belts. The wettest regions are associated with the Pacific coast, especially the western slopes of the coastal mountain ranges and the outer coasts of Vancouver Island and the Queen Charlotte Islands. East of the Coast Mountains, rain shadows create a large arid region, the interior plateau. The most extreme arid conditions are found in some of the southern interior valleys such as the Okanagan and Thompson river valleys. Other wet-dry belts are associated with the Cassiar Mountains, Rocky Mountains and Columbia Mountains (Cariboo, Monashee, Selkirk and Purcell mountains).

The vegetation of the province is predominantly coniferous forest although deciduous forest is associated with northern boreal regions and riparian habitats along rivers and lakes. Grassland and shrub-steppe habitats occur in some of the arid southern interior valleys. Grassy alpine tundra and scrubby willow-birch habitats are common in northern British Columbia and at high elevations in southern parts of the province.

A number of ecological classifications have been developed for British Columbia. In the original *Mammals of British Columbia*, Cowan and Guiguet proposed 13 terrestrial biotic areas for the province. Two more recent classifications are the B.C. Ministry of Forests' system of 14 biogeoclimatic zones (see colour fold-out map) and the 10 ecoprovinces developed by the B.C. Ministry of Environment (figure 17). The biogeoclimatic zones define areas that have relatively homogeneous climates and characteristic vegetations. In mountainous terrain, the zones are related to elevation. An ecoprovince is a broad geographical area with consistent climate and terrain. Each ecoprovince has been further subdivided into ecoregions and ecosections.

Selected References: Demarchi et al. 1990, Meidinger and Pojar 1991.

Historical Biogeography

About 16,000 to 15,000 years ago, in the late Pleistocene Period, the province was covered with thick ice from the last glaciation. Although fossils of large mammals such as bison and mammoths have been found in the province from the late Pleistocene, fossils of rodents and lagomorphs living in British Columbia before the last glaciation are almost non-existent. The only site with small mammal

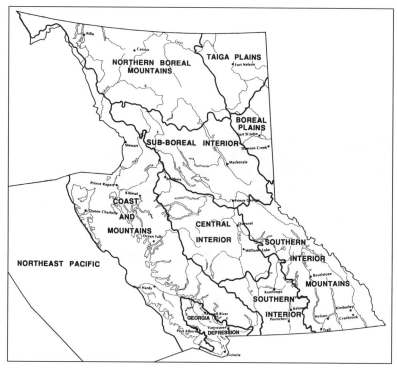

Figure 17. The 10 ecoprovinces of British Columbia.

remains dating from this period is a sea cave recently discovered at Port Eliza on the west coast of Vancouver Island. Among the some 3000 vertebrate bone and teeth fragments recovered from the cave are large numbers of voles, mostly Townsend's Vole, and a few bones of marmots, possibly the Vancouver Island Marmot (figure 18). The Port Eliza remains have been dated at 16,000 to 18,000 years ago, a time when ice was advancing over most of Vancouver Island. It appears that the environment around the cave was an open tundra with patches of trees. Subsequently, the cave was filled with deposits from glacial ice. The discovery of marmot remains at Port Eliza is interesting because it shows that this alpine mammal was living near sea level during a time when the climate was cooler.

With glacial ice covering the province from about 16,000 to 14,500 years ago, mammals would have survived the last glacial advance in ice-free areas at the southern margin of ice in the United States or in Beringia, an ice-free area in the northern Yukon

Territory and Alaska. Scientists still debate about the possibility of animals surviving in British Columbia throughout the last glacial advance. There is growing evidence that parts of the continental shelf on the Pacific coast became exposed with the lower sea levels and escaped the glacial ice. Any fossils from this period, however, would now be submerged in the ocean. Some biologists have speculated that alpine mammals such as the Vancouver Island Marmot survived on nunataks – mountain tops projecting above the thick ice. But these nunataks would have been sparsely vegetated, isolated islands surrounded by ice. It seems unlikely that marmots would have persisted for several thousand years on tiny islands in a harsh glacial landscape. As the glacial ice began to recede about 13,500 years ago, mammals recolonized the province from the northern area in Beringia, areas south of the ice and, possibly, coastal refugia. Because fossil sites with vertebrate remains from this early post-glacial period in British Columbia are rare, we know little about the rodents and lagomorphs of this period.

An exception is Charlie Lake cave, near Fort St John in northeastern B.C., which contains an extraordinary record of vertebrates dating from 10,780 years ago to the present. Rodents and lagomorphs are well represented in the deposits. The most significant finds are of jaws and teeth of the Collared Lemming (*Dicrostonyx* sp., figure 19) and Taiga Vole (*Microtus xanthognathus*), two rodents not found in the province today. An arctic species, the Collared Lemming remains are far south of its modern-day range – it may have been associated with an arctic fauna that would have flourished in British Columbia in the early period following glaciation. With the development of spruce forests about 10,000 years ago, arctic species such as the Collared Lemming would have disappeared from the

Figure 18. Vertebrate bones and teeth recovered from the cave at Port Eliza. The arrows indicate a rib from a Vancouver Island Marmot and the molar teeth of a Townsend's Vole.

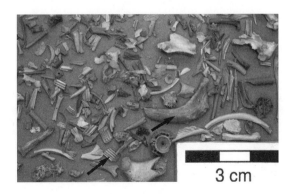

3 cm

Figure 19. The lower right jaw and molar teeth of a Collared Lemming found in Charlie Lake cave.

province. Evidently, the Taiga Vole lived in the vicinity of Charlie Lake cave throughout much of the postglacial period until as recently as 1000 years ago. Curiously, remains of this vole dating from about 6000 years ago have also been discovered recently at Bear Flats in the Peace River area of B.C. A large, distinctive vole associated with the northern boreal forest, the Taiga Vole, has not been found in B.C. in historical time. Why this vole disappeared from the province a thousand years ago is unknown. Other significant finds in Charlie Lake cave include bones of a large lagomorph that match either an Arctic Hare (*Lepus arcticus*) or White-tailed Jackrabbit and bones of ground squirrels that could be either the Columbian Ground Squirrel or Richardson's Ground Squirrel (*Spermophilus richardsoni*). Today, the only hare in the Charlie Lake region is the Snowshoe Hare, and no ground squirrels live there or anywhere in the northern Rocky Mountains.

Selected References: Driver 1988, 1998; Ward et al. 2003.

Modern Distributions

British Columbia supports the highest diversity of rodents and lagomorphs of any Canadian province (5 lagomorphs and 45 rodents), largely due to its environmental diversity and its complex glacial history. There are few striking patterns in diversity associated with latitude or elevation (see Appendix 2). The most diverse ecoprovinces for rodents are the Coast and Mountains and the Southern Interior; the northeastern ecoprovinces of Boreal Plains and Taiga Plains support the fewest species. Among the biogeoclimatic zones, the Alpine Tundra and Engelmann Spruce - Subalpine Fir zones support the greatest number of species. The high diversity in these zones reflects the occurrence of alpine specialists such the Hoary Marmot and species associated with forested ecosystems that also extend into the alpine. Rodent diversity is lowest in the biogeoclimatic zones associated with northern ecoprovinces of Spruce - Willow - Birch and Sub-Boreal Pine - Spruce.

Figure 20. An aerial view of the lower Fraser River valley looking west to the Strait of Georgia. This region supports seven species of Pacific-coast rodents found nowhere else in Canada.

Several of the ecoprovinces have distinctive rodent and lagomorph species. Parts of the Coast and Mountains ecoprovince and the lower Fraser River valley (figure 20) in the Georgia Depression ecoprovince support seven Pacific-coast rodents that occur nowhere else in Canada: Mountain Beaver, Creeping Vole, Townsend's Vole, Townsend's Chipmunk, Cascade Mantled Ground Squirrel, Douglas' Squirrel and Keen's Mouse. Most of these mammals range from northern California to southwestern British Columbia where they are associated with the lowlands and western slopes of the coastal mountain ranges. This region also supports the Vancouver Island Marmot, the only mammal species endemic to the province.

Four species confined to the Southern Interior ecoprovince (figure 21) are mammals adapted to shrub-steppe and grassland. They include two lagomorphs (Nuttall's Cottontail and White-tailed Jackrabbit) and two rodents (Great Basin Pocket Mouse and Western Harvest Mouse), all representatives of a group of mammals associated with the Great Basin and Columbia Plateau of the western United

Figure 21. The southern Okanagan Valley supports a number of grassland species, including the White-tailed Jackrabbit, Nuttall's Cottontail, Great Basin Pocket Mouse, Western Harvest Mouse and Montane Vole.

States. Grassland mammals such as the Pygmy Rabbit (*Brachylagus idahoensis*), Washington Ground Squirrel (*Spermophilus washingtoni*), Ord's Kangaroo Rat (*Dipodomys ordii*) and Sagebrush Vole (*Lemmiscus curtatus*) occur on the Columbia Plateau of Washington, but have not been found in British Columbia. Given that grasslands were more extensive during a warm, dry period 6000 to 8000 years ago, it is conceivable that some of these mammals ranged into the southern-interior grasslands of B.C. in the past.

The Northern Boreal Mountains ecoprovince (figure 22) has mammals such as the Collared Pika, Arctic Ground Squirrel, Tundra Vole and Northern Red-backed Vole that originated in Beringia in the northern Yukon and Alaska. After the last glaciation these species dispersed south from Beringia reaching their southern limits in northwestern British Columbia. Three have related and ecologically similar southern counterparts: American Pika, Columbian Ground Squirrel, Southern Red-backed Vole. The distributions of the Northern Red-backed Vole and Southern Red-backed Vole meet in a narrow zone in northern B.C. Curiously, no pikas or ground squirrels inhabit the northern Rocky Mountains, creating a large hiatus in the ranges of the Collared Pika and American Pika, and the Arctic Ground Squirrel and Columbian Ground Squirrel.

Figure 22. The Haines Triangle region of northwestern B.C. supports Beringian species such as the Collared Pika, Arctic Ground Squirrel, Tundra Vole and Northern Red-backed Vole.

One of the most intriguing aspects of small-mammal distributions in this province is their presence or absence on the numerous coastal islands (figures 23, 24, 25). The lagomorphs are not known to occur on any British Columbian island. The Snowshoe Hare and American Pika live at sea level in some areas on the adjacent coastal mainland, yet they evidently have not reached any island. The lagomorphs are either poor survivors on islands or they have poor abilities to cross sea water. Relative to the number of rodent species found on the mainland, rodents are underrepresented on the islands. For example, 20 species of rodents inhabit the Lower Mainland area, but only 6 are native to Vancouver Island, the largest island in the eastern North Pacific with an area of about 33,000 km². Graham Island (6390 km²) and Moresby Island (2615 km²), the two largest islands in the isolated Queen Charlotte Islands, support only a single rodent species, the ubiquitous Keen's Mouse. Conspicuously absent from the coastal islands are the chipmunks, ground squirrels, jumping mice and the Porcupine. Marmots are limited to Vancouver Island. Voles are also poorly represented on the islands. While eight species inhabit the south coast region (see Appendix 2, Table 4), only one, Townsend's Vole, inhabits Vancouver Island.

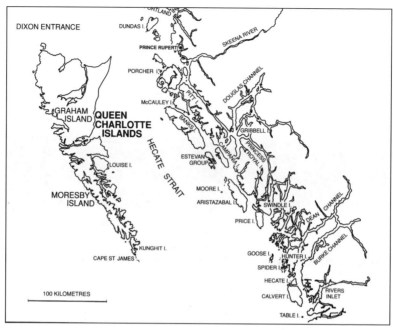

Figure 23. Islands and inlets of British Columbia's northern coast.

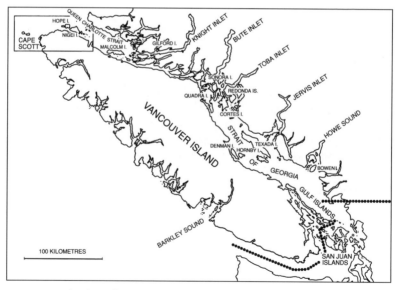

Figure 24. Islands and inlets of B.C.'s southern coast.

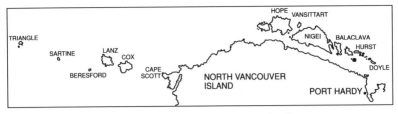

Figure 25. Islands off the north coast of Vancouver Island.

Despite the many missing species, some rodents have been successful at colonizing and surviving on islands. The Red Squirrel and Beaver are widespread on British Columbia's islands. Densely forested with coniferous trees, many of the larger coastal islands provide ideal habitat for the Red Squirrel. Capable of swimming in sea water, the Beaver has reached most of the islands near the coastal mainland with suitable wetland habitats. It is noteworthy that both of these rodents have flourished on the Queen Charlotte Islands, where they were transplanted in the 1940s.

The Deer Mouse and Keen's Mouse are by far the most widespread rodents on the province's islands. The Deer Mouse has been found on 97 islands and Keen's Mouse on some 80 islands along the coast. Their success can be attributed to their highly adaptable nature. Both species live in forest and beach habitats on islands. They also have flexible diets. On some small islands they forage in the intertidal zone, feeding on marine invertebrates such as amphipods. Keen's Mouse even feeds on seabird eggs, enabling it to survive on isolated islands with seabird colonies.

How did small rodents such as the Deer Mouse and Keen's Mouse reach these islands? Some islands may have been connected to the mainland coast at the end of the ice-age when sea levels were lower on parts of the coast. Mice could have reached these islands over land bridges and become isolated when sea levels rose about 9000 to 10,000 years ago. But this scenario only applies to a few isolated island groups, such as the Scott Islands off the north coast of Vancouver Island. It appears that most islands along the British Columbian coast have not been connected to the mainland since the ice age. In fact, in many coastal areas, sea levels were much higher immediately following the ice age than today.

Perhaps rodents reached various islands at the end of the ice age on ice bridges created by the receding glaciers. An alternate explanation is that small mammals colonized these islands more recently, crossing the various water barriers by swimming or rafting on floating logs or debris. Yet another idea is that some of these animals were transported accidentally in native canoes. Experiments by Richard Valle demonstrated that Deer Mice from Washington rarely survived more than 10 minutes swimming in water at 12°C, a temperature comparable to the summer water temperatures in British Columbia's coastal waters. Valle also found that his Deer Mice were poor at rafting – most abandoned their rafts within a few minutes and attempted to swim. The colonization of the numerous Pacific coastal islands by the Deer Mouse and Keen's Mouse remains one of the unexplained mysteries of island biogeography.

Selected References: McCabe and Cowan 1945, Valle 1962, Yensen et al. 1998.

Relations with Humans

Our view of animals is coloured by our perception of what is beneficial to human interests and by what is attractive or non-threatening. Chipmunks, marmots and pikas tend to have popular appeal because they are perceived as attractive and harmless small mammals. The Beaver and Muskrat are generally considered to be valuable fur-bearing mammals that warrant careful management. But we view most other rodents and lagomorphs with either indifference or negativity. Native mice and voles are dismissed as harmful "rats" because of their vague resemblance to the introduced Norway or Black rats. Tree squirrels are unpopular because they raid bird feeders. Many species come into conflict with humans because of their feeding activities in managed forests and agricultural lands or their occupation of human dwellings.

The most familiar rodents in human dwellings are the introduced House Mouse, Black Rat and Norway Rat. These species had a long association with humans before they were brought to the New World, and they are well adapted to living in agricultural and urban environments. Because they consume and contaminate food stores with their droppings, they are viewed as serious pests in urban and agricultural areas. They can cause structural damage by gnawing or chewing building materials and are known to transmit diseases such as salmonella and leptospirosis. Only a few species of our native rodents such as the Bushy-tailed Woodrat and Deer Mouse frequent human dwellings, and these rodents are usually found in rural areas, not in major urban centres. The Deer Mouse is a common inhabitant of cabins and cottages where it can be a nuisance because of its habit of chewing paper and mattresses for nesting material. This rodent is also known to carry the hantavirus that causes a disease know as hantavirus pulmonary syndrome (HPS). The virus can be spread by inhaling dust from the dried droppings. There have been several deaths from this disease in British Columbia over the past decade. Nevertheless, the disease is rare here, which is surprising given how common and widespread the Deer Mouse is in the province. To be safe, wear a respirator if you are cleaning or sweeping dried rodent droppings from a building. For more details on health precautions and on the disease risk from hantavirus contact your local health unit.

One area where our native rodents frequently come into conflict with human interests is in the forest industry. The Snowshoe Hare, Red Squirrel, Porcupine and voles clip off tree seedlings or girdle the bark of sapling trees (figure 26). Damage from their feeding may reduce tree regeneration and ultimately reduce productivity in managed forest stands. Voles, Deer Mice and Keen's Mice can hinder forest regeneration because they consume seeds. Considerable

Figure 26. Damage to Lodgepole Pines: (left) the Snowshoe Hare left the inner bark attached; (centre) the Red Squirrel dropped bark strips on the ground; and (right) the Porcupine marked the sapwood with its incisors.

research has been done on methods to minimize the effect of these mammals. One approach is direct control by removal or poisoning. A more holistic approach is to thin the seedling trees and provide supplemental food to divert their feeding activity.

Despite their negative image, many rodents play an important ecological role in disseminating seeds and the spores of fungi. Voles, Deer Mice, Keen's Mice and tree squirrels such as the Northern Flying Squirrel feed on the fruiting bodies of underground fungi (truffles). Recent research has shown that there is a complex relationship among trees, fungi and forest rodents. The underground fungi grow on the root tips of various trees, assisting the tree with absorbing nutrients and water from the soil. When eaten by a rodent, the spores from the fruiting body of the fungi are passed through its digestive tract unharmed. Rodent fecal pellets contain the live fungal spores as well as nutrients. By scattering their droppings on the forest floor, rodents play an important role in spreading the underground fungi.

Rodents living in agricultural areas also tend to come into conflict with human activities. Voles and Northern Pocket Gophers consume hay and cultivated crops. The burrows of ground squirrels and pocket gophers are considered a nuisance because they may reduce productivity and damage farm machinery. In fruit-growing regions of the province such as the Okanagan Valley, voles and the Northern Pocket Gopher damage fruit trees by gnawing bark or sapwood. Fruit growers reduce rodent populations in orchards by trapping or poisoning the animals and by removing the ground cover with herbicides or by mowing. Northern Pocket Gophers living in Alfalfa or hay fields are also routinely trapped or poisoned.

It would be difficult to convince a farmer with a field full of pocket gopher burrows that they can be beneficial for soil formation, but burrowing rodents bringing soil to the surface helps redistribute nutrients and their droppings, deposited near burrow entrances, may promote plant growth. It has been estimated that a population of 74 Northern Pocket Gophers living in one hectare will turn over more than 93.7 tons of soil in a year. This constant digging activity not only increases soil aeration, porosity and fertility, but it may also hasten the weathering of the small stones and cobbles. Remarkably, some Northern Pocket Gophers that were presumably sheltered in their underground burrows survived in the blast area from the volcanic eruption of Mount St Helens in the Cascade Mountains. By disturbing and mixing soils, and enhancing soil conditions, they promoted the rapid regeneration of vegetation in areas covered by volcanic ash.

Selected References: Maser et al. 1978, Sullivan and Sullivan 1982a, Sullivan et al. 1990.

Conservation and Management

Because rodents and lagomorphs are often abundant and generally perceived as nuisances or pests, the general public rarely regards them as mammals of conservation concern. Yet 6 species and 14 subspecies appear on the province's Red or Blue lists (see the table on the next page), and COSEWIC (Committee on the Status of Endangered Wildlife in Canada) shows one British Columbian lagomorph and three rodents on its national list of species at risk. Few of these rodents and lagomorphs listed provincially or nationally are at risk because of direct killing or control programs by humans. They are of conservation concern because their habitat is threatened or they occupy small, isolated areas making them vulnerable to chance effects. Ironically, the species that have been harvested by humans – the Beaver and Muskrat – do not appear on any of these lists; because these rodents have been intensively managed as fur-bearers they have received more attention by wildlife managers.

The species at highest risk is the Vancouver Island Marmot. Endemic to Vancouver Island, this is the only mammal species confined to Canada that is endangered. With a population of less than 30 in the wild, it is North America's most endangered mammal. Despite its endangered status and much research, it is still not clear why this animal is so rare. In the late 1970s, forest harvesting began at high elevations on Vancouver Island near colonies of this rodent. The Vancouver Island Marmot readily colonized clear-cut habitats in higher elevations, and there is some evidence it may have actually increased in numbers. But moving to nearby clear-cut habitats rather than adjacent mountain tops altered the normal dispersal movements of the Vancouver Island Marmot. Most of the colonies on the periphery of the range disappeared, because they depended on new immigrants to maintain their populations. As the clear-cut

**Rodents and lagomorphs considered to be at risk
in British Columbia.**

The provincial Red List includes indigenous organisms that are extirpated, endangered or threatened in British Columbia. The Blue List names those of special concern because of charateristics that make them sensitive to human activities or natural events.

Taxon	Ranking	
	B.C.	COSEWIC
Species		
White-tailed Jackrabbit (*Lepus townsendii*)	Red List	–
Nuttall's Cottontail (*Sylvilagus nuttallii*)	Blue List	Special Concern
Mountain Beaver (*Aplodontia rufa*)	*	Special Concern
Vancouver Island Marmot (*Marmota vancouverensis*)	Red List	Endangered
Great Basin Pocket Mouse (*Perognathus parvus*)	Blue List	–
Western Harvest Mouse (*Reithrodontomys megalotis*)	Blue List	Special Concern
Subspecies		
American Pika – *septentrionalis* subspecies	Red List	–
Snowshoe Hare – *washingtoni* subspecies	Red List	–
Mountain Beaver – *rufa* subspecies	Red List	–
Mountain Beaver – *rainieri* subspecies	Blue List	–
Least Chipmunk – *selkirki* subspecies	Red List	–
Least Chipmunk – *oroecetes* subspecies	Blue List	–
Red-tailed Chipmunk – *ruficaudus* subspecies	Red List	–
Red-tailed Chipmunk – *simulans* subspecies	Blue List	–
Northern Pocket Gopher – *segregatus* subspecies	Red List	–
Southern Red-backed Vole – *galei* subspecies	Blue List	–
Southern Red-backed Vole – *occidentalis* subspecies	Red List	–
Townsend's Vole – *cowani* subspecies	Red List	–
Northern Bog Lemming – *artemisiae* subspecies	Red List	–
Meadow Jumping Mouse – *alascensis* subspecies	Blue List	–

* British Columbia ranks the two subspecies separately.

areas began to regenerate with young trees, the marmots gradually abandoned their habitats. With this animal now confined to a few small scattered colonies, it is highly vulnerable to random, unpredictable events, such as a severe winter or a predator such as a Grey Wolf or Cougar concentrating its efforts on a colony.

The discovery of prehistoric bones in a number of caves far outside the current range suggests that the Vancouver Island Marmot once occupied a much larger area. The cause of this range collapse is unknown, but it could be the result of climatic changes and associated vegetational changes in subalpine-alpine areas.

An ambitious captive breeding program is now underway to save the Vancouver Island Marmot, with more than 60 marmots now held in captive breeding facilities. The intent is to restore natural populations by transplanting some of the captive animals back into their natural habitat.

The White-tailed Jackrabbit, Nuttall's Cottontail, Great Basin Pocket Mouse and Western Harvest Mouse are all species associated with shrub-steppe and grassland habitats in the southern interior. The White-tailed Jackrabbit may be gone from the province, as there is no recent evidence of a breeding population. Early fruit farmers killed many, but the loss of large tracts of native grassland in the southern Okanagan Valley had the greatest effect on this large hare. Populations of the Western Harvest Mouse, the Great Basin Pocket Mouse and Nuttall's Cottontail still persist, although the status of Great Basin Pocket Mouse in the Thompson River valley and the north Okanagan is unknown. Mostly associated with arid grassland or shrub-steppe habitats, these species are most vulnerable to habitat loss from urban growth and the conversion of natural areas to irrigated agricultural land such as orchards and vineyards (see figure 21). The recent establishment of several large, protected areas in the southern Okanagan-Similkameen valleys is a positive step for their conservation.

The *washingtoni* subspecies of the Snowshoe Hare, *rufa* subspecies of the Mountain Beaver and the *occidentalis* subspecies of the Southern Red-backed Vole are all races associated with the Lower Mainland on the south side of the Fraser River valley. This area experienced major changes in the last century when settlers converted forests to agricultural land, then again during the past 50 years with the urban growth of Vancouver and adjacent municipalities (see figure 20).

Some of the other subspecies listed in the table are considered to be potentially at risk simply because they have small isolated ranges and occur in small population densities – two life-history traits that make them vulnerable to extinction from random events. The *selkirki* subspecies of the Least Chipmunk, for example, is limited to about 30 km^2 in two alpine areas in the Purcell Mountains. The taxonomic validity of many of the listed subspecies is questionable, as they were described decades ago from a few specimens. Modern taxonomic studies must be done to verify that these are valid races that warrant treatment as unique populations for conservation and management (see Taxonomy, page 36).

Another conservation concern is the introduction of two alien lagomorphs (European Rabbit and Eastern Cottontail) and five alien rodents (Eastern Grey Squirrel, Eastern Fox Squirrel, Black Rat, Norway Rat and House Mouse) to the province. These introduced species may have a negative impact on the province's native flora and fauna, including its native rodents and lagomorphs. Although the loss of forested habitat was the major cause of the disappearance of the Lower Mainland subspecies of the Snowshoe Hare (*L. a. washingtoni*), the arrival of the Eastern Cottontail in the late 1940s probably accelerated its decline through direct competition. Introduced Black Rats and Norway Rats on the Queen Charlotte Islands appear to be responsible for the disappearance of the native Keen's Mouse from Langara and Bischof islands. Feral populations of the House Mouse are now established in grasslands of the southern Similkameen and Okanagan valleys. The impact of this mouse on the Great Basin Pocket Mouse and Western Harvest Mouse, two native rodents of conservation concern, is unknown. Similarly, the effect of Eastern Grey Squirrels introduced to the Lower Mainland and Vancouver Island on the native Douglas' Squirrel and Red Squirrel is unknown.

Selected References: Cannings et al. 1999, Carl and Guiguet 1972, Hafner et al 1998.

Studying Rodents and Lagomorphs

Because they are secretive and often nocturnal, most rodents and lagomorphs are challenging to study in the wild. It is no coincidence that some of the best researched species are the marmots and ground squirrels. Active outside their burrows during the daylight hours, these large, conspicuous rodents are easy to study by direct observation with a pair of binoculars; many live in colonies or family groups, making them ideal subjects for behavioural studies. For most other species, biologists have to rely on indirect methods such as observing signs, trapping or tracking with radio transmitters.

Various traps are available for capturing rodents and lagomorphs. Early museum collectors relied on kill traps such as the snap trap for collecting mice and voles. Modern versions of the mouse trap

Figure 27. This Tomahawk live trap contains a captured Snowshoe Hare.

Figure 28. Sherman live traps.

(Museum Specials) are still used in some inventory work. Specialized kill traps have been designed for taking rodents such as the Northern Pocket Gopher. The disadvantage of kill traps is that they are nonselective and potentially harmful to the populations of rare or endangered species. They are also not very useful for population studies because removing animals affects the natural population structure.

Live traps are available for capturing all sizes of rodents and lagomorphs. The Havahart and Tomahawk companies produce wire-mesh live traps (figure 27) designed to capture mammals from marmot-size to pika-size. The doors at each end of the trap are connected to a treadle that releases the doors when an animal steps on it. For smaller rodents, such as mice and voles, most researchers prefer Sherman, Longworth or Little Critter Trap aluminum-box live traps, which all have a door that is closed by a treadle device. The Sherman trap (figure 28) comes in various sizes; one model is a folding design that is easy to transport in the field. Although effective for the House Mouse and New World mice (Deer Mouse,

Keen's Mouse and Western Harvest Mouse), the Sherman is less effective for capturing voles. The Longworth trap (figure 29) has a tunnel section and a larger nest box that clip together. It is a heavy trap to transport, but its nest box ensures a high survival rate. Also, voles seem to readily enter the narrow tunnel section. Cotton batting is often added to provide nesting material. The

Figure 29. Little Critter or Longworth type live trap.

Figure 30. Weighing a Red Squirrel in a net bag.

bait depends on the species – peanut butter, rolled oats, fruits or sunflower seeds are effective. Another type of trap is the pitfall trap consisting of a plastic bucket or ice-cream container set flush with the ground. Although sometimes used as a kill trap, the pitfall can become a live trap if food and nesting material are added. Pitfalls tend to capture species not easily captured in conventional live traps. Highly specialized traps have been designed for some species, such as a trap constructed from plastic tubing for the live capture of pocket gophers.

The placement of traps and the trapping schedule depends on the objectives of the researcher. For determining simple presence-absence of species in an area or habitat, a combination of live and pitfall traps set in transecting lines may be sufficient. Studies of relative or absolute abundance and day-to-day movements require long-term trapping, with traps set in a grid pattern and spaced at a standard distance. Researchers determine the sex, age and weight (figure 30) of captured animals, then tag or mark them so that individuals can be recognized when recaptured. The province's Resources Inventory Committee (RIC) has standardized procedures, with instruction manuals, for taking inventories of various kinds of rodents and lagomorphs. For more information on RIC standards contact the British Columbia Ministry of Sustainable Resource Management. Before undertaking any field work with small mammals you should check with public health agencies for precautions on how to protect yourself from potential diseases such as hantavirus or bubonic plague. Be aware that native rodents and lagomorphs are protected under the British Columbia *Wildlife Act.* Furbearers such as Beaver and Muskrat and endangered species listed by the province or COSEWIC have additional protection. Before taking voucher specimens or even live captures of any species, contact a regional office of the Ministry of Water, Land and Air Protection about permits.

Several new technological developments have revolutionized the study of rodents and lagomorphs. One is a small radio transmitter that can be placed on a mammal as small as a Deer Mouse. By tracking an animal with a receiver, researchers can collect far more detailed information on its daily movements than they can from recaptures in traps. Radio-tracking also reveals the location of hidden den sites underground or in trees. Some radio transmitters record and transmit physiological information such as body temperature. Another new tool is DNA technology. By taking blood or other tissue from a captured animal, it is possible to determine the animal's DNA sequences. This has provided invaluable information in behavioural studies where researchers are interested in determining the genetic relationship of animals in a colony or establishing which males mate with which females.

Naturalists can also make important contributions to our understanding of rodents and lagomorphs, even if they do not have the elaborate traps and equipment that biologists use. We know little about the distribution and basic natural history of some of the province's small mammals, and inventories have not yet been done for vast areas of the province. If you see a rare, threatened or endangered species, or a species that appears to be outside its provincial range, as shown in this book's species accounts, carefully record your observations and report them to the regional wildlife office. Record the date, location and circumstances, and locate your observation on a topographic map. Better yet, if you have a GPS device, record the coordinates.

If you are not scientifically inclined, you may wish to simply observe rodents and lagomorphs in their natural environment for enjoyment. In summer, alpine-subalpine areas in some of our provincial and national parks can be very rewarding for wildlife viewing. Diurnal mammals, such as pikas, marmots and chipmunks, can be spotted easily in this open landscape. One summer afternoon in the Columbia Mountains, while sitting quietly in a talus field for several hours, I saw an American Pika, Hoary Marmot, Golden-mantled Ground Squirrel and Yellow-pine Chipmunk, as well as an Ermine hunting pikas in the rocks. Identifying and following mammals tracks in the snow is also an effective method for detecting the presence of some species in an area.

Selected References: Murie 1974, Wilson et al. 1996.

Taxonomy and Nomenclature

The modern system of classification, developed in 1758 by the Swedish botanist Carl Linnaeus, uses a hierarchy of taxonomic categories (e.g., class, order, family, genus, species) with the species as the basic unit. Although the definition of a species is somewhat contentious, most biologists consider a species to be a group of interbreeding populations. Each species has a unique scientific name or binomial consisting of the genus followed by the species name. By convention, the binomial is written in italics with the genus capitalized and the species name in lower case. For example, the scientific name for Townsend's Vole is *Microtus townsendii*. Closely related species that share similar traits are usually grouped in the same genus. Scientists recognize about 60 species in the rodent genus *Microtus*; 7 are found in British Columbia. Some species have distinct geographic races or subspecies, formally recognized by taxonomists with a trinomial. For example, *Microtus townsendii cowani* is a race of Townsend's Vole restricted to Triangle Island off the north coast of Vancouver Island. Species are grouped into higher taxonomic categories based on their presumed relationships. Taxonomists group the 2021 known species of rodents (the mammalian order Rodentia) into 28 families, with 8 families represented in British Columbia.

The scientific names for species and higher taxonomic categories used in this book are based on *Mammal Species of the World* (Wilson and Reeder 1993), published by the Smithsonian Institute and the American Society of Mammalogists, and the *Revised Checklist of North American Mammals North of Mexico, 2003* (Baker et al. 2003). Subspecies are mostly based on *The Mammals of British Columbia: A Taxonomic Catalogue* (Nagorsen 1990). Consult these publications for detailed information on the taxonomy of British Columbian species, including citations to subspecies descriptions. There is no universally accepted list of mammalian common names equivalent to the American Ornithologists Union checklist of bird names. Common names used in this book are from *The Vertebrates of British Columbia: Scientific and English Names* (Sustainable Resource Management 2002). This checklist was an attempt to establish standard common names for the province's vertebrates. Be aware, however, that the common names for many mammals are somewhat contrived and rarely used by scientists. Moreover, several species have other common names that are widely used. For example, the Snowshoe Hare (*Lepus americanus*) is also called the Varying Hare or Snowshoe Rabbit.

Taxonomy is a dynamic science and there have been important changes in taxonomy and nomenclature since the original *Mammals of British Columbia* was last published in 1965. The three jumping mice (*Zapus*) are now classified in the family Dipodidae (formerly placed in the family Zapodidae); rats, mice, voles and lemmings classified in the families Cricetidae and Muridae are now treated as three subfamilies (Arvicolinae, Murinae and Sigmodontinae) in the family Muridae. The North American chipmunks are now classified in the genus *Tamias* rather than *Eutamias* (but the generic classification of the chipmunks is unresolved, as results from a recent study assessing DNA suggest that the western North American chipmunks should be classified in the genus *Neotamias*).

Some of the most extensive changes at the species level since the original *Mammals of British Columbia*, involve the genus *Peromyscus*. Cowan and Guiguet recognized three species in the province: the Sitka Mouse (*Peromyscus sitkensis*), Columbian Mouse (*Peromyscus oreas*) and the Deer Mouse (*Peromyscus maniculatus*). Recent genetic studies by Ira Greenbaum and colleagues have revealed that the Sitka Mouse and Columbian Mouse are a single coastal species, now classified as Keen's Mouse (*Peromyscus keeni*). Their research also has demonstrated that the Deer Mouse and Keen's Mouse co-occur over much of Vancouver Island and the southern coastal mainland where they can even be found at the same locality. The distribution map in *The Mammals of British Columbia* handbook does not reflect the existence of two coexisting species in southwestern B.C. Other deviations from the original handbook include my treating the Lower Mainland population of red-backed vole as a subspecies (*occidentalis*) of the Southern Red-backed Vole (*Clethrionomys gapperi*) rather than a separate species, and all populations of the Heather Vole as a single species (*Phenacomys intermedius*) rather than splitting them into two species: *P. intermedius* and *P. ungava*. See the Taxonomy sections in those species accounts for more details.

Powerful new taxonomic tools, particularly the use of DNA analysis, are providing new insights into mammalian taxonomy and will undoubtedly result in more taxonomic changes at the species and subspecies levels. Recent DNA research, for example, has revealed the existence of several strongly divergent genetic lineages in the Northern Flying Squirrel (*Glaucomys sabrinus*) and Yellow-pine Chipmunk (*Tamias amoenus*). These species may actually be several distinct species. But the Red Squirrel (*Tamiasciurus hudsonicus*) and Douglas' Squirrel (*T. douglasii*) show minimal divergence in their DNA – they may be a single species. More genetic research is needed to resolve the taxonomy of these rodents, especially in areas where these related forms come into contact.

Many of the subspecies listed in the Taxonomy section of the species accounts are based on descriptions from 50 to 100 years ago. Few of these subspecies have been evaluated with modern taxonomic methods. For the few species where their DNA has been analysed, the patterns of genetic variation are often inconsistent with the traditional subspecies. But DNA studies evaluating the two subspecies of the Red-tailed Chipmunk (*Tamias ruficaudus*) have shown that the two named races are genetically distinct. Clarifying the validity of these named subspecies is particularly critical for species such as the American Pika (*Ochotona princeps*), Snowshoe Hare (*Lepus americanus*), Mountain Beaver (*Aplodontia rufa*), Townsend's Vole (*Microtus townsendii*) and Least Chipmunk (*Tamias minimus*). One or more of their subspecies currently appear on the province's Red or Blue lists of mammals potentially at risk.

Selected References: Arbogast et al. 2001, Baker et al. 2003, Hogan et al. 1993, Sustainable Resource Management 2002, Nagorsen 1990, Wilson and Reeder 1983.

CHECKLIST OF SPECIES

This is a list of the scientific and common names for the 45 native and 7 introduced rodents and lagomorphs that occur in the province of British Columbia. Orders and families are ordered according to their generally accepted phylogenetic arrangement. Genera and species within a genus are ordered alphabetically by their scientific names. The only introduced species (I) on the list are those that have established feral populations in the province.

Order Lagomorpha: Lagomorphs

Family Ochotonidae: Pikas
  Ochotona collaris (Nelson)   Collared Pika
  Ochotona princeps (Richardson)   American Pika

Family Leporidae: Hares and Rabbits
  Lepus americanus Erxleben   Snowshoe Hare
  Lepus townsendii Bachman   White-tailed Jackrabbit
  Oryctolagus cuniculus (Linnaeus)   European Rabbit (I)
  Sylvilagus floridanus (J.A. Allen) Eastern Cottontail (I)
  Sylvilagus nuttallii (Bachman)   Nuttall's Cottontail

Order Rodentia: Rodents

Family Aplodontiidae: Mountain Beavers
  Aplodontia rufa (Rafinesque)   Mountain Beaver

Family Sciuridae: Squirrels

Glaucomys sabrinus (Shaw)	Northern Flying Squirrel
Marmota caligata (Eschscholtz)	Hoary Marmot
Marmota flaviventris (Audubon and Bachman)	Yellow-bellied Marmot
Marmota monax (Linnaeus)	Woodchuck
Marmota vancouverensis Swarth	Vancouver Island Marmot
Sciurus carolinensis Gmelin	Eastern Grey Squirrel (I)
Sciurus niger Linnaeus	Eastern Fox Squirrel (I)
Spermophilus columbianus (Ord)	Columbian Ground Squirrel
Spermophilus lateralis (Say)	Golden-mantled Ground Squirrel
Spermophilus parryii (Richardson)	Arctic Ground Squirrel
Spermophilus saturatus (Rhoads)	Cascade Mantled Ground Squirrel
Tamias amoenus J.A. Allen	Yellow-pine Chipmunk
Tamias minimus Bachman	Least Chipmunk
Tamias ruficaudus (A.H. Howell)	Red-tailed Chipmunk
Tamias townsendii Bachman	Townsend's Chipmunk
Tamiasciurus douglasii (Bachman)	Douglas' Squirrel
Tamiasciurus hudsonicus (Erxleben)	Red Squirrel

Family Geomyidae: Pocket Gophers

Thomomys talpoides (Richardson)	Northern Pocket Gopher

Family Heteromyidae: Heteromyids

Perognathus parvus (Peale)	Great Basin Pocket Mouse

Family Castoridae: Beavers

Castor canadensis Kuhl	Beaver

Family Muridae: Rats, Mice, Voles and Lemmings
Subfamily Arvicolinae: Voles, Lemmings and Muskrat

Clethrionomys gapperi (Vigors)	Southern Red-backed Vole
Clethrionomys rutilus (Pallas)	Northern Red-backed Vole
Lemmus trimucronatus (Richardson)	Brown Lemming
Microtus longicaudus (Merriam)	Long-tailed Vole
Microtus montanus (Peale)	Montane Vole
Microtus oeconomus (Pallas)	Tundra Vole
Microtus oregoni (Bachman)	Creeping Vole

Microtus pennsylvanicus (Ord)	Meadow Vole
Microtus richardsoni (DeKay)	Water Vole
Microtus townsendii (Bachman)	Townsend's Vole
Ondatra zibethicus (Linnaeus)	Muskrat
Phenacomys intermedius Merriam	Heather Vole
Synaptomys borealis (Richardson)	Northern Bog Lemming

Subfamily Murinae: Old World Rats and Mice

Mus musculus Linnaeus	House Mouse (I)
Rattus norvegicus (Berkenhout)	Norway Rat (I)
Rattus rattus (Linnaeus)	Black Rat (I)

Subfamily Sigmodontinae: New World Rats and Mice

Neotoma cinerea (Ord)	Bushy-tailed Woodrat
Peromyscus keeni (Rhoads)	Keen's Mouse
Peromyscus maniculatus (Wagner)	Deer Mouse
Reithrodontomys megalotis (Baird)	Western Harvest Mouse

Family Dipodidae: Jumping Mice and Jerboas

Zapus hudsonius (Zimmermann)	Meadow Jumping Mouse
Zapus princeps J. A. Allen	Western Jumping Mouse
Zapus trinotatus Rhoads	Pacific Jumping Mouse

Family Erethizontidae: New World Porcupines

Erethizon dorsatum (Linnaeus)	Porcupine

IDENTIFICATION KEYS

The keys are designed to identify animals in the hand or voucher specimens – they are not intended as a field guide for identifying sightings or signs such as tracks. The first key is for mammalian orders, which directs you to the appropriate key for identifying your animal. There are three independent keys: one for whole animals, another for skulls and a specialized key to identify chipmunk genital bones (the most reliable way to identify some chipmunks).

All of the keys are dichotomous with diagnostic characters arranged into couplets; each couplet offers two mutually exclusive choices labelled "a" or "b". Begin with couplet number one and select either "a" or "b". This will give you a species name or direct you to another couplet in the key. By systematically working through the various steps, you will eventually arrive at an identification. Consult the species account to verify that your identification is consistent with the species' description and range map.

An introductory section at the beginning of each key summarizes the important diagnostic traits and the pitfalls in identifying species of that group. I tried to avoid subjective characteristics (e.g., slightly smaller than or slightly darker than) and instead emphasized present or absent traits, absolute size differences, and absolute colour or pelage pattern differences. Although some species of rodents and lagomorphs have restricted distributions in the province, I deliberately avoided geography in the keys (e.g., found only in the Okanagan Valley), because accurate locality information may be lacking for some individuals. Moreover, we do not know the precise range for many small mammals in the province and excluding a species on the assumption that it is absent from an area could be misleading. Similarly, designing separate keys for different geographic regions of the province was not practical for a book designed to cover the entire province.

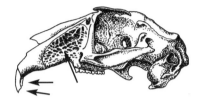 

Figure 31.   Figure 32.

Key to Orders

Whole Animals
1a.   Tail like a powder puff . . . . .Lagomorphs
1b.   Tail not like a powder puff . . . . .Rodents

Skulls
1a.   Two pairs of upper incisors; side of rostrum extensively fenestrated
        (figure 31) . . . . .Lagomorphs
1b.   One pair of upper incisors; side of rostrum not extensively fenestrated
        (figure 32) . . . . .Rodents

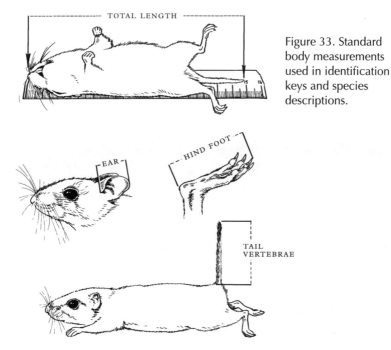

Figure 33. Standard body measurements used in identification keys and species descriptions.

Key to Whole Animals

This key is intended for use on restrained or anaesthetized animals, dead animals or museum study skins. You will require a millimetre rule or callipers for body measurements (figure 33) and a metric scale (in grams). A magnifying device is essential for examining dental traits such as the presence of grooves on the incisors of rodents. A dissecting microscope is ideal but, if you are working in difficult field conditions, a portable hand lens with a battery-powered light will suffice. You can expose the teeth by pushing up the lips and gently opening the mouth.

Lagomorphs: Hares, Rabbits and Pikas

Pikas are easily distinguished from hares and rabbits by their small size and short ears. In most areas of the province the only species of hare or rabbit is the Snowshoe Hare and identification is straightforward. But identification can be problematic in areas where this hare co-occurs with cottontails (e.g., lower Fraser River or Okanagan valleys) or regions where the Eastern Cottontail coexists with feral populations of the European Rabbit (e.g., southern Vancouver Island). Although the Snowshoe Hare generally turns white in winter, animals in summer pelage resemble the colour of cottontails. Positive identification of problematic or immature animals may require voucher specimens.

1a. Ears short (< 40 mm) and rounded; no visible external tail 2
1b. Ears long (> 40 mm) and elongated; visible external tail 3

2a. Dorsal pelage grey with an indistinct pale grey collar on neck and shoulders; ventral pelage whitish . . .Collared Pika (*Ochotona collaris*)
2b. Dorsal pelage brown with no distinct grey collar on neck or shoulders; ventral pelage grey . . . .Common Pika (*Ochotona princeps*)

3a. No nape on the back of head (figure 34) .4
3b. Rufous or brown nape on back of head (figure 35) 5

4a. Ear length > 100 mm . . . . .White-tailed Jackrabbit (*Lepus townsendii*)
4b. Ear length < 100 mm . . . . .Snowshoe Hare (*Lepus americanus*)

5a. Ear length usually > 80 mm; total length > 490 mm
     .European Rabbit (*Oryctolagus cuniculus*)
5b. Ear length usually < 80 mm; total length < 490 mm 6

6a. Pelage dull grey; weight < 700 g; ear length < 60 mm
     . . . . .Nuttall's Cottontail (*Sylvilagus nuttallii*)
6b. Pelage brownish; weight > 700 g; ear length > 60 mm
     . . . . .Eastern Cottontail (*Sylvilagus floridanus*)

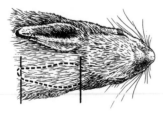

Figure 34.

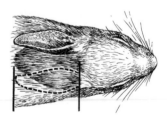

Figure 35.

Rodents

Diagnostic external traits for identifying this diverse group include pelage colour and markings, relative size of the tail, hind foot length, and head and body length (see figure 33 for body measurements). Some species are exceedingly difficult, if not impossible, to identify confidently from external characters; positive identification of many of the arvicoline rodents (voles and lemmings) for example, may require an examination of skulls for various cranial or dental characters. In some parts of the province, an examination of the male or female genital bones is necessary to identify chipmunks (*Tamias*). Note that immature animals can be problematic to identify from criteria based on absolute size; some pelage traits (e.g., the median reddish stripe of red-backed voles) may be indistinct in immature individuals.

1a.   Upper part of body covered with long stiff quills
. . . . .Porcupine (*Erethizon dorsatum*)
1b.   Upper part of body not covered with long stiff quills 2

2a.   Hind foot webbed 3
2b.   Hind foot not webbed 4

3a.   Broad tail flattened dorsal-ventrally; head and body length > 400 mm
. . . . .Beaver (*Castor canadensis*)
3b.   Tail flattened laterally, not broad; head and body length < 400 mm
. . . . .Muskrat (*Ondatra zibethicus*)

4a.   Tail with long hairs and bushy (figure 36). 5
4b.   Tail without long hairs and not bushy (figure 37). 22

5a.   Vibrissae extending well beyond the ears to shoulders; underside of
tail white; five digits on front foot
. . . . .Bushy-tailed Woodrat (*Neotoma cinerea*)
5b.   Vibrissae not extending well beyond the ears; underside of tail not
white; four digits on front foot 6

6a.   Tail flat in appearance; skin on side of body extends between front
and hind legs to form a gliding membrane
. . . . .Northern Flying Squirrel (*Glaucomys sabrinus*)

Figure 36.

Figure 37.

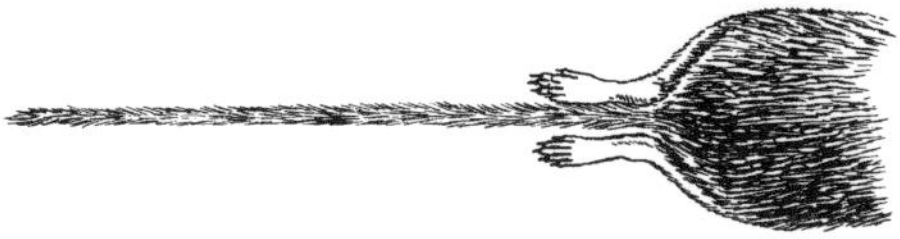

6b.   Tail not flat; no gliding membrane present on side of body 7

7a.   Light and dark longitudinal stripes present on back 8
7b.   No stripes present on back 13

8a.   Light and dark stripes present on side of head; body stripes consist of
      five black and four light stripes 9
8b.   No stripes on the side of head; body stripes consist of four black and
      two light stripes 12

9a.   The four pale body stripes indistinct; hairs on dorsal surface of tail
      have white or grey tips . . .Townsend's Chipmunk (*Tamias townsendii*)
9b.   The four pale body stripes distinct; hairs on dorsal surface of tail have
      pale yellow to rufous tips 10

10a. Total length usually < 195 mm; belly fur whitish-grey
      ....Least Chipmunk (*Tamias minimus*)★
10b. Total length usually > 195 mm; belly fur buff or whitish-grey 11

11a. Belly fur usually buff; underside of tail orange
      ....Yellow-pine Chipmunk (*Tamias amoenus*)★
11b. Belly fur usually whitish; underside of tail rufous
      ....Red-tailed Chipmunk (*Tamias ruficaudus*)★

★ Because size and belly fur colour vary geographically among subspecies, in some
areas of the province these three chipmunk species are impossible to distinguish
based on these traits. They can only be positively identified from skulls or genital-
bone morphology (see Key to Chipmunk Genital Bones, page 69).

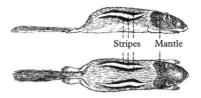

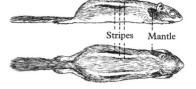

Figure 38. Golden-mantled Ground Squirrel (*Spermophilus lateralis*).

Figure 39. Cascade Mantled Ground Squirrel (*Spermophilus saturatus*).

12a. Prominent black stripes bordering the two light stripes on the back; conspicuous mantle of brown or reddish hairs on neck and shoulders (figure 38) . . .Golden-mantled Ground Squirrel (*Spermophilus lateralis*)

12b. Black stripes bordering the two light stripes on back indistinct or absent; mantle on neck and shoulders inconspicuous or absent (figure 39) . . . . .Cascade Mantled Ground Squirrel (*Spermophilus saturatus*)

13a. Dorsal pelage spotted or mottled 14
13b. Dorsal pelage not spotted or mottled 15

14a. Side of neck tawny brown; ventral side of tail predominately reddish with black tip; hind legs and dorsal surface of feet grey to yellowish-brown . . . . .Arctic Ground Squirrel (*Spermophilus parryii*)

14b. Side of neck greyish; ventral side of tail a mix of black and white; hind legs and dorsal surface of hind feet reddish-brown
. . . . .Columbian Ground Squirrel (*Spermophilus columbianus*)

15a. Body heavy, stout; tail length < 50% of head and body length 16
15b. Body not heavy, stout; tail length about equal to head and body length 19

16a. Posterior pad on sole of hind foot oval (figure 40) 17
16b. Posterior pad on sole of hind foot circular (figure 41) 18

Figure 40.

Figure 41.

17a. Dorsal surface of feet dark brown; side of neck grizzled brown; neck
not distinct from side or back . . . . . . .Woodchuck (*Marmota monax*)

17b. Dorsal surface of feet yellowish to buff; side of neck uniform buff to
yellow; neck paler than side or back
. . . . .Yellow-bellied Marmot (*Marmota flaviventris*)

18a. Dorsal pelage grizzled with a mix of white and black; individual guard
hairs bicoloured or tricoloured . . .Hoary Marmot (*Marmots caligata*)

18b. Dorsal pelage light to chocolate brown, not grizzled; individual guard
hairs unicoloured
. . . . .Vancouver Island Marmot (*Marmota vancouverensis*)

19a. Head and body length > 220 mm; tail vertebrae length > 190 mm;
hind foot length > 60 mm 20

19b. Head and body length < 190 mm; tail vertebrae length < 190 mm;
hind foot length < 60 mm 21

20a. Dorsal pelage rust grey; ventral pelage yellow or orange; tips of tail
hairs and tufts behind ears reddish-orange
. . . . .Eastern Fox Squirrel (*Sciurus niger*)

20b. Dorsal pelage grey with whitish ventral pelage; tips of tail hairs and
tufts of hairs behind ears whitish – or melanic with black or dark
brown pelage . . . . .Eastern Grey Squirrel (*Sciurus carolinensis*)

21a. Ventral pelage predominately yellow or orange
. . . . .Douglas' Squirrel (*Tamiasciurus douglasii*)

21b. Ventral pelage predominately white or grey
. . . . .Red Squirrel (*Tamiasciurus hudsonicus*)

22a. External cheek pouches present on each side of the mouth (figure 42)
. 23

22b. No external cheek pouches present 24

Figure 42.

Figure 43. Body form of a typical "mouse-rat" rodent.

Figure 44. Body form of a "vole-lemming" rodent.

23a. Front feet have long claws (figure 42); tail no more than 50% of the length of head and body
.............Northern Pocket Gopher (*Thomomys talpoides*)
23b. Front feet without long claws, tail longer than head and body
.... Great Basin Pocket Mouse (*Perognathus parvus*)

24a. Tail markedly shorter than hind foot length and not readily visible; long claws on front feet ....Mountain Beaver (*Aplodontia rufa*)
24b. Tail equal length or longer than hind foot and readily visible; front feet without long claws 25

25a. Conspicuous eyes and ears, ears not concealed in fur; long tail; sleek body (figure 43) .... 26 (mice and rats)
25b. Inconspicuous eyes and ears, ears concealed in fur; tail short; body stocky (figure 44) ....34 (voles and lemmings)

Figure 45.

Figure 46.

Figure 47. Elongated hind legs and feet of jumping mice (*Zapus*).

Figure 48.

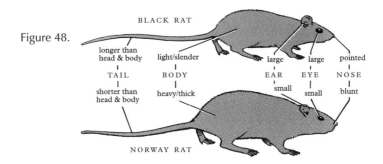

BLACK RAT

longer than head & body — light/slender — large — large — pointed

TAIL — BODY — EAR — EYE — NOSE

shorter than head & body — heavy/thick — small — small — blunt

NORWAY RAT

26a. Tail nearly naked, scaly, annulations readily visible (figure 45) 27
26b. Tail hairy, scales and annulations not readily visible (figure 46) . . .32

27a. Hind legs and hind feet greatly elongated (figure 47) 28
27b. Hind legs and hind feet not elongated 30

28a. Ventral pelage typically pure white; usually no buff patches on chest,
     throat or belly . . . . .Western Jumping Mouse (*Zapus princeps*)
28b. Ventral pelage white, typically with buff wash; buff patches often on
     chest, throat or belly 29

29a. Total length usually > 230 mm; tail vertebrae length usually
     > 140 mm . . . . .Pacific Jumping Mouse (*Zapus trinotatus*)[*]
29b. Total length usually < 230 mm; tail vertebrae length usually
     < 140 mm . . . . .Meadow Jumping Mouse (*Zapus hudsonius*)[*]

30a. Hind foot length < 20 mm; weight < 30 g
     . . . . .House Mouse (*Mus musculus*)
30b. Hind foot length > 20 mm; weight > 30 g 31

31a. Tail longer than head and body length; snout pointed; rump light and
     slender (figure 48) . . . . .Black Rat (*Rattus rattus*)
31b. Tail less than head and body length, snout blunt, rump thick and
     heavy (figure 48) . . . . .Norway Rat (*Rattus norvegicus*)

---

[*] The jumping mice show considerable geographic variation in size and ventral fur colour. In areas of the province where two species co-occur, positive identification requires a skull (see Key to Skulls).

Figure 49.

Figure 50.

32a. Total length < 150 mm; weight < 15 g, distinct groove on anterior
face of upper incisors (figure 49)
. . . . .Western Harvest Mouse (*Reithrodontomys megalotis*)
32b. Total length > 150 mm; weight > 15 g; upper incisors have no groove
(figure 50) 33

33a. Tail vertebrae length < 98 mm . .Deer Mouse (*Peromyscus maniculatus*)★
33b. Tail vertebrae length > 98 mm . . . .Keen's Mouse (*Peromyscus keeni*)★

34a. Dorsal pelage with broad chestnut or rufous median stripe 35
34b. Dorsal pelage lacks median stripe 36

35a. Tail densely furred, with yellow on the underside
. . . . .Northern Red-backed Vole (*Clethrionomys rutilus*)
35b. Tail sparsely furred, with white or grey on the underside
. . . . .Southern Red-backed Vole (*Clethrionomys gapperi*)

36a. Tail vertebrae length < 50 mm 37
36b. Tail vertebrae length > 50 mm 43

37a. Tail roughly the same length as the hind foot; tail vertebrae length less
than 20% of total length 38

★ Tail length will only discriminate adults. This trait is not completely reliable as there
is some overlap among the two species. Moreover, the diagnostic measurement of
98 mm is derived from a few island and mainland specimens from southwestern
British Columbia identified from genetic criteria. The reliability of this diagnostic
measurement for discriminating the two species of *Peromyscus* in other regions needs
to be assessed.

37b. Tail distinctly longer than hind foot; tail vertebrae length more than 20% of total length 39

38a. Tail sparsely furred; dorsal pelage short and grey or dull brown; upper incisor has a groove (figure 49)
.... Northern Bog Lemming (*Synaptomys borealis*)
38b. Tail densely furred; dorsal pelage long and orange to brown; upper incisors not grooved (figure 50)
.... Brown Lemming (*Lemmus trimucronatus*)

39a. Fur short and velvety; eye diameter < 3 mm
.... Creeping Vole (*Microtus oregoni*)
39b. Fur not short and velvety; eye diameter > 3 mm 40

40a. Tail thin, not thicker at base, and sparsely furred; body fur soft and fine .... Heather Vole (*Phenacomys intermedius*)
40b. Tail not thin, with a thicker base, and covered in hairs, body fur coarse 41

41a. Pelage grizzled grey-brown or "salt-and-pepper"; tops of hind feet whitish to grey .... Montane Vole (*Microtus montanus*)
41b. Pelage brown with no "salt-and-pepper" pattern; tops of hind feet dark brown to nearly black 42

42a. Dorsal pelage has tinges of yellow on flanks and rump; underparts pale yellow; tail strongly bicoloured . . . Tundra Vole (*Microtus oeconomus*)*
42b. Dorsal pelage lacks yellow tinge on flanks and rump; underparts grey; tail not strongly bicoloured . . Meadow Vole (*Microtus pennsylvanicus*)*

43a. Hind foot usually > 26 mm; body weight usually > 100 g
.... Water Vole (*Microtus richardsoni*)
43b Hind foot length usually < 26 mm; body weight < 100g 44

44a. Tail vertebrae length usually 33% or more of the total length
.... Long-tailed Vole (*Microtus longicaudus*)
44b. Tail vertebrae length usually less than 33% of the total length
.... Townsend's Vole (*Microtus townsendii*)

---

* Positive identification of these two species may require an examination of the morphology of the second upper molar in the skull (see Key to Skulls).

Key to Skulls

This key relies on cranial and dental characters taken from clean skulls or skulls recovered from raptor pellets. You will require a dissecting microscope to examine teeth, and needle-nosed callipers for taking cranial and dental measurements.

Lagomorphs: Hares, Rabbits and Pikas

Important cranial traits for identifying species in this order include the fenestration of the rostrum, the supraorbital processes, distinctness of the interparietal bone and cranial size. Basilar length and maxillary toothrow length are measures for skull size (figure 51). Although basilar length is a good index of skull size, it cannot be measured from broken skulls and may be influenced by age. Maxillary toothrow length can usually be taken from broken skulls and it is less affected by age variation than basilar length.

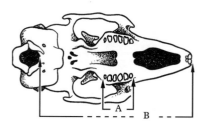

Figure 51. Key measurements on a lagomorph skull: A, maxillary toothrow length; B, basilar length.

1a. Skull lacks supraorbital processes (figure 52); rostrum with single oval fenestration (figure 53) . . . . .Collared Pika (*Ochotona collaris*)
. . . . .or American Pika (*Ochotona princeps*)
1b   Skull has supraorbital processes (figure 54); rostrum with lattice-like fenestration (figure 55) 2

2a. Interparietal bone indistinct, fused to parietal bones; supraorbital processes broad and wing-like (figure 56) .3
2b. Interparietal bone distinct, not fused to parietal bones; supraorbital bone narrow and strap-like (figure 57) 4

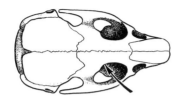

Figure 52.

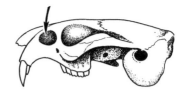

Figure 53.

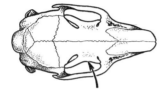

Figure 54.

Figure 55.

3a.   Basilar length > 69.0 mm; maxillary toothrow > 16.0 mm
       .White-tailed Jackrabbit (*Lepus townsendii*)
3b.   Basilar length < 69.0 mm; maxillary toothrow length < 16.0 mm
       . . . . .Snowshoe Hare (*Lepus americanus*)

4a.   Tip of posterior extension of supraorbital process free of the
       braincase . . . . .European Rabbit (*Oryctolagus cuniculus*)
4b.   Tip of posterior extension of supraorbital process touches the
       braincase 5

5a.   Basilar length > 57.0 mm; maxillary toothrow length > 14.0 mm
       .Eastern Cottontail (*Sylvilagus floridanus*)
5b.   Basilar length < 57.0 mm; maxillary toothrow length < 14.0 mm
       . . . . .Nuttall's Cottontail (*Sylvilagus nuttallii*)

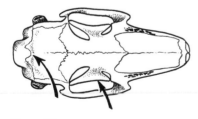

Figure 56.

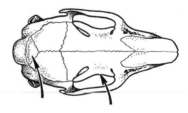

Figure 57.

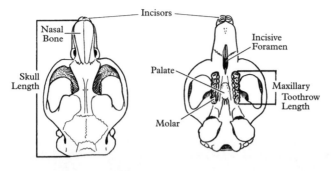

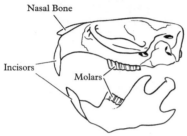

Figure 58. Rodent skull (three views) showing key measurements.

Rodents

Important cranial traits for identifying rodents include size (skull and maxillary toothrow lengths shown in figure 58), the morphology of the crowns of the molar teeth (figure 59), and the size, shape and position of the infraorbital opening. It may not be possible to use size to identify immature animals where the skull is not completely developed. The cusp patterns that distinguish the Old World rats and the New World rats and mice are not discernible in old animals with worn teeth. Enamel patterns on the molar teeth of most species of arvicoline rodents (voles and lemmings) are not affected by tooth wear.

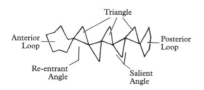

Figure 59. Nomenclature of an arvicoline rodent's lower molar.

Figure 60.

Figure 61.

Figure 62.

Figure 63.

1a	Skull has large and conspicuous infraorbital opening (figure 60) . . .2
1b.	Skull has small and inconspicuous infraorbital opening (figure 61) .25

2a.	Infraorbital opening on the side of the rostrum (figure 62)
	. . . . .Great Basin Pocket Mouse (*Perognathus parvus*)
2b.	Infraorbital opening not on side of rostrum 3

3a.	Infraorbital opening exceeding the foramen magnum in size, round in outline . . . . .Porcupine (*Erethizon dorsatum*)
3b.	Infraorbital opening not exceeding the foramen magnum in size, oval or V-shaped in outline 4

4a.	Infraorbital opening oval in outline (figure 60) 5
4b.	Infraorbital opening V-shaped in outline (figure 63) 7

5a.	First upper cheek tooth (premolar) has a distinct crescent fold at the base of the main cusp (figure 64)★
	. . . . .Pacific Jumping Mouse (*Zapus trinotatus*)
5b.	First upper cheek tooth (premolar) lacks a distinct crescent fold at the base of the main cusp (figure 65) 6

★ This trait tends to be variable and it is not discernible in animals with worn teeth. For animals taken from areas of the Cascade Mountains, where this species overlaps with the range of the Western Jumping Mouse (*Zapus princeps*), skulls should be submitted to an expert for verification.

Figure 64.

Figure 65.

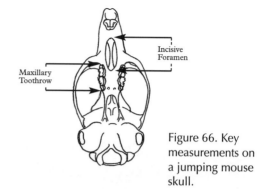

Figure 66. Key measurements on a jumping mouse skull.

6a. Incisive foramen length (figure 66) < 4.2 mm; maxillary toothrow length < 3.7 mm . . . . . . .Meadow Jumping Mouse (*Zapus hudsonius*)

6b. Incisive foramen length (figure 66) > 4.2 mm; maxillary toothrow length > 3.7 mm . . . . .Western Jumping Mouse (*Zapus princeps*)

7a. Crowns of molars have triangles or prisms of dentine surrounded by enamel (figures 67 and 68) 8

7b. Molar crowns have cusps but no triangles or prisms (figure 69) . . .21

8a. Zygomatic plate extending anteriorly from zygomatic process of maxilla (figure 70), molars with prisms not arranged in alternating triangles (figure 68) . . . . .Bushy-tailed Woodrat (*Neotoma cinerea*)

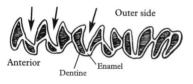

Figure 67.

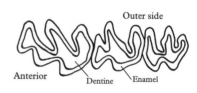

Figure 68.

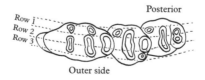

Figure 69.

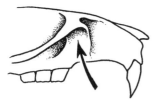 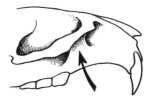

Figure 70.   Figure 71.

8b.   Zygomatic plate not extending anteriorly from zygomatic process of
      maxilla (figure 71); molars with prisms arranged in alternating
      triangles (figure 67) 9

9a.   Maxillary toothrow length > 14.0 mm . .Muskrat (*Ondatra zibethicus*)
9b.   Maxillary toothrow length < 14.0 mm 10

10a. Re-entrant angles deeper on outer side of upper molars; first and
      second molar re-entrant angles extend to inner border of tooth
      (figure 67) 11
10b. Re-entrant angles not deeper on outer side of upper molars; first and
      second molar re-entrant angles do not extend to inner border of tooth
      (figure 72) 12

11a. Mandibular molars lack closed triangles on outer sides (figure 73)
      . . . . .Northern Bog Lemming (*Synaptomys borealis*)
11b. Closed triangles present on outer sides of mandibular molars
      (figure 74) . . . . .Brown Lemming (*Lemmus trimucronatus*)

Figure 72.

Figure 73.   Figure 74.

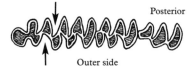

Figure 75.   Figure 76.

12a. On mandibular molars inner re-entrant angles deeper than outer
      re-entrant angles (figure 75) . .Heather Vole (*Phenacomys intermedius*)
12b. On mandibular molars inner and outer re-entrant angles about equal
      in depth (figure 76) 13

13a   Bony palate terminates as a thin transverse shelf extending between
      the last molars (figure 77) 14
13b. Posterior palate terminates as a sloping median ridge between the last
      molars, bordered by lateral pits (figure 78) 15

14a. Posterior edge of palate usually completely fused (figure 79)⋆
      . . . . .Southern Red-backed Vole (*Clethrionomys gapperi*)
14b. Posterior edge of palate usually incompletely fused (figure 80)⋆
      . . . . .Northern Red-backed Vole (*Clethrionomys rutilus*)

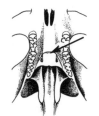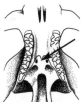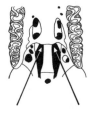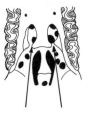

Figure 77.   Figure 78.   Figure 79.   Figure 80.

⋆ Development of the palate is related to growth and age. Some old Northern Red-
backed Voles (*Clethrionomys rutilus*) may have palates with their posterior edge completely fused and a few immature Southern Red-backed Voles (*C. gapperi*) may have
palates with their posterior edge incompletely fused. Skulls taken from areas of overlap among the two species should be examined by an expert.

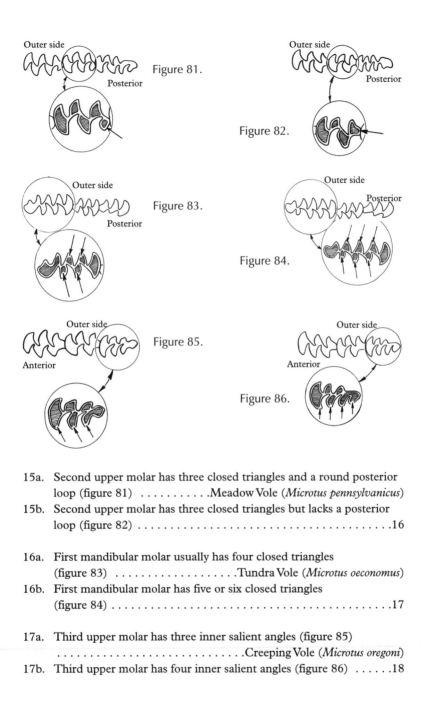

Figure 81.

Figure 82.

Figure 83.

Figure 84.

Figure 85.

Figure 86.

15a. Second upper molar has three closed triangles and a round posterior
      loop (figure 81) . . . . .Meadow Vole (*Microtus pennsylvanicus*)
15b. Second upper molar has three closed triangles but lacks a posterior
      loop (figure 82) 16

16a. First mandibular molar usually has four closed triangles
      (figure 83) . . . . .Tundra Vole (*Microtus oeconomus*)
16b. First mandibular molar has five or six closed triangles
      (figure 84) 17

17a. Third upper molar has three inner salient angles (figure 85)
      . . . . .Creeping Vole (*Microtus oregoni*)
17b. Third upper molar has four inner salient angles (figure 86) 18

Figure 87.   Figure 88.   Figure 89.

18a. When skull is viewed dorsally, upper incisors protrude extensively beyond the nasal bones (figure 87); incisive foramina taper to narrow slits posteriorly (figure 90) . . . . .Water Vole (*Microtus richardsoni*)

18a. When skull is viewed dorsally, upper incisors hidden or protrude only slightly beyond the nasal bones (figures 88 and 89); incisive foramina not tapered to narrow slits posteriorly (figures 91 and 92) 19

19a. When skull is viewed dorsally, upper incisors hidden by the nasal bones (figure 88) . . . . .Long-tailed Vole (*Microtus longicaudus*)

19b. When skull is viewed dorsally, upper incisors extend slightly beyond the nasal bones (figure 89) 20

20a. Incisive foramina taper gradually in the posterior region* (figure 91) . . . . .Townsend's Vole (*Microtus townsendii*)

20b. Incisive foramina taper abruptly in the posterior region* (figure 92) . . . . .Montane Vole (*Microtus montanus*)

Figure 90.   Figure 91.   Figure 92.

---

* Because the protrusion of the incisors and shape of the incisive foramina vary with age, these traits are not completely reliable. Townsend's Vole (*Microtus townsendii*) and the Montane Vole (*M. montanus*) are widely separated in their ranges in British Columbia. But the Long-tailed Vole (*M. longicaudus*) co-occurs with Townsend's Vole in the lower Fraser River valley and the Montane Vole in the interior grasslands. Discriminating the skulls of these voles from these regions is difficult and identifications should be verified by an expert.

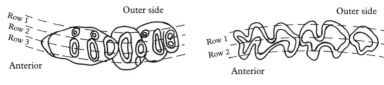

Figure 93.   Figure 94.

21a. Cusps of first two upper molars arranged in three longitudinal rows
     (figure 93) 22
21b. Cusps of first two upper molars arranged in two longitudinal rows
     (figure 94) 24

22a. Upper incisor notched in side view (figure 95); maxillary toothrow
     length < 6.0 mm; greatest length of skull < 25.0 mm
     . . . . .House Mouse (*Mus musculus*)
22b. Upper incisor not notched in side view; maxillary toothrow length >
     6.0 mm; greatest length of skull > 25.0 mm 23

23a. Braincase rectangular; temporal ridges straight and nearly parallel
     (figure 96) .Norway Rat (*Rattus norvegicus*)
23b. Braincase rounded; temporal ridges curved (figure 97)
     . . . . .Black Rat (*Rattus rattus*)

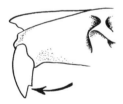

Figure 95.

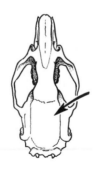   

Figure 96.   Figure 97.

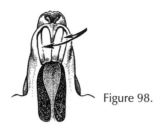

Figure 98.

Figure 99.

24a. Upper incisor has a distinct groove on anterior face (figure 98);
maxillary toothrow length < 3.8 mm; skull length < 23.0 mm
. . . . .Western Harvest Mouse (*Reithrodontomys megalotis*)

24b. Upper incisor lacks a groove on anterior face (figure 99); maxillary
toothrow length > 3.8 mm; skull length > 23.0 mm
. . . . .Keen's Mouse (*Peromyscus keeni*)★
. . . . .or Deer Mouse (*Peromyscus maniculatus*)★

25a. Upper cheek-teeth have a distinct projection on outer side; auditory
bullae flask shaped (figure 100) . .Mountain Beaver (*Aplodontia rufa*)

25b. Upper cheek teeth lack a distinct projection on outer side; auditory
bullae not flask shaped 26

26a. Upper and lower premolars indented on the side forming an "8"
shape (figure 101) . . .Northern Pocket Gopher (*Thomomys talpoides*)

26b. Upper and lower premolars not "8" shaped. 27

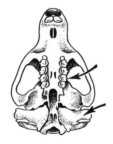

Figure 100.

Figure 101.

★ Keens' Mouse (*Peromyscus keeni*) and the Deer Mouse (*P. maniculatus*) cannot be
discriminated by any single cranial measurement. For specimens taken from coastal
areas where the two species co-occur, skulls should be submitted to an expert for
verification.

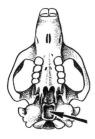

Figure 102.　　　Figure 103.　　　Figure 105.

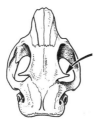

Figure 104.　　　　　Figure 106.

27a. Prominent depression in basioccipital region (figure 102); width of
first upper cheek tooth > 9.5 mm . . . . .Beaver (*Castor canadensis*)
27b. No prominent depression in basioccipital region; width of first upper
cheek tooth < 9.5 mm. 28

28a. Dorsal profile of skull straight (figure 103); interorbital region with a
shallow depression; postorbital processes at right angles to main axis
of skull (figure 104). 29
28b. Dorsal profile of skull convex (figure 105); no shallow depression in
interorbital region; postorbital processes directed obliquely backwards
(figure 106) 32

29a. Length of maxillary toothrow > 21.2 mm 30
29b. Length of maxillary toothrow < 21.2 mm 31

30a. Posterior border of nasal bones forms a V-shaped notch (figure 107)
. . . . .Vancouver Island Marmot (*Marmota vancouverensis*)
30b. Posterior border of nasal bones squared or arched (figure 108)
. . . . .Hoary Marmot (*Marmota caligata*)

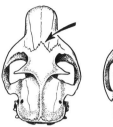 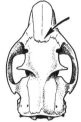 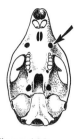 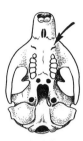

Figure 107.   Figure 108.   Figure 109.   Figure 110.

31a. Upper toothrows parallel; posterior border of nasal bones forms a
    V-shaped notch (figure 107) . . . . .Woodchuck (*Marmota monax*)
31b. Upper toothrows diverge anteriorly; posterior border of nasal bones
    forms an arch (figure 108)
    . . . . .Yellow-bellied Marmot (*Marmota flaviventris*)

32a. Infraorbital opening simple and rounded piercing the zygomatic
    plate (figure 109); skull length < 40.0 mm, maxillary toothrow
    length < 7.0 mm 33
32b. Infraorbital opening modified as a canal that passes between the
    zygomatic plate and the side of the rostrum (figure 110); skull
    length > 40.0 mm; maxillary toothrow length > 7.0 mm 35

33a. Skull length > 37.0 mm; maxillary toothrow length > 6.0 mm
    . . . . .Townsend's Chipmunk (*Tamias townsendii*)
33b. Skull length < 37.0 mm; maxillary toothrow length < 6.0 mm 34

34a. Skull length usually < 32.5 mm; mandibular length < 17.5 mm[1]
    . . . . .Least Chipmunk (*Tamias minimus*)
34b. Skull length usually > 32.5 mm; mandibular length > 17.5 mm.
    . . . . .Red-tailed Chipmunk (*Tamias ruficaudus*)[2]
    . . . . .or Yellow-pine Chipmunk (*Tamias amoenus*)[2]

---

[1]. These diagnostic measurements will discriminate most Least Chipmunks (*Tamias minimus*) from the Rocky and Columbia mountains of southern British Columbia. But in northern areas of the province, where this species is larger and co-occurs with the Yellow-pine Chipmunk (*T. amoenus*), the two species can only be positively identified from genital bone morphology (see Key to Chipmunk Genital Bones, page 69).

[2]. In southeastern British Columbia, where the Red-tailed Chipmunk (*T. ruficaudus*) and Yellow-pine Chipmunk (*T. amoenus*) co-occur, the two species overlap in size and

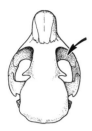

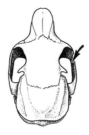

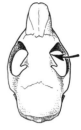

Figure 111.   Figure 112.   Figure 113.

35a. Zygomatic arches converge anteriorly and are flat horizontally
     (figure 111). 36
35b. Zygomatic arches parallel and not flat (figure 112). 37

36a. Skull length > 49.0 mm; maxillary toothrow > 10.0 mm
     . . . . .Arctic Ground Squirrel (*Spermophilus parryii*)[3]
     . . . . . . .or Columbian Ground Squirrel (*Spermophilus columbianus*)[3]
36b. Skull length < 49.0 mm; maxillary toothrow < 10.0 mm
     . . . . . .Cascade Mantled Ground Squirrel (*Spermophilus saturatus*)[4]
     . . . . . .or Golden-mantled Ground Squirrel (*Spermophilus lateralis*)[4]

37a. Interorbital region indented with a V-shaped notch (figure 113)
     . . . . .Northern Flying Squirrel (*Glaucomys sabrinus*)
37b. Interorbital region not indented with a V-shaped notch 38

cannot be distinguished from any single skull measurement. Positive identification is
based on genital bone morphology or examination of skins and skulls by an expert
(see Key to Chipmunk Genital Bones).

*3. Although the Columbian Ground Squirrel (*Spermophilus columbianus*) and Arctic
ground Squirrel (*S. parryii*) show minor differences in their pelage, their skulls cannot
be discriminated by any single measurement. But distributions of the two species in
B.C. are widely separate.

*4. Although the Golden-mantled Ground Squirrel (*S. lateralis*) and Cascade
Mantled Ground Squirrel (*S. saturatus*) differ in their pelage colour and stripes, their
skulls cannot be discriminated by any single measurement. But distributions of the
two species in B.C. are separate.

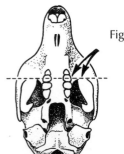

Figure 114.

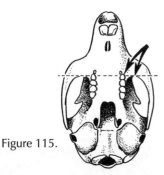

Figure 115.

38a. Skull length > 55.0 mm; maxillary toothrow > 10.0 mm; anterior ventral border of orbit opposite the first upper molar (figure 114) 39

38b. Skull length < 55.0 mm; maxillary toothrow < 10.0 mm; anterior ventral border of orbit opposite the second upper premolar (figure 115) . . . . .Red Squirrel (*Tamiasciurus hudsonicus*)*
. . . . .or Douglas Squirrel (*Tamiasciurus douglasii*)*

39a. Usually two upper premolars
. . . . .Eastern Grey Squirrel (*Sciurus carolinensis*)

39b. Usually one upper premolar
. . . . .Eastern Fox Squirrel (*Sciurus carolinensis*)

---

* Although these species generally differ in ventral pelage colour, their skulls cannot be discriminated by any single measurement. Specimens taken from areas on the eastern slopes of the Cascade and Coast mountains where the two species co-occur should be submitted to an expert for verification.

Key to Chipmunk Genital Bones

Because species converge in pelage colour and size, chipmunks are difficult or impossible to identify from pelage or even skull morphology in areas of the province where two or more species co-occur. An important diagnostic trait for discriminating chipmunk species is the morphology of the male and female genital bones (baculum and baubellum). Imbedded in the genitalia, these structures have to be specially prepared by clearing the surrounding tissue with potassium hydroxide and applying Alazarin red, a stain that turns bone deep red. Preparation techniques are given in the publications of Dallas Sutton and John White. Because they are minute, the genital bones have to be viewed with a microscope and measured with an ocular micrometer.

Figure 116. Baculum measurements: A, shaft length; B, tip height; C, keel height.

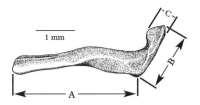

Male Genital Bone (Baculum) (figure 116)

1a.   Shaft length > 3.20 mm; keel height > 0.45 mm
. . . . .Red-tailed Chipmunk (*Tamias ruficaudus*)
1b.   Shaft length < 3.20 mm; keel height < 0.45 mm 2

2a.   Ratio of the tip length to shaft length > 30% 3
2b.   Ratio of the tip length to shaft length < 30%
. . . . .Least Chipmunk (*Tamias minimus*)

3a   Tip length > 1.0 mm . . . .Townsend's Chipmunk (*Tamias townsendii*)
3b.   Tip length < 1.0 mm . . . . .Yellow-pine Chipmunk (*Tamias amoenus*)

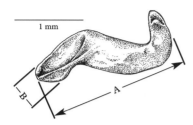

Figure 117. Baubellum measurements: A, greatest length; B, tip height.

Female Genital Bone (Baubellum) (figure 117)

1a.   Greatest length > 2.0 mm; tip height > 0.40 mm
. . . . Red-tailed Chipmunk (*Tamias ruficaudus*)
1b.   Greatest length < 2.0 mm; tip height < 0.40 mm 2

2a.   Base U-shaped and usually tapered sharply near the proximal end
(figure 118) . . . . .Townsend's Chipmunk (*Tamias townsendii*)*
. . . . .or Least Chipmunk (*Tamias minimus*)*
2b.   Base not U-shaped and usually not tapered sharply near the proximal
end (figure 119) . . . . .Yellow-pine Chipmunk (*Tamias amoenus*)

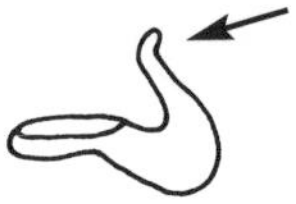

Figure 118.

Figure 119.

Selected References: Allard and Greenbaum 1988; Glass 1951; Maser
and Storm 1970; Nagorsen 2002; Sutton 1982; White 1951, 1953.

---

* Baubella of the Least Chipmunk (*Tamias minimus*) and Townsend's Chipmunk
(*T. townsendii*) overlap in all measurements and are similar in general shape. But these
two chipmunks are widely separated in their range in British Columbia and are easily
discriminated by skull size and pelage colour.

SPECIES ACCOUNTS

This book provides a detailed species account for each of the province's 7 lagomorph and 45 rodent species. The accounts follow the order in the checklist. Each one has an illustration of the species and its skull. These figures show general features; diagnostic features important for identification are illustrated in the identification keys. Information in the species accounts is divided into six categories:

Other Common Names is a list of alternative English common names.

Description is a concise description, including measurements, dental formula and a comparison with similar species that may be confused in identification. I based my descriptions of fur colour, body measurements and weights on adult museum specimens and live animals from British Columbia.

All linear measurements are in millimetres (mm); the weight is in grams (g) to the nearest tenth for animals weighing less than a kilogram and in kilograms (kg) to the nearest hundredth for animals weighing more. The values given are the mean, range (in parentheses), and sample size (number of individuals measured). For example, the mean total length of the Snowshoe Hare is 443 mm, based on measurements obtained from 139 animals with total lengths ranging from 388 to 530 mm – these figures are written as "443 (388-530) n=139". Because weight varies with season and reproductive condition, there is considerable variation associated with the mean for each species.

The dental formula describes the number of teeth in one side of the head. The first number is for the upper jaw and the second for the lower jaw. For example, incisors 2/1, canines 0/0, premolars 3/2, molars 3/3 for the Snowshoe Hare indicates two upper and one lower incisor, no upper and lower canines, three upper and two lower premolars, and three upper and lower molars.

Natural History includes habitat, elevational range, movements, population estimates, food habits, behaviour and reproduction. Wherever possible, I used information from studies done on British Columbian populations or populations from adjacent areas of Canada and the United States. I obtained considerable habitat, elevational and reproductive data from museum specimens collected in the province.

Range is a general description of the overall distribution and a detailed description of the provincial range. It includes a range map that is based on an extensive review of all the known locality records from the province. The database for maps consisted of about 30,000 museum specimens and 3000 observational records. Maps with symbols representing actual locality records were generated by computer using Universal Transverse Mercator Grid (UTM) co-ordinates. Provincial base maps (Albers projection) are all 1:2,000,000 scale. For species with very localized ranges in the province, maps cover only the localized area to provide maximum detail.

Taxonomy lists subspecies found in the province and discusses any major changes in nomenclature from *The Mammals of British Columbia* (Cowan and Guiguet 1965). Subspecies are based on *The Mammals of British Columbia: A Taxonomic Catalogue* (Nagorsen 1990).

Remarks gives interesting details about the species in British Columbia, including conservation status, and identifies areas for future study.

Selected References lists important publications on the species. It is not intended to be a comprehensive review of all literature – the emphasis is on studies done in British Columbia and adjacent areas. I cite unpublished reports only if they contain significant original information.

Order Lagomorpha

The lagomorphs (rabbits, hares and pikas) have two pairs of upper incisors, separated from the cheek teeth by a space, and a fenestrated rostrum; they have no canine teeth. There are 70 species worldwide; 5 are found in British Columbia.

Family Ochotonidae: Pikas

Pikas are small lagomorphs with short, nearly circular ears; the tail is not visible and the hind limbs are only slightly longer than front limbs. The skull is flat with a short narrow rostrum that has a single oval fenestration (figure 53). Two species inhabit British Columbia.

Collared Pika   *Ochotona collaris*

Other Common Names: None.

Description
The dorsal pelage of the Collared Pika is dull grey with some buff fur on the head and back. An indistinct grey collar is present on the neck and shoulders. The ventral pelage is white.

Measurements:
total length: 197 (170-217) n=18
tail vertebrae: 16 (12-14) n=6
hind foot: 32 (28-34) n=18
ear: 22 (20-22) n=7
weight: 146.7 g (123.0-173.0) n=7

Dental Formula:
incisors: 2/1
canines: 0/0
premolars: 3/2
molars: 3/3

Identification:
The only similar mammal is the American Pika (*Ochotona princeps*). Its distribution is restricted to southern and central British Columbia and the two species do not overlap in their geographic ranges. The Collared Pika is distin-

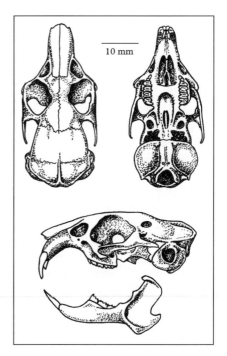

10 mm

guished from the American Pika by its pale dorsal pelage, distinct grey collar and white undersides, but the two species overlap extensively in their skull and dental measurements.

Natural History

The Collared Pika is usually associated with talus slopes and rock slides in mountainous terrain. Ideal habitat seems to be at the edge of rock slides in close proximity to abundant plant material for food. In Alaska, Robert Rausch found a colony living under scattered rocks in a forested valley more than 180 metres from the nearest rock slide. The few records from British Columbia with elevational data range from 700 to 1220 metres above sea level.

Solitary and territorial, the Collared Pika aggressively defends its food piles from other pikas. It frequently emits alarm calls that, presumably, identify its territory to others. Harold Broadbooks estimated the population density in his study area at Denali National Park Alaska to be 6.4 to 7.2 animals per hectare. He observed that the distances between individual Collared Pikas ranged from about 30 to 90 metres. A sedentary mammal, the Collared Pika rarely moves more than 90 metres from its den site. Its home-range size is about 700 m^2, although the area it defends is somewhat smaller. A Collared Pika's territory contains two essential items for survival: hay piles and a den. The den is a nest lined with grass located in an opening in the rocks that is inaccessible to most predators.

Strictly diurnal the Collared Pika spends most of its time feeding and gathering plant material for hay piles. By late summer the construction of hay piles reaches a peak, the animal spending its time gathering plants for its winter food supply. Because they provide the main food source throughout winter, these hay piles are essential for the overwinter survival. The Collared Pika remains active throughout winter under the snow where it is protected from the extreme northern weather conditions.

This mammal is well camouflaged with its greyish coat, which matches the colour of the rocks. It may be difficult to see among the rocks except when it is running, but the distinct "eek" or "ank" alarm call is sure to give away its presence. Other signs of this pika are its distinctive hay piles stored under rocks and its tiny spherical fecal pellets, about a third the size of those made by the Snowshoe Hare.

The diet includes various plants that grow in or near rock slides. Plants found in hay piles of the Collared Pika include the leaves of

willow and birch, various grasses, sedges, mountain avens, Partridgefoot, Crowberry, blueberry, heather plants, Sitka Burnet, Broad-leaved Willowherb, and ferns. Dried fecal pellets of the Hoary Marmot and Ermine have also been recovered from hay piles.

There are no breeding data for British Columbian populations, but in Alaska and the Yukon Territory the breeding season is from May to July. Pregnant animals have been found in late May and June; nursing females have been observed from mid June to mid July. Females bear two to six young after a gestation period of about 30 days. Newborns are blind and naked. Robert Rausch captured a pregnant female that was still nursing on June 13, suggesting that females can produce two litters per year. Collared Pikas do not breed in the summer of their birth.

Potential predators include various raptors and small mammalian carnivores such as the Red Fox, Marten and Ermine. With a slim elongated body, the Ermine is especially adept at capturing Collared Pikas among the rock debris of talus slides.

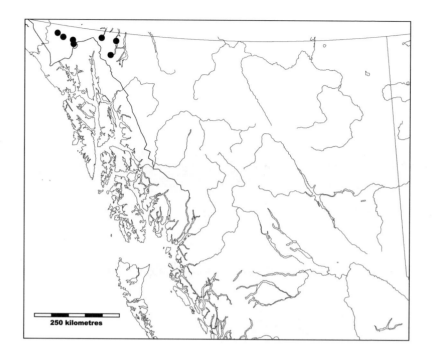

250 kilometres

Range

The Collared Pika inhabits Alaska, the Yukon Territory, the western Northwest Territories and a very restricted area in extreme northwestern British Columbia. There are only 12 substantiated locality records from B.C. Most are from the Haines Triangle area: Carmine Mountain, Shini Creek and Sediments Creek in Tatshenshini Provincial Park; and the Chilkat Pass, Mile 60, Mount Glave, Seltat Peak, Stonehouse Creek and Three Guardsmen Lake, all sites near the Haines Road. Other locality records include Bennett City, the White Mountains and Ben-My-Chree near Tagish Lake. Cowan and Guiguet listed specimen records from Teslin Lake and Atlin Lake in the original *Mammals of British Columbia*, but I could not substantiate their records.

Taxonomy

No subspecies are recognized. Although the Collared Pika was originally described in 1893 as a distinct species, some taxonomists considered the two North American pikas (*Ochotona princeps* and *O. collaris*) to be conspecific; yet others suggested that the North

American pikas and the Old World pika (*O. alpina*) are representatives of a single species. Recent research, such as the study by Marla Weston, has demonstrated that the Collared Pika in northwestern North America is a distinct species.

Conservation Status

Known from only a few isolated populations, the Collared Pika appears to be rare and sparsely distributed in northern British Columbia. Nevertheless, more field inventory is needed in the remote northwestern corner of the province to determine its precise distributional area. This species is easily identified and I encourage naturalists, hikers or hunters to report any observations. Currently, the British Columbian populations are not of conservation concern and the species is not listed by the province.

Remarks

The geographic ranges of the Collared Pika and American Pika in British Columbia are separated by a hiatus of nearly 800 km. More surveys may extend the known ranges of these species, but sufficient field work has been done to conclude that there is a large area in northern B.C. that lacks pikas. Ian McTaggart Cowan suggested that the Peace and Skeena rivers act as lowland barriers separating the two species. Nevertheless, these barriers do not account for the southern limits of the range of the Collared Pika. This species is curiously absent from the northern Rocky Mountains, Cassiar Mountains and northern Coast Mountains, all mountains north of the Peace and Skeena rivers that have potential habitat for pikas. Isolated in northern Alaska and the Yukon Territory during the last glaciation, it is possible that the Collared Pika is still expanding its range southward.

Selected References: Broadbooks 1965, Cowan 1954, MacDonald and Jones 1987, Rausch 1962, Weston 1981.

American Pika   *Ochotona princeps*

Other Common Names: Common Pika, Rocky Mountain Pika.

Description

The nine subspecies of American Pika found in British Columbia demonstrate considerable variation in dorsal fur colour, from pale grey or brown in interior forms to blackish-brown in coastal races. The species lacks a distinct collar; its undersides are whitish with a buff wash.

Measurements:
total length: 178 (154-207)
  n=456
tail vertebrae: 15 (13-18)
  n=192
hind foot: 31 (26-36)
  n=463
ear: 21 (19-25) 65
weight: 162.3 g (142.5-
  190.4) n=18

Dental Formula:
incisors: 2/1
canines: 0/0
premolars: 3/2
molars: 3/3

Identification:
The only similar mammal in B.C. is the Collared Pika (*Ochotana collaris*); see that account (page 73) for identification traits.

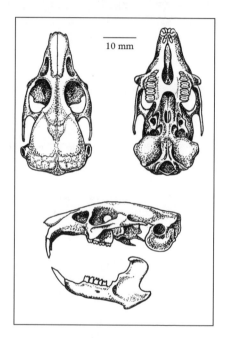

10 mm

Natural History

The American Pika is restricted to talus slopes and piles of broken rock debris in close proximity to meadows or clearings with food plants. It also exploits human-made rocky debris associated with road sides, railroad beds and mine tailings. In British Columbia, the elevational range is from sea level to about 2500 metres. The lower elevational limits of the American Pika appear to be a function of its inability to tolerate higher temperatures. On the coast where the climate is cool and maritime, the American Pika may live at sea level.

The presence of American Pikas at sea level was first reported by the naturalist Hamilton Mack Laing in the 1930s, when he discovered populations at the mouth of Bute Inlet and on both sides of the Dean River near Kimsquit at the northern limits of this species' range. In southern interior valleys where summer temperatures are much warmer than coastal areas, I have not observed this animal below 600 metres elevation.

Population densities have been estimated at 2.2 to 10 animals per hectare; the home-range size range is from about 600 to 3300 m². Males and females have similar home-range sizes, but young-of-the-year maintain smaller home ranges than adults. The American Pika defends a territory about half the size of its home range. Adjacent home ranges are usually occupied by animals of the opposite sex, a pattern that is evidently maintained over many generations with new occupants being the same sex as the replaced occupant. Home ranges of adjacent pairs of the opposite sex overlap to some extent in summer, but individuals of the same sex maintain mutually exclusive home ranges. The American Pika defends its territory by chasing away or fighting intruders. The most aggressive encounters are between animals of the same sex or unfamiliar pikas attempting to establish territories. Other behaviours also play a role in maintaining territories, such as scent marking by rubbing rocks with the cheek glands and the emission of short "ank" calls. These pikas also produce "ank" calls when predators come near.

Pikas are remarkably sedentary animals. Removal experiments have shown that the American Pika can move up to three kilometres between talus patches, but such long-distance movements are rare. In California, Andrew Smith found that for low-elevation populations living in mine tailings, a distance of only 300 metres prevented colonization. After an adult pika has managed to establish a territory it will remain there for life unless it is unable to find a mate. Most young animals remain close to their birth site. After weaning, some young-of-the-year disperse, but most establish home ranges that overlap the home range of the parents. This is a temporary measure and, unless a vacant territory becomes available, they will move beyond their birth site. Their sedentary nature results in a high incidence of inbreeding among pikas and accounts for the low genetic variation demonstrated by most populations.

The American Pika locates its den or nest under or among rocks that are inaccessible to most predators. Although it does not excavate burrows, it will dig shallow depressions under rocks to enlarge den sites, and in winter it excavates tunnels in the snow. The diet includes a wide range of graminoids, forbs, woody shrubs and conifer boughs. This species demonstrates two distinct feeding strategies: grazing directly on vegetation near rock slides, and collecting and storing plants in distinct hay piles for later use in the winter and early spring. Grazing pikas stay within a few metres of the protective cover of talus, but those collecting plants for hay piles travel up to 400 metres from talus areas. The types of plants they harvest varies with these activities. Short grasses and flower heads are heavily grazed, but forbs with flowering stalks, tall grasses or sedges and small shrubs are the predominant plants in hay piles. The plants selected for winter storage in hay piles tend to be larger with a higher nutrient content.

American Pikas graze throughout the year. Even in autumn and early winter, when green plants are no longer available, they will eat lichens growing on rocks and the bark of Engelmann Spruce and Subalpine Fir; these foods may also be available under the snow. In contrast, haying is seasonal, from late June to late September. Within its defended territory, a pika constructs one or two hay piles under rocks or on the surface. Although there has been some controversy as to how important these hay piles are for winter survival, a recent study by Denis Dearing showed that they are of sufficient size to provide the major food source throughout winter. A popular idea is that American Pikas cure or dry the hay on the surface before they store it. Nevertheless, researchers have noted the presence of mouldy vegetation in hay piles.

Breeding data are not available for British Columbian populations. In the Rocky Mountains of Alberta, the breeding season extends from May to August, with mating beginning when there is still snow cover. Females produce two litters, the second conceived only days after the first litter is born. Young of the first litter are born in early June, young of the second in July. The timing of the first litter is closely linked to the first appearance of food plants with spring snowmelt. Because of the high energy demands of nursing their first litter, many female American Pikas are not successful at rearing a second. The gestation period is 30 days; litter size is typically 3 (range: 1 to 5).

Weighing only 10 to 12 grams, newborn American Pikas are undeveloped with closed eyes and sparse hair cover. But they grow rapidly – by nine days their eyes open and by three weeks they can survive on their own. At four weeks of age, the young are independent of their mother and siblings. They reach adult weight by about three months. American Pikas do not breed in the summer of their birth, but reach sexual maturity in the spring after their birth. This species can survive up to seven years in the wild, surprisingly long for a small mammal. Mortality is highest during the first year and after five years of age.

Mammalian predators include the Ermine, Long-tailed Weasel, Marten, Bobcat, Lynx, Red Fox and Coyote. Weasels are probably the most effective predator because they can readily enter the narrow crevices and small openings among rock piles where the American Pika has its den. When young disperse, moving between patches of talus across open meadows or forest in search of a vacant territory, they are especially vulnerable to mammalian predators, such as the Coyote, and raptors, such as the Golden Eagle, Northern Goshawk and Red-tailed Hawk.

Range

The American Pika inhabits the mountainous regions of North America from central British Columbia south to California and New Mexico. Because of its association with talus and rock debris, its range is fragmented into a number of discontinuous populations in the coastal and interior mountain ranges across the southern and central parts of the province. Northern limits of the range are Kismquit in the Coast Mountains, the Itcha and Ilgachuz mountains

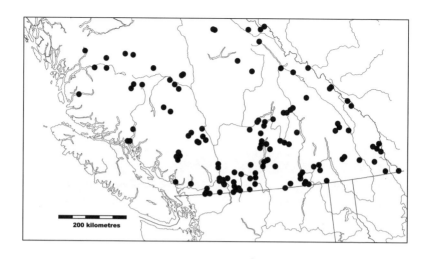

200 kilometres

in the Chilcotin region, Bowron Lakes Provincial Park in the Cariboo Mountains and Mount Robson Provincial Park in the Rocky Mountains. Although there have been a few reports of pikas from the Ootsa Lake and the Bulkley Mountains in central B.C., these observations have not been substantiated. The American Pika is not found on any British Columbian island.

Taxonomy

Thirty-six subspecies have been described; nine occur in British Columbia:

Ochotona princeps brooksi Howell – restricted to B.C. where it is known from the plateau area south of Shuswap Lake.

Ochotona princeps brunnescens Howell – western slopes of the coastal mountain ranges from northern California to B.C. where it extends as far north as Garibaldi Provincial Park.

Ochotona princeps cuppes Bangs – northwestern United States and the Columbia Mountains in B.C.

Ochotona princeps fenisex (Lord) – eastern slopes of the coastal mountain ranges from the Columbia River in Washington to the east slope of the Cascade and Coast mountains in the province from Kimsquit to the United States border.

Ochotona princeps littoralis Cowan – restricted to the province where it inhabits the west slope of the Coast Mountains from Bella Coola to Bute Inlet.

Ochotona princeps lutescens Howell – the Rocky Mountains of Alberta. Ian McTaggart Cowan speculated that it may extend as far west as Kicking Horse Pass and Yoho National Park. The status of this race in B.C.'s Rocky Mountains needs to be verified.

Ochotona princeps princeps (Richardson) – the Rocky Mountains in Montana, Idaho, Alberta and B.C. where it ranges from Mount Robson and the Yellowhead Pass south to the United States border.

Ochotona princeps saturatus Cowan – restricted to B.C. where it inhabits the Cariboo Mountain Range.

Ochotona princeps septentrionalis Cowan and Racey – described from only two specimens from the Itcha Mountains of B.C. In the nearby Ilgachuz Mountains, I have observed pikas that presumably belong to this race.

Conservation Status

The subspecies *O. p. septentrionalis* is on the provincial Red List. Confined to the Itcha and Ilgachuz mountains, this race appears to be isolated by extensive lowland forest from the nearest American Pika populations in the Coast Mountains. Nevertheless, no threats are known and much of its range falls within the boundaries of a protected area – Itcha Ilgachuz Provincial Park. Because it was described from only two specimens, the taxonomic validity of this race needs to be assessed. Genetic samples of *O. p. septentrionalis* analysed by David Hafner and Robert Sullivan showed little divergence from the coastal race *O. p. littoralis.*

Remarks

The northern limits of the geographic range of this species in British Columbia are poorly known and I encourage naturalists to report any observations from central B.C. Some readily accessible areas where you can find the American Pika are taluses and rockslides adjacent to highways, such as the Hope Slide east of Hope on Highway 3, and old abandoned railroad beds converted to trails, such as the Kettle Valley Railway.

Selected References: Cowan 1954, Dearing 1997, Hafner and Sullivan 1995, Smith 1987, Smith and Weston 1990.

Family Leporidae: Hares and Rabbits

Hares and rabbits have long ears and long hind feet. The skull is arched with prominent supraorbital processes; the rostrum is long and wide with lattice-like fenestrations. There are three native and two introduced species in the province.

Snowshoe Hare

Snowshoe Hare   *Lepus americanus*

Other Common Names: Snowshoe Rabbit, Varying Hare, Varying Rabbit.

Description

The Snowshoe Hare is a medium-sized hare with long ears and large hind feet. In summer its dorsal pelage is brown, but the dorsal surface of the hind feet often remains white. The ventral pelage is whitish. The ears are black-tipped, the outer margins edged with white. In winter it typically has white dorsal and ventral pelage with some brown hairs on the head and hind feet; the ear tips are black, contrasting sharply with the white body fur. In spring and autumn, the pelage is a mixture of white and brown. In the lower Fraser River valley, where Snowshoe Hares remain brown throughout winter, their winter pelage resembles the summer pelage. The skull has an indistinct interparietal bone (figure 56) and the posterior extension of the supraorbital processes are usually free of the braincase.

Measurements:
total length: 443 (388-530) n=139
tail vertebrae: 39 (20-61) n=116
hind foot: 135 (117-155) n=133
ear: 72 (64-95) n=49
weight: 1.34 kg (0.89-1.92) n=12

Dental Formula:
incisors: 2/1
canines: 0/0
premolars: 3/2
molars: 3/3

Identification:
Throughout most of the province this is the only species of hare or rabbit. However, where its distribution overlaps with the White-tailed Jackrabbit (*Lepus townsendii*) and Nuttall's

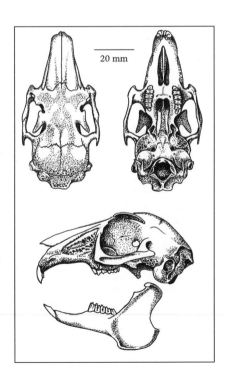

20 mm

Cottontail (*Sylvilagus nuttallii*) in the Okanagan and Similkameen valleys and with the introduced Eastern Cottontail (*S. floridanus*) in the Fraser River valley, identification could be problematic, particularly in summer when the Snowshoe Hare is brown.

The White-tailed Jackrabbit is much larger (ear > 100 mm, basilar length of skull > 69.0 mm, maxillary toothrow length > 16.0 mm) and has a silver-grey winter pelage. The cottontails (*S. floridanus* and *S. nuttallii*) have smaller hind feet (< 115 mm), a rufous or brown nape on the back of their head, brown pelage in winter and a distinct interparietal bone in the skull.

Natural History

The Snowshoe Hare lives in forests and riparian thickets with a dense understorey of shrubs and young trees. Elevational records in British Columbia range from sea level on the coast to 2200 metres in the Cascade and Columbia mountains. It occupies virtually all the province's biogeoclimatic zones – tracks have even been observed in winter on windswept alpine ridges with sparse shrub cover. Generally, the Snowshoe Hare is most abundant in lowland forests and in early successional stages that follow fires or forest harvesting.

Throughout most of its range, including northern B.C., the Snowshoe Hare demonstrates regular population cycles, with peaks occurring every 10 years. During these peaks, population densities may reach 14 to 136 animals per hectare in spring and 22 to 177 animals per hectare in autumn. In southern parts of the range, including southern B.C., where populations appear to be non-cyclic, the densities are much lower, rarely exceeding 20 animals per hectare. The cause of the regular ten-year cycles has long intrigued naturalists and biologists. Long-term studies by Lloyd Keith and colleagues in Alberta and Charles Krebs and colleagues in the Yukon Territory have largely resolved this curious biological phenomenon. The ten-year cycle is associated with predictable changes in reproduction and survival rates. The underlying cause of these changes is largely mortality from predators and food shortages. In southern parts of the range, where the Snowshoe Hare occurs in low population densities, it does not undergo population cycles. This may result from suitable habitats being discontinuous and a greater diversity of predators. In these regions, high predation prevents Snowshoe Hares from dispersing to marginal habitats and building up in population.

Home-range size varies with season and the population cycle, but it rarely exceeds 10 ha. Most estimates of home-range size are from 1.6 to 10.2 ha. Individual home ranges of both sexes overlap extensively, with little evidence of territories. Snowshoe Hares can travel up to 10 km when dispersing. Thomas O' Farrell recovered a tagged animal about 20 km outside his study area in Alaska. Most active at night, this species depends on concealment to escape predators. Its usual tactic when threatened is to freeze and remain motionless. In summer, the Snowshoe Hare spends the day concealed in resting sites in dense thickets or grass cover. In winter, it hides during the day under snow-covered branches of trees and shrubs or in shallow depressions made in the snow. This hare will also use abandoned Woodchuck burrows and hollow logs for shelter. Although it does not excavate burrows in soil, it will dig holes in the snow. Other than a shallow depression in the ground, females do not make a nest for their newborn young. The Snowshoe Hare travels on familiar runways, which appear as a network of worn paths in its habitat. In winter, it packs down the snow in its conspicuous trails, forming compacted runways that provide vital escape routes from predators.

Summer diet consists of grasses, sedges and forbs. In winter the Snowshoe Hare browses on the stems and branches of woody plants, especially the small terminal twigs less than 1.5 cm in diameter that are more nutritious and easily digested than larger twigs. If food is scarce, this hare will clip off larger twigs or girdle the bark of large stems; it can reach stems up to 60 cm above the ground or snow level. A Snowshoe Hare requires about 300 grams of woody browse daily to support its nutritional and energy requirements. Its favourite winter foods are deciduous trees and shrubs, such as Trembling Aspen, Paper Birch, willows, alders, maples, blueberry and Labrador tea, but it will also eat conifers such as spruce, larch and pine. Snowshoe Hares also dig craters (15 to 30 cm deep) in the snow to reach dried plants such as hedysarum and mountain avens. Breeding in a Snowshoe Hare population is synchronized – the females in a local area will come into estrus and conceive their first litter at the same time. The onset of breeding varies geographically and even from year-to-year within a population. Males become sexually active in March or April. In western North America females conceive their first litter from mid March to early May. Females can produce two to four litters in a year. The number of litters produced by a female depends on the date of first breeding. The gestation

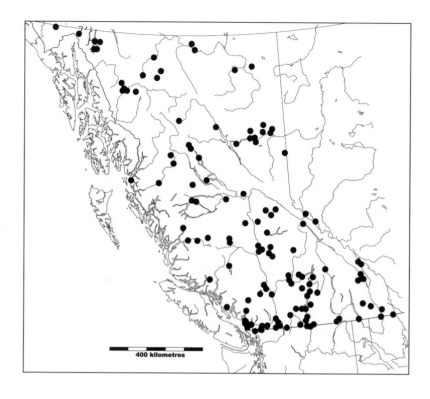

400 kilometres

period is about 36 days. Litter sizes vary greatly but average two to five young.

Newborn Snowshoe Hares are well developed with a body covering of fur, open eyes and weighing 40 to 96 grams. They remain together for only a few days, then scatter from their birth site to occupy separate individual hiding places. They assemble for about 10 minutes each evening to be nursed by their mother, then disperse to their hiding places. Females appear to nurse their young only once a day; except for nursing, females play no role in raising their young. Most young Snowshoe Hares are weaned by 25 to 28 days, but those in the last litter of the breeding season may be nursed up to 56 days. Although most Snowshoe Hares do not breed until the spring following their birth, a few females from the first litter evidently breed in the summer of their birth. Mortality mostly from predation is high, especially among young animals. During a population decline as few as 6 to 11% of the young will survive to their first winter.

The Snowshoe Hare is one of the most important prey species in the northern boreal forest – many mammalian and avian predators rely on it as a food source. Few predators are more dependent on the Snowshoe Hare than the Lynx, whose year-to-year population fluctuations closely track those of the hare. Other mammalian predators include the Bobcat, Coyote, Ermine, Grey Wolf, Marten, Fisher and Red Fox. The Golden Eagle, Great Horned Owl and Northern Goshawk are major avian predators.

Range
The Snowshoe Hare is found throughout the boreal regions of North America. It ranges across the entire mainland of the province but is absent from all coastal islands. North of Burrard Inlet, Snowshoe Hares appear to be rare and spottily distributed on the coastal mainland.

Taxonomy
Fifteen subspecies are recognized, with seven occurring in British Columbia:

Lepus americanus bairdii Hayden – the Rocky Mountains of the United States, extreme southwestern Alberta and, according to Cowan and Guiguet, the southern Kootenay region (Elko, Newgate) of extreme southeastern B.C.

Lepus americanus cascadensis Nelson – the coastal mountain ranges of Washington and B.C. as far north as Jervis Inlet.

Lepus americanus columbiensis Rhoads – the Rocky Mountains of Alberta, north-central Washington and southern B.C., where it occurs in the Okanagan Highlands and the Columbia and Rocky mountains.

Lepus americanus dalli Merriam – Alaska, the Yukon Territory, the Northwest Territories and northern B.C. Although Cowan and Guiguet in the original *Mammals of British Columbia* classified populations in northern B.C. as *L. a. macfarlani*, this race is now considered to be inseparable from *L. a. dalli*.

Lepus americanus pallidus Cowan – restricted to the province where it inhabits the Interior Plateau and the coastal region from Jervis Inlet to Prince Rupert.

Lepus americanus pineus Dalquest – northeastern Washington, northwestern Idaho and possibly the Kootenay River valley at Creston and the Columbia River valley at Trail. The status of this race in B.C. is not clear.

Lepus americanus washingtoni Baird – a small coastal form that remains brown in winter found west of the Cascade Mountains in Washington and Oregon and the lower Fraser River valley in B.C. Cowan and Guiguet in the original *Mammals of British Columbia* considered this subspecies restricted to the south side of the Fraser River. But, there are records from the north side of the Fraser River south of Burrard Inlet. In my 1983 study, I found historical (1895–1947) museum specimens in brown winter pelage collected from Lulu Island, Hastings and Point Grey. Hares in the University of British Columbia Research Forest at Haney appear to remain brown in winter, and a recent road kill found at Mission was in brown winter pelage. These new findings suggest that Snowshoe Hares in the foothills of the Coast Mountains on the north side of the Fraser River are *L. a. washingtoni*.

Conservation Status

The coastal subspecies *L. a. washingtoni* is on the provincial Red List. Historical museum specimens suggest that it was once widespread throughout the Lower Mainland in areas now heavily disturbed by urbanization. On the north side of the Fraser River, Clark Streator of the United States Biological Survey, collected a specimen in Hastings in 1895 and there are museum specimens collected in the 1940s from Point Grey near the University of British Columbia. On the south side of the Fraser River, a series of museum specimens were taken in 1947 by Ken Racey from Huntingdon near Abbotsford. Only two confirmed records exist for the Lower Mainland area in the past 30 years – road kills at Burnaby Lake in 1970 and at Mission in 1997. Populations presumably survive in the foothills of the Coast Mountains and the lowlands of the Chilliwack River valley. Loss of forest from agriculture and urban development as well as the appearance of the introduced Eastern Cottontail in the 1950s contributed to the demise of the Snowshoe Hare in the lower Fraser River valley. A thorough field survey throughout the Lower Mainland and a taxonomic study of Snowshoe Hares from the north side of the Fraser River and foothills of the Cascade Mountains is needed to resolve the conservation status of *L. a. washingtoni* in British Columbia.

Remarks

The Snowshoe Hare demonstrates three remarkable adaptations for surviving in the harsh winter environment of Canada's northern boreal forests: its large, heavily furred hind feet enable it to move on top of deep snow; the long guard hairs of the winter pelage effectively insulate the animal from the cold; and the white winter pelage provides camouflage against a snowy background, essential for a mammal that depends on concealment to escape predators. The value of fur colour for camouflage is further demonstrated by the brown winter pelage of hares living in lowland areas of the Pacific coast, where rain is more common than snow in winter.

In northern regions of the province, the Snowshoe Hare was an important food source for aboriginal people, especially in winters when large mammals were scarce. The pelts were cut into long strips and woven into cloaks and blankets or used as trim in fur clothing.

Selected References: Keith 1990, Krebs et al. 1995, Nagorsen 1983, O' Farrell 1965, Pease et al. 1979, Wolff 1978.

White-tailed Jackrabbit *Lepus townsendii*

Other Common Names: Townsend's Jackrabbit.

Description

Our largest hare, the White-tailed Jackrabbit has long hind legs and very long ears. The summer pelage consists of pale brown dorsal fur mixed with light grey and white undersides. The ears are tipped with dark brown or black and the tail is white. The few available museum specimens in winter pelage from British Columbia have a silvery-white dorsal pelage that is washed with brown rather than pure white; their undersides are brown. The skull is large with an indistinct interparietal bone, a broad rostrum and supraoccipital processes that usually touch the braincase.

Measurements:

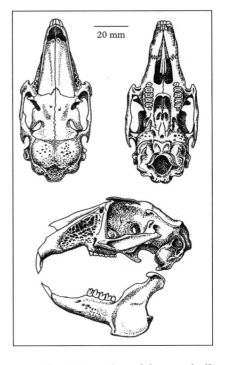

20 mm

total length: 564 (510-602) n=11

tail vertebrae: 90 (75-113) n=11

hind foot: 150 (137-171) n=11

ear: 120 (109-137) n=3

weight: none available for British Columbia

Dental Formula:

incisors: 2/1

canines: 0/0

premolars: 3/2

molars: 3/3

Identification:

Where the White-tailed Jackrabbit co-occurs with Nuttall's Cottontail (*Silvilagus nuttallii*) and the Snowshoe Hare (*Lepus americanus*) in southern British Columbia, it can be readily identified by its larger size, especially the longer ears (> 100 mm) and larger skull (basilar length > 69.0 mm, maxillary toothrow length > 16.0 mm). The winter pelage of the White-tailed Jackrabbit is dirty silver-grey, opposed to the nearly pure white winter fur of the Snowshoe Hare.

Natural History

Information on the biology of this species is largely based on studies done in the western United States. The only data available for the British Columbian population comes from historical museum specimens and observations from the late 1920s recorded in the unpublished field notes of Hamilton Mack Laing, a noted naturalist who collected mammals in the south Okanagan region. The few historical records from British Columbia suggest that the White-tailed Jackrabbit was associated with low-elevation grasslands (below 600 metres) with Bluebunch Wheatgrass, Big Sage and Antelope-Bush. Populations may reach 44 animals per square kilometre in prairie habitats of the Great Plains, but densities are much lower in Cordilleran populations. For example, Gordon Rogowitz and Michael Wolfe reported densities of about 7 animals per square kilometre in Wyoming. There are no data on the historical populations in B.C., but Hamilton Mack Laing described this hare as abundant on the benches above Keremeos in 1928, and uncommon but widespread in the Osoyoos Lake region in 1929.

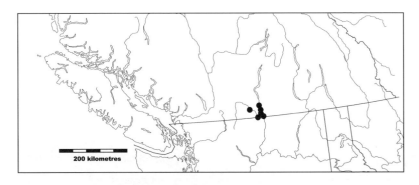

This is a highly mobile species that can cover great distances. Gordon Rogowitz tracked animals with radio transmitters and found that most White-tailed Jackrabbits moved 0.4 to 1.0 km during their daily activity periods. One female moved 2.2 km in an hour. Running White-tailed Jackrabbits have been clocked at speeds of about 55 km per hour. Strictly nocturnal, White-tailed Jackrabbits are active from about an hour after sunset until sunrise. They spend the daylight hours resting in shallow depressions at the base of shrubs or rocks. In the Osoyoos Lake area, Hamilton Mack Laing found that resting sites were always situated at the base of Big Sage shrubs. Although the White-tailed Jackrabbit generally does not excavate burrows, it will construct short snow tunnels for shelter in winter. It's diet consists of shrubs, forbs and grasses, primarily woody shrubs in winter and forbs in summer.

In the western United States, the breeding season begins in late February or early March. In some regions, females may produce as many as four litters per year, but northern populations probably have fewer litters. The gestation period is about 42 days; females typically have five to seven young. The young are born well developed with open eyes and covered in fur. They become independent by about two months. Breeding data for the B.C. population are scanty. Two females collected on May 23 and 24 were pregnant, each with six large fetuses. A female taken August 4 was nursing. In the Okanagan, Hamilton Mack Laing observed that by late May young from the first litter were well-developed and many females were carrying well developed fetuses of their second litter; by early August females were nursing their second litters. Although there are reports of White-tailed Jackrabbits living up to eight years, most do not survive a year in the wild and few survive beyond two years. Predators include the Coyote, Bobcat, Cougar and large raptors.

Range

The White-tailed Jackrabbit is found throughout the Great Plains and dry intermontane regions of western Canada and the United States. In British Columbia, where it is at the northern periphery of its range, it inhabits the southern Okanagan and Similkameen valleys. There are historical museum specimens and sightings from Osoyoos Lake, Testalinden Creek near Oliver, White Lake, Chopaka and Keremeos.

Taxonomy

Two subspecies are recognized; one occurs in British Columbia:

Lepus townsendii townsendii Bachman – west of the continental divide in the United States; in Canada this race is restricted to B.C.

Conservation Status

The White-tailed Jackrabbit appears on the provincial Red List. But with no evidence of a breeding population, the species probably should be considered extirpated from British Columbia. The few sporadic recent sightings may be animals from Washington. The population of White-tailed Jackrabbits in B.C. suffered from the effects of livestock grazing and the loss of shrub-steppe habitat to orchards, vineyards and urban growth. Because of its tendency to eat young fruit trees, this hare was regarded as a pest. Hamilton Mack Laing reported that it was shot on sight by fruit farmers in the Okanagan Valley in the late 1920s. Except for the extensive grassland in the Richter Pass area that borders Washington, the remaining natural grassland habitats in the Okanagan and Similkameen valleys are probably too small and fragmented to support viable populations of this hare.

Remarks

The last museum specimen from British Columbia was collected in 1957 at Osoyoos. The most recent confirmed sighting was at Chopaka in 1981, although there have been several more recent unconfirmed sightings from the southern Okanagan Valley.

Selected References: Lim 1987, MacCracken and Hansen 1984, Rogowitz 1997, Rogowitz and Wolfe 1991.

European Rabbit　　*Oryctolagus cuniculus*

Other Common Names: Domestic Rabbit, Old World Rabbit
(and many others for the different domestic breeds).

Description

The European Rabbit is a medium to large rabbit with variable fur
colour and markings depending on the breed. In British Columbia,
wild populations generally contain a mix of melanic, tan, piebald
and brown animals. The brown wild form has a prominent, rufous
nape at the base of the head. The skull has a distinct interparietal
bone (figure 57) and the posterior extension of the supraorbital
processes is free of the braincase.

Measurements:
　total length: 569 (500-620) n=4
　tail vertebrae: 70 (55-85) n=4
　hind foot: 112 (102-121) n=5
　ear: 116 (84-152) n=5
　weight: 3.26 kg (2.9-3.95) n=4

Dental Formula:
　incisors: 2/1
　canines: 0/0
　premolars: 3/2
　molars: 3/3

Identification:

Most domestic breeds are
easily distinguished from our
wild rabbits or hares by either
their distinctive pelage colour
and markings or by their
unusual body proportions
such as elongated ears.
Nevertheless, the brown form
of the European Rabbit could
be confused with the Snowshoe Hare (*Lepus americanus*)
or Eastern Cottontail (*Silvilagus floridanus*). The European Rabbit is larger (total
length > 490 mm), more
heavily built, and has longer

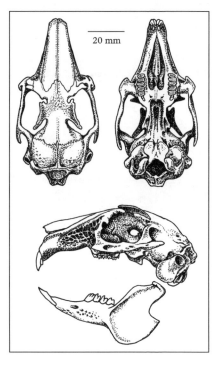

20 mm

ears (> 80 mm) than a cottontail. The brownish nape and heavier build discriminates the European Rabbit from a Snowshoe Hare in summer pelage. The skull of the European Rabbit differs from that of the Snowshoe Hare by its distinct interparietal bone. The Eastern Cottontail skull also has a distinct interparietal bone, but the tip of the posterior extension of the supraorbital processes are fused to the braincase.

Natural History

No research has been done on wild populations of this species in the province. Therefore, most of the natural history data given in this account are based on studies done on European populations or the research by Stevens and Weisbrod on a population on San Juan Island, Washington, an island with habitat and climate similar to that of southeastern Vancouver Island. In British Columbia, feral populations occur on Vancouver Island and Triangle Island. On Vancouver Island, it lives in meadows, fields, agricultural land, shrub thickets and open forest. Two essential habitat features appear to be suitable soil for excavating burrows and shrub thickets or rock cover that provide concealment from predators when the rabbit is feeding. In contrast, the habitat on Triangle Island is considerably harsher. This small, isolated island (1.07 km², 46 km from Vancouver Island) is windswept and treeless, supporting a dense vegetation cover of various grasses and sedges, Salmonberry, Salal, Pacific Crab Apple, Tufted Hairgrass, and Lady Fern. European Rabbits burrow in areas with well developed soil among nesting seabird colonies.

On San Juan Island, populations may exceed 200 animals per hectare when they reach the annual peak in July; by late winter the population declines to 3 to 45 animals per hectare. Evidently the San Juan population is stable from year to year. A relatively sedentary mammal, the European Rabbit has a small home range (0.3 to 3.0 ha), although animals may move up to four kilometres in response to habitat disturbances.

Unlike our other hares and rabbits, which are solitary, the European Rabbit is highly social and strongly territorial. It digs complex burrow systems or warrens that have a number of entrances with interconnecting tunnels that can be as deep as 3 metres and extend 45 metres underground. A warren system is usually occupied by a unique social group consisting of two to eight animals. A well-defined hierarchy, established by fighting bouts, exists within the group, with a dominant male and dominant female doing much of the breeding. The dominant animals have access to abundant food and the best burrows where they are well protected from predators. Subordinate females are often forced into peripheral burrows where they are more vulnerable to predators. Some European Rabbits are solitary, living on the edge of the warren systems. Territorial boundaries among groups and the social status of an individual rabbit are marked with complex odours or pheromones. Urine, for example, is an important marker that identifies the sexual and group status of an individual. European Rabbits have well developed anal glands that produce a pheromone that is scattered around the territory in the fecal pellets. Other glands located under the chin produce a secretion used to mark territory by rubbing on various objects in the territory and at the burrow entrance.

Grasses and various forbs are the predominant food, but the European Rabbit will browse the small branches of shrubs and girdle bark from trees and shrubs. European Rabbits at the hospital grounds in Victoria eat the succulent parts of Himalayan Blackberry, Common Snowberry and wild rose.

On San Juan Island, the breeding season begins in early March, and females produce three litters; the first litter emerges from the burrows in early May. Breeding may be earlier on southern Vancouver Island, because I have observed small rabbits (about one-third adult size) above ground as early as mid February. Litter size in the San Juan Island population is large averaging 7.3 young (range: 5 to 9). The gestation period for the European Rabbit is 28 to 33 days. The newborn young are naked, blind and weigh 40 to 50 grams. They appear at the burrow entrances at about 18 days of age; by 21 to 25 days they leave their nest. Young-of-the-year probably do not breed until the following spring. The European Rabbit may survive up to nine years. Nevertheless, survival rates are low for this species with as few as five per cent of the young surviving beyond their first year in some populations. On San Juan Island, raptors are the major predators – 20 species of hawks and owls have been observed in the vicinity of the rabbit colonies. In the urban and agricultural areas of southern Vancouver Island, Domestic Cats are a major predator of young European Rabbits.

Range

The original range of this rabbit was probably southern France, the Iberian Peninsula of Spain and north Africa. In historical times, feral populations were introduced to all the continents and many oceanic islands. In British Columbia, there have been many introductions, either from deliberate releases of pets or escapes from rabbit breeders. Feral populations currently exist on southern Vancouver Island in the Victoria region, and on Triangle Island, the outermost of the Scott Islands off the northern coast of Vancouver Island.

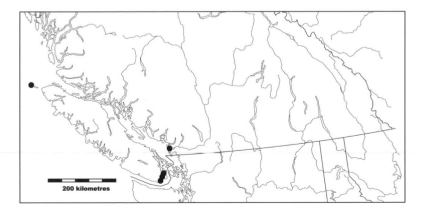

200 kilometres

Taxonomy
No subspecies are recognized. Feral populations on Vancouver Island are a complex mix of various domesticated breeds. The European Rabbits inhabiting Triangle Island are all the wild brown colour but the source of this population is unknown.

Conservation Status
An introduced animal, this species is not of conservation concern except for its potential impact on native species. Triangle Island is one of the province's most important sites for breeding seabirds, supporting about 1.25 million colonial seabirds. The impact of the European Rabbit on these colonies has not been studied.

Remarks
The European Rabbit was presumably brought to British Columbia as a domestic animal by the first European settlers. Releases or escapes over the past century have resulted in a number of feral populations, but few persisted. According to Clifford Carl and Charles Guiguet who reviewed some of the history of these introductions, the European Rabbit was released on various islands off the southeast coast of Vancouver Island: Bare, Chatham, James, Strongtide, Piers and Sidney. None survive on these islands today. Similarly, several releases on the Queen Charlotte Islands did not persist, although a few European Rabbits are found on northern Graham Island. The Triangle Island population, which has persisted for nearly a century, originated from European Rabbits brought by lighthouse keepers in the early 1900s (though their source remains unknown). Over the years, many escapes and releases occurred on Vancouver Island, but wild populations exist today only around Victoria and outlying agricultural areas. The best-known populations inhabit the grounds of the Victoria General Hospital and the University of Victoria. They appear to be maintained by the ongoing release of unwanted pets.

Domesticated in AD 600 to 1000, the European Rabbit now comprises 66 recognized breeds. Some of the exotic breeds show little resemblance to the original wild form. White or albino rabbits are used as laboratory animals.

Selected References: Carl and Guiguet 1972, Stevens and Weisbrod 1981.

Eastern Cottontail   *Sylvilagus floridanus*

Other Common Names: None.

Description

The Eastern Cottontail is a slim, medium-sized rabbit with a brown or tan dorsal pelage and a conspicuous rufous nape on the back of the head. The ventral pelage is white with a grey base. The ears are black-tipped with some white on their outer edges. The tail is brown or grey on top and completely white underneath. The skull is characterized by a distinct interparietal bone (figure 57) and supraoccipital processes that touch the braincase.

Measurements:
total length: 423 (370-449) n=31
tail vertebrae: 51 (35-70) n=31
hind foot: 99 (90-110) n=31
ear: 68 (61-76) n=14
weight: 1.33 kg (1.11-1.72) n=32

Dental Formula:
incisors: 2/1
canines: 0/0
premolars: 3/2
molars: 3/3

Identification:
In the lower Fraser River valley, the Eastern Cottontail superficially resembles the coastal subspecies of the Snowshoe Hare (*Lepus americanus washingtoni*), a race that remains brown in winter. On some parts of Vancouver Island the Eastern Cottontail could be confused with brown forms of the European Rabbit (*Oryctolagus cuniculus*). See the accounts for those species (pages 85 and 96) for diagnostic traits.

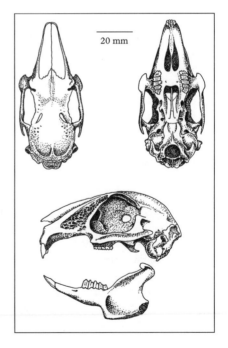

20 mm

Natural History

In its native range, the Eastern Cottontail lives in farmland, open woods and fence rows bordering cultivated fields. In the lower Fraser River valley of British Columbia, it is associated with agricultural land, suburbs and urban green belts. It occupies similar habitats on Vancouver Island, where it can be found even in the urban gardens of Victoria; it also inhabits coastal forests on Vancouver Island. The Eastern Cottontail seems to be most common in edge habitats that border highways, roads, fields and railways. Dense shrub thickets of wild rose, Scotch Broom, Gorse and Himalayan Blackberry provide cover from predators. Confined to elevations below 500 metres, the Eastern Cottontail is absent from the higher mountains of Vancouver Island.

Populations reach a peak in late summer or early autumn with recruitment of the young. No data exist on population densities in B.C., but in the mid-western United States autumn densities can reach 10 animals per hectare. Male Eastern Cottontails have larger home ranges than females, and they tend to increase their home-range size during the peak of the breeding season. Females are more sedentary, especially during the breeding season when they remain close to the nest with their young. During the breeding season, male home ranges overlap considerably, but females maintain separate areas. Although there is little evidence that this rabbit is strongly territorial, males have a well-developed dominance hierarchy that minimizes fighting.

Eastern Cottontails are most active at dawn and dusk. They spend most of their active period foraging. During the day, this rabbit rests in dense thickets or under woody brush piles. Although females construct elaborate nests for their young consisting of a shallow depression lined with grass, leaves and hair, they provide little maternal care. The female only associates with her young when nursing. Although Eastern Cottontails do not excavate burrows, in eastern North America, especially in winter, they use the abandoned burrows made by Woodchucks. Because there are no comparable burrowing mammals in the Lower Mainland or on Vancouver Island, burrows are not available to B.C. populations. When threatened, this species demonstrates two escape behaviours: it flushes from its resting site with rapid ziz-zag movements running to the nearest cover, a strategy that presumably startles or confuses a predator; or it slinks away keeping its body close to the ground. The Eastern Cottontail eats forbs, grasses and woody browse of shrubs

and trees. Woody browse is the main food in winter, and herbaceous plants in summer.

The Eastern Cottontail is prolific with a lengthy breeding season. Females can produce as many as eight litters in a single breeding season. The most comprehensive breeding data for the American Pacific Northwest comes from a study done on a population in western Oregon by D.E. Trethewey and B.J. Verts. They found that the males were in breeding condition from January to August. Females bred from January or February to September producing five to eight litters per year. In the B.C. populations, being farther north, the breeding season would likely begin later and be shorter in duration, but this needs to be verified from field studies. The gestation period is about 26 to 29 days. Litter size is highly variable – in Oregon it ranged from 3 to 8 with an average of 5.6. Both males and females are capable of breeding in the summer of their birth. At birth, Eastern Cottontails weigh about 25 to 30 grams. They are pink, essentially naked and have closed eyes. Within 24 hours they develop a sparse covering of light hairs. By 6 or 7 days the eyes open; at about 14 days they leave the nest. The potential life span is about 10 years but the survival rates are low with few individuals surviving 4 years in the wild. The many predators include: the Great Horned Owl, Golden Eagle, Bald Eagle, Northern Goshawk, Bobcat, Grey Wolf, Cougar, Coyote, Marten and Ermine. Domestic Cats in urban areas also take a toll on the Eastern Cottontail.

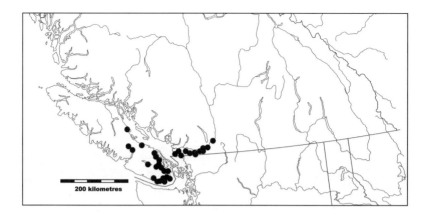

Range
Widely distributed throughout eastern and central North America, Middle America and northern South America, the Eastern Cottontail was introduced to northwestern North America. Two introduced populations occur in British Columbia: a lower Fraser River valley population that is derived from introductions to western Washington and a population introduced to Vancouver Island. The Fraser-valley population occurs south of the river where it ranges as far east as Hope. The Vancouver Island population has now spread from the Victoria region as far north as Courtenay and west to Port Renfrew and the Port Alberni Valley.

Taxonomy
Victor Diersing concluded that the British Columbian populations are derived from two different eastern subspecies. He found that Vancouver Island specimens are similar to *Sylvilagus floridanus mearnsi* from southern Ontario; specimens from the lower Fraser River population resemble *S. f. alacer* from Kansas and Missouri. Because several subspecies were introduced into western Washington, the source of the British Columbian mainland population, it is possible that the lower Fraser River population is derived from several subspecies. Whatever their taxonomic affinities, Eastern Cottontails inhabiting the lower Fraser River valley are noticeably paler and smaller than the Vancouver Island population.

Sylvilagus floridanus alacer (Bangs) – Texas, Oklahoma, Missouri, Arkansas and Louisiana. Individuals from this race were introduced into western Washington and subsequently spread into the lower Fraser River valley of southwestern B.C.

Sylvilagus floridanus mearnsi (J.A. Allen) – southern Quebec, Ontario, and the eastern and north-central United States. Introduced to Vancouver Island.

Conservation Status

An introduced species, the British Columbian populations are not of conservation concern except for their potential impact on native species.

Remarks

Unfortunately, the spread of this introduced rabbit in British Columbia has not been well documented. Eastern Cottontails first appeared in the lower Fraser River valley around Huntington near the Washington border about 1952. It has now spread throughout much of the Lower Mainland south of the Fraser River. The Vancouver Island population first appeared about 1964 after the release of animals in Metchosin, near Victoria. Eastern Cottontails now occupy much of southeastern Vancouver Island. Given the slow rate of spread of the introduced populations in Oregon, this species is probably still expanding its range on Vancouver Island. It only appeared in the Courtenay and Port Alberni areas within the past few years. With no native hares or rabbits on Vancouver Island, the Eastern Cottontail seems to be filling an empty ecological niche. Remarkably no research has been done in B.C. on this alien species' general life history or its ecological impact on our native flora and fauna.

Selected References: Carl and Guiguet 1972, Chapman et. al 1980, Diersing 1978, Racey 1953, Trethewey and Verts 1971.

Nuttall's Cottontail   *Sylvilagus nuttallii*

Other Common Names: Mountain Cottontail.

Description

Nuttall's Cottontail is our smallest rabbit. Its dorsal pelage is pale brown with grey on the sides and rump. The nape is pale brown; the ears are tipped with black. The tail is white on the dorsal surface and grey underneath. The ventral pelage is white. The skull is small with a distinct interparietal bone (figure 57) and supraoccipital processes that touch the braincase.

Measurements:
total length: 319 (263-363) n=9
tail vertebrae: 33 (24-44) n=9
hind foot: 85 (75-90) n=9
ear: 57 (54-58) n=4
weight: 495.4 g (341.9-777.5) n=4

Dental Formula:
incisors: 2/1
canines: 0/0
premolars: 3/2
molars: 3/3

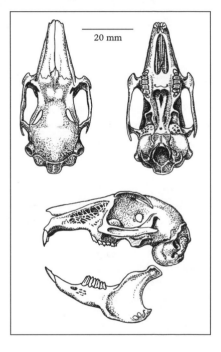

20 mm

Identification:
In British Columbia, the distributions of the Eastern Cottontail (*Silvilagus floridanus*) and Nuttall's Cottontail (*S. nuttallii*) do not contact each other. Nevertheless, Nuttall's Cottontail can be discriminated by its smaller size (weight < 700 g, ear length < 60 mm), duller fur colour and smaller skull (basilar length < 57 mm, maxillary toothrow length < 14 mm). A Snowshoe Hare (*Lepus americanus*) in summer pelage could be confused with a Nuttall's Cottontail, but the hare lacks the brown nape, is much larger (hind foot > 115 mm, ear > 60 mm); its skull is also larger (basilar length > 57 mm, maxillary toothrow length > 14 mm), and it has an indistinct interparietal bone and supraoccipital processes that are free of the braincase.

Natural History

Nuttall's Cottontail is confined to Big Sagebrush and Antelope-Bush habitats in British Columbia. Two essential features of its habitat are rock crevices and a shrub cover of 30% or more. It rarely inhabits cultivated fields or orchards. In B.C., this species ranges from 320 to 1200 metres above sea level, but is generally confined to the valley bottoms and low benches with most occurrences below 800 metres. Thomas Sullivan estimated population densities of 0.2 to 0.4 animals per hectare for his study site at Summerland. With the recruitment of young, populations peak in August then decline throughout autumn and winter. Year-to-year fluctuations in population numbers have been attributed to variations in the summer drought. Dry summers where there is little succulent food available, may produce a shorter breeding season and poor survival of young animals. Home-range size and movements have not been documented for this species.

Nuttall's Cottontail is solitary, with most interactions between individuals confined to the breeding season. It has two distinct daily activity peaks: one in the second hour after sunset and the other about an hour before sunrise. Nuttall's Cottontail spends much of its active period feeding. In Oregon, this rabbit has been observed climbing into the lower branches of juniper shrubs in early morning hours. Because this climbing behaviour occurs only in July and

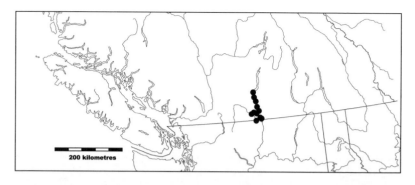

August during the summer drought, it is suspected that Nuttall's Cottontail is pursuing water droplets condensed on the boughs and succulent foliage of this shrub. During the day, Nuttall's Cottontail rests at the base of shrubs or in the crevices of rocky outcrops. A nest with young found under a pine tree in California was a cup-like structure lined with fur and grass and covered with grass, fur and small sticks. Nuttall's Cottontail eats grasses (such as Bluebunch Wheatgrass, Needle-and-thread Grass, Cheatgrass), forbs and shrubs (particularly Big Sage and Common Juniper). Grasses are generally most important in summer, and woody browse from shrubs is the main winter food.

The most detailed studies on reproduction in Nuttall's Cottontail were done in Oregon by B.J. Verts and colleagues. They found that the breeding season extends from January to the end of July. The gestation period is about 26 to 28 days. Embryo counts range from 1 to 6 with an average of 4.6. Females generally produce four litters per year, although a few animals may have five. The first litter is born in mid March. Males do not breed during the breeding season of their birth, but a few females may breed in the summer of their birth. The limited breeding data available for the B.C. population suggest that the breeding season begins in March and ends in July. In the original *Mammals of British Columbia*, Cowan and Guiguet reported that females have three litters per year with the first litter born in April. This appears to be the only breeding information available for the B.C. population. Although there are no descriptions of newborns, they are presumably small, blind and covered with fine fur, similar to other cottontail species. Population turnover is high and few animals survive over winter. Potential predators include the Coyote, Bobcat, Cougar and various raptors, particularly the Great Horned Owl.

Range

Nuttall's Cottontail is associated with the grasslands and intermontane regions of the southwestern United States, southern Alberta and Saskatchewan and south-central British Columbia. In B.C., the species is confined to the southern Okanagan and Similkameen valleys where it ranges as far north as Keremeos, Summerland and the south end of Okanagan Mountain Provincial Park.

Taxonomy

Is his unpublished taxonomic revision, Victor Diersing recognized two subspecies; one occurs in British Columbia:

Sylvilagus nuttallii nuttallii (Bachman) – across the Cordillera of the western United States (Washington, Idaho, Montana, Oregon, California, Nevada, Arizona, Utah) and southern B.C.

Conservation Status

Appearing on the province's Blue List, Nuttall's Cottontail is protected from hunting. Nationally, COSEWIC designates the British Columbian subspecies (*S. n. nuttallii*) as Special Concern. Although its populations appear to be stable, it is considered to be sensitive because it has a small distributional area and sagebrush-steppe habitat has declined with agricultural and urban development.

Remarks

The first record of Nuttall's Cottontail in British Columbia was in 1939 when James Hatter collected a specimen on Anarchist Mountain near Osoyoos. In the early 1900s, naturalists and museum collectors intensively surveyed the southern Okanagan Valley, and they likely would have detected Nuttall's Cottontail had it lived there. Its appearance in 1939, therefore, represents range expansion, presumably from Washington. Since then, Nuttall's Cottontail has slowly spread throughout the low-elevation shrub-steppe grasslands of the southern Okanagan and Similkameen valleys. By 1951, it was described as common throughout the benches on the eastern side of Osoyoos Lake. With little suitable habitat available in the northern Similkameen and Okanagan valleys, this rabbit now occupies most of its potential range in the province.

Selected References: Carter et al. 1993, Chapman 1975, Cowan and Hatter 1940, Diersing 1978, Guiguet 1952, Powers and Verts 1971, Sullivan et al. 1989.

Order Rodentia

Rodents show a variety of forms, with the anatomy of some species highly modified for specializations associated with locomotion. The skull is characterized by a single pair of upper incisors, no canines, the incisors separated from cheek teeth by a space, and the rostrum lacking fenestrations. All British Columbian rodents have rooted cheek teeth, except for some species of voles, the Mountain Beaver (*Aplodontia rufa*) and Northern Pocket Gopher (*Thomomys talpoides*), which have ever-growing cheek teeth.

Family Aplodontiidae: Mountain Beavers

This family has only one living species, the Mountain Beaver (*Aplodontia rufa*). It is considered the most primitive living rodent.

Mountain Beaver

Mountain Beaver   *Aplodontia rufa*

Other Common Names: Boomer, Sewellel.

Description

The Mountain Beaver is a thickset rodent with short limbs, an indistinct stubby tail that is furred, small eyes and ears, long stiff vibrissae and long claws on the front and hind feet. The coarse dorsal pelage consists of scattered long black-tipped guard hairs and shorter brown hairs. Some individuals have a white spot below the ear. The ventral pelage is paler, ranging from light tan to grey. Many specimens from British Columbia have prominent white patches on their chest, abdomen or flanks. The skull is flat and triangular in outline, with a flask-shaped auditory bullae. There are no postorbital processes. Adults have molars with distinct projections on their cheek side.

Measurements:
  total length: 349 (260-470) n=88
  tail vertebrae: 27 (17-47) n=72
  hind foot: 57 (48-68) n=77
  ear: 20 (16-22) n=5
  weight: 1.05 kg (0.67-1.30) n=7

Dental Formula:
  incisors: 1/1
  canines: 0/0
  premolars: 2/1
  molars: 3/3

Identification:
Although the Mountain Beaver is a distinctive mammal, naturalists and wildlife biologists sometimes confuse it with young marmots. The Mountain Beaver has a much shorter tail and longer claws on its front feet. The skull's flask-shaped auditory bullae and the projections on the molar teeth are unique among B.C. rodents.

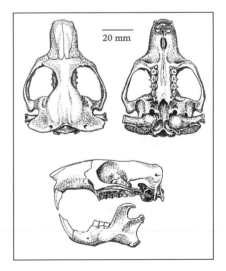

20 mm

Natural History

The Mountain Beaver inhabits benchlands, foothills and mid slopes of mountains where it occupies forested areas with a dense cover of young trees, shrubs and woody debris on the forest floor. In British Columbia it ranges from near sea level in the lower Fraser River valley to 1925 metres in the Cascade Mountains. Because it has a primitive kidney that cannot produce concentrated urine, the Mountain Beaver needs to ingest a large amount of water – this may explain its preference for cool, moist sites with succulent vegetation.

In the lowlands of the Fraser River valley it inhabits second growth forests of Red Alder, Western Hemlock and Douglas-fir with a shrub cover of Vine Maple, Thimbleberry and raspberries. The most comprehensive habitat data for the province are from Les Gyug's research on the eastern slopes of the Cascade Mountains where the highest population densities are in the upper-elevation forests of Engelmann Spruce, Subalpine Fir and Amabilis Fir. Important habitat features are deep soils for excavating tunnels and burrows, food plants within close proximity, a cool, moist microclimate, and good drainage of the burrow systems. Most burrows in the Cascade Mountains are located between 1200 and 1600 metres elevation on moderate slopes near small streams or seepage areas, usually on the cool north sides of gullies. Burrows are never located in the floodplains of large streams or rivers or among cascading streams flowing over rock faces or canyons.

Population densities are usually about 4 to 8 animals per hectare, although populations of 15 to 20 per hectare have been reported in Oregon. Les Gyug estimated that ideal habitats in the Cascade Mountains of B.C. support 4.4 to 5.8 nest sites per hectare, though these habitats tend to be quite localized and small – less than 2.5 ha. In habitats of moderate quality, Gyug found that the numbers of nest sites were only 0.5 to 0.7 per hectare.

A fossorial rodent, the Mountain Beaver rarely ventures far from its burrow. Estimates of home-range size are from 0.04 to 0.70 ha, with females occupying smaller home ranges than males. Paul Martin used radio transmitters to track animals in western Washington over a period of 12 months and found that they moved only 37 to 43 metres from their nests. But young animals move up to 560 metres when they disperse from their birth sites.

This rodent constructs an elaborate burrow system that has an underground nest chamber, nearby dead-end tunnels used for refuse and food storage, toilet chambers filled with fecal pellets, and a number of tunnels that connect with openings to the surface. Evidently, most entrances are plugged – only a few have fresh soil from recent excavating. The tunnels are 15 to 46 cm below the surface. The nest chamber, which is about 30 to 150 cm below ground and about 50 to 60 cm in diameter, has an outer layer of coarse ferns or twigs and a soft inner layer of dried leaves such as Salal. Although the tunnels and burrows may be wet, the nest chambers are always located above the water table where they remain dry. Surface runways or trails are often visible in the vicinity of burrows. Mountain Beaver burrows provide important habitat for other vertebrates, including small mammals such as moles, voles, mice, Snowshoe Hares, Western Spotted Skunks, Minks, Ermines and Long-tailed Weasels.

Although its burrows tend to be concentrated in areas where there is ideal habitat, Mountain Beavers are not colonial. Adult males and females live separately; nest sites are occupied by only one adult animal. An animal will occupy the same nest site for long periods, up to 44 months. Mountain Beavers defend their nest sites, although individuals may overlap in their home-range movements. They use a strong musky odour to mark their burrow and the area around it – presumably, this scent plays a role in communication or territorial behaviour. Because the Mountain Beaver's sight and hearing are poorly developed, it relies mostly on smell and touch for finding its way above and below ground. Primarily nocturnal, this animal is most active several hours after sunset. Much of the time above ground is spent feeding and gathering food. It cuts and piles plants on top of logs or on the ground near burrow entrances, and leaves them there to wilt before transporting them underground for consumption.

The Mountain Beaver eats a variety of plants. In coastal regions, it eats Bracken Fern, Sword Fern and Salal leaves throughout year, and various forbs and grasses in summer. Summer food caches collected in the Cascade Mountains by Les Gyug contained mostly Sitka Valerian, horsetails, twisted stalks, Thimbleberry, Cow-parsnip, Indian Hellebore and Black Huckleberry; Fireweed was also common in food caches, especially in clear-cut habitats. In some coastal regions, Red Alder leaves are a major food in late summer.

The Mountain Beaver does not hibernate, but remains active throughout the winter. Its winter diet, especially in high elevations with snow cover, shifts to the bark and foliage of coniferous trees, especially Douglas-fir, and the bark of various hardwood trees and

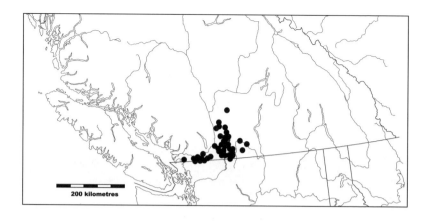

200 kilometres

shrubs. In the Sierra Nevada Mountains of California, James O'Brien found that winter food caches were predominately conifer branches and leaves. The winter diet in British Columbia is unknown. In his study area in the Cascade Mountains, Les Gyug observed many White-flowered Rhododendrons that were heavily clipped and he speculated that this plant was a major winter food. The Mountain Beaver can climb up to six metres into trees where it will clip off twigs and transport them to the ground. Because it girdles bark and clips the seedlings of young coniferous trees, this rodent is considered a major forest pest species in some areas of the United States. Considerable research has been done on various techniques to control populations and damage. In B.C., damage from the Mountain Beaver appears to be rather localized and insignificant.

Breeding data are lacking for the B.C. population. In the Puget Sound area of Washington, most females are pregnant from late February to March with births in late March or early April, although pregnant individuals have been found as late as May 3. They produce only one litter a year. The gestation period is 28 to 30 days. Litter size ranges from two to four, with most females producing three young. At birth, the young are naked and blind; at two days they weigh about 20 grams. By two weeks their body is covered with hair. At about four weeks the incisor teeth erupt and by 45 days the eyes are open. They begin to eat solid food by six to eight weeks. Females do not breed until their second year. In the wild, Mountain Beavers may survive five or six years.

Predators include the Golden Eagle, owls, Bobcat, Coyote and Cougar. Small mammalian predators such as the Mink, Long-tailed

Weasel, Ermine and Western Spotted Skunk may prey on young Mountain Beavers inside their burrows.

Range

The Mountain Beaver ranges from the Sierra Nevada Mountains and Pacific coast of California to southwestern British Columbia. Here it is restricted to the Cascade Range and lower Fraser River valley as far west as Sumas Mountain and the Langley region, and east to the eastern slopes of the Cascades. There are historical records as far north as Canford in the Nicola Valley. The species has not been found north of the Fraser River.

Taxonomy

Seven subspecies are recognized; two occur in British Columbia. The two B.C. races appear to intergrade in the foothills of the Cascade Mountains. Demonstrating only minor differences in size and no differences in pelage colour, the validity of these subspecies needs to be assessed with genetic studies.

Aplodontia rufa rainieri Merriam – the Cascade Mountains of Washington and B.C. as far west as Hope.

Aplodontia rufa rufa (Rafinesque) – along the coast from California to B.C.'s lower Fraser River valley, where it ranges as far east as the Chilliwack region.

Conservation Status

Both subspecies are considered at risk in British Columbia with *Aplodontia rufa rufa* on the provincial Red List and *A. r. rainieri* on the Blue List. Nevertheless, given the questionable taxonomy of these subspecies, it may be inappropriate to treat them as separate populations for conservation and management. The Mountain Beaver was designated nationally by COSEWIC as Special Concern in 1999. Les Gyug concluded that this rodent is vulnerable to habitat loss from urban development in the lower Fraser River valley and habitat destruction from forest harvesting. Although Mountain Beavers are common in clear-cuts, these habitats are not ideal. Bulldozers and excavators used in forest harvesting compact the soil and collapse burrows. Another impact from forestry operations results from draining wet areas for replanting.

Remarks

The Mountain Beaver's burrows are easy to identify (see figure 8), but few naturalists or biologists, including me, have observed this animal in the wild. Les Gyug interviewed forestry workers and hunters in his study area and found that most had no idea what animal created the conspicuous burrows and runways. Although I have received a number of reports of extralimital records over the years – some as far north as Glenora on the Stikine River and Pine Pass in the northern Rocky Mountains – any that I could verify from photographs or specimens were young marmots. Aboriginal groups in the Puget Sound area hunted the Mountain Beaver for food and clothing. They made robes from its fur and chisel tools from its incisors.

Selected References: Carraway and Verts 1993, Gyug 2000, Martin 1971, O' Brien 1988.

Family Sciuridae: Squirrels

Typical squirrels have a bushy tail, dense fur, four toes on the front feet and five toes on the hind feet. Some species have internal cheek pouches. The skull has pronounced postorbital processes, a wide interorbital region and a small infraorbital opening. There are 15 native and 2 introduced species in British Columbia.

Northern Flying Squirrel
Glaucomys sabrinus

Other Common Names: None.

Description

The Northern Flying Squirrel is a medium-sized squirrel with a broad flat tail, large eyes and a fold of skin that forms a gliding membrane extending along each side between the wrist and ankle. The soft, silky dorsal pelage varies from grey-brown to rich brown.

The dorsal guard hairs are dark grey with brown tips. The undersides are pale with hairs that are grey at the base and whitish or tan at the tips. A distinct black lateral line separates the dorsal and ventral fur. The flat tail tends to be darker at the tip. The conspicuous dark eyes are bordered by a black ring. The ears are large and round with a sparse covering of hair. The skull has a distinct V-shaped notch in the interorbital region (figure 113), an inflated braincase and postorbital processes that taper to a sharp point.

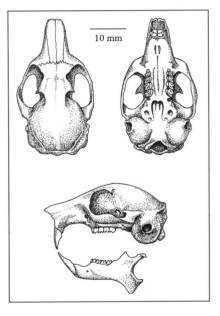

10 mm

Measurements:
  total length: 322 (298-344) n=51
  tail vertebrae: 144 (130-164) n=51
  hind foot: 42 (39-46) n=50
  ear: 21 (17-25) n=50
  weight: 155.5 (124.0-200.6) n=58

Dental Formula:
  incisors: 1/1
  canines: 0/0
  premolars: 2/1
  molars: 3/3

Identification:
This species cannot be confused with any other squirrel in British Columbia.

Natural History

Associated with forests, the Northern Flying Squirrel has two major habitat requirements: tree cavities for den sites, and an abundant supply of lichens and truffles for food. Although some biologists and naturalists have suspected that this species is an old-growth-forest specialist, studies done in the coastal forests of the western United States and the boreal forests of British Columbia and Alberta have revealed that the Northern Flying Squirrel inhabits both young and old forests, although some researchers have reported higher population densities in older forests. In B.C., the elevational range is from sea level on the coast to 2130 metres in the Columbia Mountains.

Estimates of population density range from about 0.12 to 5.00 animals per hectare. A number of estimates of home-range size have been calculated by tracking movements with radio transmitters. Males generally have larger home ranges than females. In northern B.C., Laine Cotton and Katherine Parker estimated average home-range sizes of 1.4 ha for females and 3.7 ha for males, based on distances moved between den sites. Estimates of average home-range size in Oregon, which included the areas used for feeding, varied from 3.9 to 5.9 ha. This species often switches den sites moving several hundred metres between them.

The Northern Flying Squirrel usually makes its den above ground in trees, although dens have been found in stumps and fallen trees. Dens may be woodpecker holes or natural cavities in dead or live trees, nests in tree branches made of twigs, leaves, lichen, moss, or bark, or nests in clumps of branches created by witches' broom. Tree species used for dens in northwestern North America include Douglas-fir, Grand Fir, Engelmann Spruce, Western Hemlock, Lodgepole Pine, White Spruce, Subalpine Fir and Trembling Aspen. The characteristics of these trees are highly variable, depending on the age of the forest and the geographic region; but in general, Northern Flying Squirrels select the broader, taller, older trees in a forest and they use live trees more often than dead trees. In coastal Washington and Oregon, Andrew Carey and colleagues found that females mostly used cavities for breeding nests, but the few breeding nests described by Ian McTaggart Cowan in B.C. were all located outside of cavities. Most of these external nests were constructed of

twigs, roots and bark, but a breeding nest found near Newgate in the Kootenay region was constructed entirely from Old Man's Beard lichen. This squirrel also uses bird houses and it can be attracted to artificial nesting boxes.

The Northern Flying Squirrel is nocturnal and rarely leaves its nest in daytime. Laine Cotton and Katherine Parker found variable activity patterns among Northern Flying Squirrels in northern B.C. Some animals had two activity peaks: one within two hours after dusk and another ending a few hours before dawn. Other animals were most active in the middle of the night. This species remains active throughout the winter. Although it shortens its activity period in cold weather, it will leave its den at night for one or two hours, even when temperatures fall below -20°C. The social behaviour of the Northern Flying Squirrel is largely unknown. Adults will share a den throughout the year. In winter this is presumably a strategy to conserve heat, but in other seasons, these aggregations may have some behavioural significance.

The Northern Flying Squirrel can glide up to 50 metres from tree to tree, though the average length of a glide is about 20 metres. To launch a glide from a tree, it stretches out its front and hind limbs to extend the gliding membrane and then leaps into the air. As it descends, the tail acts as a rudder to help steer; muscles in the gliding membrane can alter the shape or contour of the gliding surface. Just before landing on a tree trunk or branch, it will shift to a vertical position. This species generally descends to the forest floor to forage for food. On the ground, where it is vulnerable to predators, the Northern Flying Squirrel moves by awkward jumps.

No diet studies have been done in British Columbia. In Oregon and Washington, the principal food is fungi, particularly truffles. This squirrel eats more than 40 species of truffles, excavating their fruiting bodies from under the soil, and leaving distinct holes or pits in the forest floor. In coastal regions, where there is little snow cover, truffles provide a reliable food source throughout the year. In regions where truffles are not accessible in winter because of deep snow, the Northern Flying Squirrel switches its diet to arboreal lichens such as Edible Horsehair. It also eats mushrooms, as well as berries, maple seeds, the male flowers of deciduous and coniferous trees, carrion, birds' eggs, and insects. Although the Northern Flying Squirrel will steal fungi stored from the middens of the Red Squirrel, the extent that it caches its own food stores is unknown. Some researchers speculate that lichens used as insulation material in nests also provide a food source during inclement winter weather.

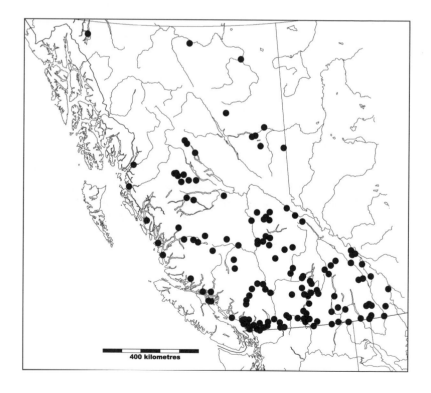

Mating occurs from late March to May. After a gestation period of 37 to 42 days, females produce two to four young. At one day of age, Northern Flying Squirrels weigh about six grams and are about 70 mm long. They are naked with closed eyes. Their eyes open at 26 to 32 days; by 47 days they begin to eat solid food. The young begin to glide at about three months. The scanty reproductive data for British Columbia is based on museum specimens and the few observations reported by Ian McTaggart Cowan. Two pregnant females each with two embryos were taken on May 5 and 6; three females with two or three newborns were taken between May 14 and 31. Females probably produce only one litter each year in British Columbia. Because it is nocturnal, this species is especially vulnerable to avian predators such as the Barn Owl, Great Horned Owl and Spotted Owl. It has been estimated in the northwestern United States that one Spotted Owl can eat as many as 260 Northern Flying Squirrels per year. Despite its gliding abilities and arboreal nests, this squirrel is also taken by arboreal mammalian predators such as the Ermine, Long-tailed Weasel and Marten. While feeding on the ground, the Northern Flying Squirrel is vulnerable to predators such as the Bobcat, Cougar, Coyote and Grey Wolf.

Range

The Northern Flying Squirrel is broadly distributed across Alaska, Canada and boreal-montane habitats in the conterminous United States. It occupies the entire mainland of British Columbia. The distribution of this squirrel on the coastal islands of B.C. is poorly documented. There are records from Campbell and Princess Royal, two islands on the central coast, and from Cortes, Quadra and Stuart islands off the northeastern coast of Vancouver Island. The absence of this species from Vancouver Island, the largest island in the eastern North Pacific, is a biogeographic curiosity, because the Northern Flying Squirrel population on nearby Quadra Island is only 700 metres away, across Discovery Passage.

Taxonomy

Twenty-five subspecies are recognized; seven occur in British Columbia, all defined by relatively minor differences in fur colour and size. Recent DNA studies by Brian Arbogast have revealed that populations associated with the Pacific coast region of the United States and extreme southwestern B.C. are strongly differentiated from populations in other parts of the range.

Glaucomys sabrinus alpinus (Richardson) – western Alberta, the southern Yukon Territory, and most of northern and central B.C. May be indistinguishable from the boreal race *G. sabrinus sabrinus*.

Glaucomys sabrinus columbiensis Howell – extreme northern Washington and the southern dry interior valleys (Okanagan, Nicola and Similkameen) of B.C.

Glaucomys sabrinus fuliginosus (Rhoads) – from the Cascade Mountains of Oregon and Washington to the southern Coast Mountains in British Columbia.

Glaucomys sabrinus latipes Howell – southeastern Alberta, northwestern Montana, northern Idaho, northeastern Washington, and the Selkirk, Purcell and southern Rocky mountain ranges in B.C. as far north as Glacier National Park.

Glaucomys sabrinus oregonensis (Bachman) – from the coastal lowlands of Oregon to Loughborough Inlet of coastal B.C. and a few southern coastal islands (Cortes, Quadra and Stuart).

Glaucomys sabrinus reductus Cowan – restricted to B.C. where it ranges from the central coast (including Campbell Island) and Coast Mountains as far east as Quesnel and north to Ootsa Lake.

Glaucomys sabrinus zaphaeus (Osgood) – southeastern Alaskan mainland and the northwestern coastal region of B.C. including Princess Royal Island.

Conservation Status

The British Columbian populations are not of conservation concern. But taxonomic and ecological research is needed on the populations inhabiting the coastal islands of British Columbia. The Prince of Wales subspecies (*G. sabrinus griseifrons*) restricted to the islands of the Alexander Archipelago in southeastern Alaska is listed as Endangered by the IUCN. In the United States, this species has attracted considerable attention from conservation biologists because of its possible dependence on old-growth forests and its role as a major prey item of the endangered Spotted Owl.

Remarks

Because it is nocturnal, shy, secretive and arboreal, the Northern Flying Squirrel is a challenging small mammal to study. With new technology, especially modern radio-tracking techniques, biologists have learned a great deal in the past few decades about the role of this squirrel in the forests of western North America. It plays an important role in dispersing the spores of truffles. Unaffected by digestive enzymes, the fungal spores pass through the gut of the squirrel and are scattered over the forest floor in its droppings. It is also suspected that this species may assist in the spread of lichens by inadvertently leaving lichen fragments on the forest floor or in tree branches while transporting lichens to its nests.

Selected References: Arbogast 1999; Carey et al. 1997; Cotton and Parker 2000a, 2000b; Cowan 1936; Maser et al. 1985; Wells-Gosling and Heaney 1984.

Hoary Marmot   *Marmota caligata*

Other Common Names: Rockchuck, Whistler.

Description

The Hoary Marmot is our largest marmot. It has a distinctive black-and-white pelage. Although the fur colour varies geographically among the five subspecies found in the province, the dorsal pelage is usually white to grey with a dark wash in the pale shoulder region, giving it a mantled appearance. It has a white nose, a black or blackish-brown cap on the head that extends to the side of the face, and a white patch between the eyes on the snout. The dorsal surface of the hind feet are black or dark brown. The short bushy tail varies from blackish-brown to pale brown. Melanic individuals occur in some northern populations. The posterior pad on the sole of the hind foot is circular (figure 41). The skull is straight in dorsal profile with a distinct depression in the interorbital region. The postorbital processes are oriented at right angles to the main axis of the skull; the posterior border of the nasal bones are square or arched.

Measurements:
total length: 673 (505-800) n=102
tail vertebrae: 199 (140-246) n=106
hind foot: 94 (78-114) n=101
ear: 26 (19-32) n=11
weight: 4.59 kg (3.09-5.59) n=6

Dental Formula:
incisors: 1/1
canines: 0/0
premolars: 2/1
molars: 3/3

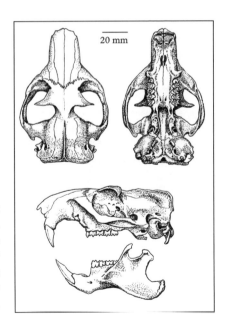

20 mm

Identification:
The Yellow-bellied Marmot (*Marmota flaviventris*) and Woodchuck (*M. monax*) are smaller, lack the pale hoary colour pattern and have smaller skulls (maxillary toothrow length < 21.2 mm).

Melanic Hoary Marmots can be distinguished from melanic Yellow-bellied Marmots or Woodchucks by the circular posterior pad on the sole of its hind foot; the others have oval pads. Its distribution does not overlap with the Vancouver Island Marmot (*M. vancouverensis*), which can be discriminated by its dark brown to nearly black dorsal pelage, and the V-shaped notch on the posterior border of the nasal bones in its skull (figure 108).

Natural History

The Hoary Marmot is associated with subalpine and alpine areas, where it inhabits meadows and open forests of Subalpine Fir or Subalpine Larch in close vicinity to rock slides and large boulders. The elevational range in British Columbia typically extends from 1250 to 2450 metres, although dispersing Hoary Marmots can be found at lower elevations. The only density estimate for this marmot is 0.36 to 0.54 animals per hectare for an Alaskan population studied by Warren Holmes. Estimates of home-range size are 11.0 to 15.6 ha, with feeding activity restricted to smaller areas of 8.9 to 10.0 ha. Except when dispersing to new colony sites, Hoary Marmots are relatively sedentary; for example, breeding females rarely move more than 30 metres from their burrows when feeding.

The Hoary Marmot is a colonial species that lives in family groups. In Alaska, Warren Holmes found that colonies consisted of an adult male, an adult female and their young. In more moderate environments such as the Cascade Mountains, Hoary Marmot colonies have two adult females with one adult male. Two-year-old animals disperse from their colony of birth after spring emergence if their mother has a new litter. Members of a colony use a common home range and often share the same burrow. Typical social interactions among colony members include play fighting, nose-greetings, chases and mutual grooming. Adjacent Hoary Marmot colonies have distinct territories and their members rarely interact with those of neighbouring colonies; they usually chase away intruders from adjacent colonies (often the dispersing two-year-olds).

Seven vocalizations have been described for the Hoary Marmot. The most familiar is a high pitched whistle emitted while the animal stands upright on its hind legs. This call is an alarm signalling the presence of an intruder or predator. Other marmots in the colony will respond by immediately surveying the area for danger or running to the nearest burrow.

A diurnal mammal, the Hoary Marmot is most active in early morning and late afternoon with a midday lull. Except on hot days, animals usually remain above ground during midday. Much of the time above ground is spent resting on rocks or feeding. In late August, with the approach of winter hibernation, Hoary Marmots spend less and less time above ground often remaining in their burrows for two or three consecutive days.

Hoary Marmots construct three types of burrows: sleeping burrows, refuge burrows and hibernacula. One or more colony members may share a summer sleeping burrow at night, and a colony will maintain four to seven separate sleeping burrows. These elaborate burrows, located in talus or under large boulders, are at least 3.5 metres long and have three to five entrances. In contrast, refuge burrows are simple structures that provide temporary escape from predators. Strategically scattered under rocks throughout the colony's feeding area, they have a single entrance and seldom exceed 1.5 metres in length. A single hibernaculum, usually located in a vegetated hillside, is shared by all members of a colony for hibernation throughout winter. The hibernaculum is essential for the survival of the colony. It not only provides shelter during the severe winter hibernation period, but it is strategically located in close proximity to areas where there is early snow melt and an accessible food supply in early spring.

Hoary Marmots hibernate for about eight months of the year. There are no data available for the hibernation period in British

Columbia. But in Alaska, the Hoary Marmot emerges from hibernation in early May when there is still deep snow covering the hibernaculum. Emergence appears to be synchronized among members of a colony, because they all emerge within a few days after the first animal appears above ground. Throughout the summer these marmots steadily gain weight until they enter hibernation in late September or early October. The diet includes graminoids, forbs, willows, lichens and moss. Hoary Marmots in Alaska eat mostly sedges, grasses, vetches and fleabanes. In B.C.'s Cascade Mountains, David Gray found that Hoary Marmots mainly eat Western Anemone, paintbrush, Yellow Glacier-lily, lupines, betony and False Indian Hellebore.

Mating occurs in early May during the first two weeks after animals emerge from hibernation. Females usually produce litters of three young after a gestation period of about 30 days. After nursing for 25 to 30 days, the young appear above ground. The young-of-the-year generally emerge in July, although in Manning Provincial Park, David Gray did not observe young until August 1. Females produce only one litter per year and breed in alternate years, although they have the potential to breed every year. In some populations litters may be spaced three or even four years apart. The Hoary Marmot reaches sexual maturity when three years old. Predators include the Golden Eagle, Swainson's Hawk, Northern Goshawk, Lynx, Wolverine, Cougar, Coyote, Grey Wolf, Black Bear and Grizzly Bear.

Range

An inhabitant of alpine areas, the Hoary Marmot ranges across the western Cordillera from Alaska, the Yukon Territory and the extreme western Northwest Territories to Washington, northern Idaho and northern Montana. It occupies higher elevations throughout the entire mainland of British Columbia except for lowland boreal forests in the northeast and low elevation grasslands in the dry interior. This species is not known to occur on any island in B.C.

Taxonomy

Variation in size and pelage colour among populations is substantial. Eight subspecies are recognized; five are found in the province:

Marmota caligata caligata (Eschscholtz) – Alaska south of Brooks Range, the Yukon Territory, the Mackenzie River valley in the Northwest Territories and the northern and central interior of the

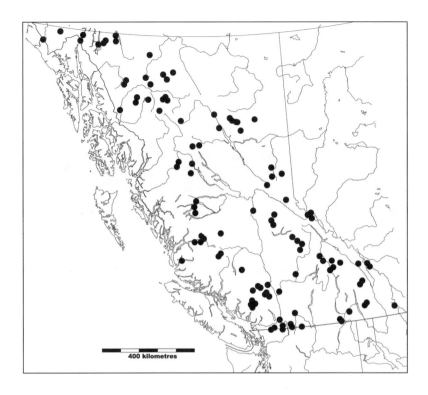

400 kilometres

province as far south as Barkerville and Mount Robson. This sub-species includes populations classified as the subspecies *M. c. oxytona* in the original *Mammals of British Columbia*.

Marmota caligata cascadensis Howell – on the coast from the Cascade Range of Washington north to Garibaldi Provincial Park.

Marmota caligata okanagana (King) – extreme southwestern Alberta and the Monashee, Selkirk, Purcell and Rocky mountains in southeastern B.C. Some authorities treat populations from Tornado Pass, Monarch Mountain and Farrow Pass northwest of Mount Assiniboine as *M. c. nivaria*.

Marmota caligata oxytona Hollister – the southeastern Yukon Territory, the southwestern Northwest Territories and part of the Rocky Mountains in Alberta.

Marmota caligata raceyi Anderson – restricted to B.C. where it inhabits mountains in the Chilcotin Plateau and the Coast Mountains as far north as Kitimat and Ootsa Lake.

Conservation Status

The British Columbian populations are not of conservation concern. There have been anecdotal accounts of population declines in some parts of the province.

Remarks

With its large size, attractive markings and piercing whistle-like alarm calls, the Hoary Marmot is one of the most conspicuous and familiar mammals in the high country of British Columbia. Whistle Mountain and Whistler Glacier on the western edge of Garibaldi Provincial Park are named after this animal. Aboriginal groups prized this marmot, using its pelts for robes which were valued for their soft warm fur and attractive markings. Meat and fat from the Hoary Marmot provided an important food resource.

Selected References: Gray 1967; Holmes 1979, 1984; Taulman 1977.

Yellow-bellied Marmot
Marmota flaviventris
Other Common Names: Groundhog.

Description
The Yellow-bellied Marmot is a small marmot with brown dorsal pelage grizzled with tan and grey. The ventral pelage is tan to yellow. The head is black with a white or tan area in front of the eyes. The side of the neck is a uniform pale yellow or tan that contrasts with the flanks and back. The dorsal surface of the hind feet is yellow to tan. In contrast to the Woodchuck, melanic Yellow-bellied Marmots are rare in British Columbian populations. The posterior pad on the sole of the hind foot is oval (figure 40). The skull is straight in dorsal profile with a distinct depression in the interorbital region; the postorbital processes are positioned at right angles to main axis of skull. The posterior border of nasal bones forms an arch and the upper toothrows diverge anteriorly.

Measurements:
total length: 519 (415-660) n=44
tail vertebrae: 147 (103-184) n=46
hind foot: 72 (63-83) n=45
ear: 27 (23-34) n=11
weight: 2.34 kg (1.80-3.27) n=8

Dental Formula:
incisors: 1/1
canines: 0/0
premolars: 2/1
molars: 3/3

Identification:
To discriminate from the Hoary Marmot (*Marmota caligata*), see that account (page 124). The Yellow-bellied Marmot's range is mostly separate from that of the Woodchuck (*M. monax*), but the two species may co-occur in the Chilcotin and northern

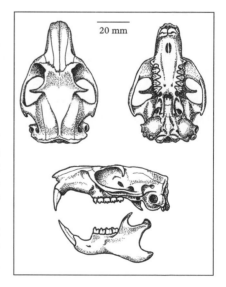

20 mm

Fraser River valley region. The Woodchuck lacks the white markings on the face and light patches on the side of the neck, its ventral fur is reddish-brown and the dorsal surface of its feet are dark brown. The Woodchuck's skull has parallel upper toothrows and the posterior border of the nasal bones form a V-shaped notch.

Natural History

The Yellow-bellied Marmot inhabits valleys and lower slopes of mountains where it is associated with open grassland, clearings and disturbed habitats such as road or railroad embankments. Its burrows are under rocks, talus and logs, or beneath human-made equivalents such as buildings, old foundations, lumber piles, and exposed road or railroad embankments. One of the more unusual human-made habitats was a cemetery at Kelowna where Yellow-bellied Marmots burrowed under the headstones. Although an alpine species throughout much of the western United States, in British Columbia this marmot is restricted to low elevations with most colonies below 1300 metres. Where its range overlaps with that of the Hoary Marmot, the two marmots are clearly separated by elevation, with the Hoary Marmot confined to alpine habitats. Nevertheless, solitary Yellow-bellied Marmots, presumably dispersing from colonies, have been observed at high elevations. For example, a solitary Yellow-bellied Marmot was observed in a Hoary Marmot colony on Blackwell Peak in Manning Provincial Park. The only alpine colony of Yellow-bellied Marmots that I have observed in B.C. was on the Cornwall Hills west of Cache Creek in the Clear Range at about 1980 metres.

Detailed studies by Ken Armitage and his colleagues have revealed that Yellow-bellied Marmots are colonial, living in social groups that consist of one or more adult females, their related kin and an adult male. Influenced by the abundance and distribution of females, the home-range size of males is quite variable, with estimates ranging from 0.09 to 1.90 ha. Although rather sedentary, on rare occasions adult males with established home ranges will make long-distance excursions of more than a kilometre. Within its home range, a male maintains a smaller territory that it defends against other males. In large tracts of continuous habitat, a male will defend two or three females, but in areas with small patches of habitat, it may defend only one female. Mothers and their offspring (both young-of-the-year and yearlings) live within the male's territory where they occupy home ranges of 0.13 to 1.02 ha. Yellow-bellied Marmots disperse from their birth site when they are one year old. Most dispersing animals move less than 4 km from their birth site, but Ken Armitage has observed long distance movements up to 15.4 km by males and 6.4 km by females.

This marmot hibernates for seven or eight months in the alpine areas of the United States, emerging from hibernation in late April or early May. Adult males are the first to appear above ground in the spring, followed by adult females, yearling males and yearling females. No hibernation data are available for lowland populations in B.C., where it would be expected to have a shorter hibernation period. Leo Couch reported that spring emergence was around mid March in northern Washington. The marmots steadily gain weight throughout the summer, growing 11 to 40 grams per day. They begin to enter hibernation in late August and by mid September the entire colony is underground.

Yellow-bellied Marmots construct three types of burrows: a home, an escape from predators and a hibernaculum – although they may use the same burrow for all of these functions. The most important requirements for a burrow are suitable soil for digging, a steep slope, and rocks or tree roots at the entrance for support. A steep slope provides drainage and gives the marmot a clear view of its surrounding area and any intruders or predators. Rocks or tree roots not only support the burrow entrance, but prevent predators such as the Badger from excavating the burrow. A typical burrow system has a main passageway, several lateral tunnels and a nest chamber. Home burrows may be occupied by as many as 12 marmots, but usually 1 to 5. They hibernate in groups of 2 to 11, sharing the same hibernaculum. Before the occupants enter hibernation, they seal the entrance of the hibernaculum with a mix of feces and dry grass.

This is a diurnal species, active above ground mostly in the morning and late afternoon. Yellow-bellied Marmots spend most of their time above ground resting or sunning themselves on rocks or foraging for food in close proximity to their burrows (usually within 25 metres). Activity falls off noticeably when temperatures exceed 20°C, presumably to avoid heat stress.

Yellow-bellied Marmots produce whistles, screams and clicking or chattering sounds with their teeth. Researchers have categorized their whistles into six distinct types based on their context, intensity and the interval between each whistle. Most whistles function as alarm or threat calls given in the presence of intruders or predators. The diet includes forbs, grasses and sedges. Cultivated crops such as Alfalfa are eaten in agricultural areas. Leo Couch reported that this species climbs small fruit trees to feed on ripe fruits.

This species displays a number of complex social behaviours when mating. Adult males tend to be dominant over other individuals in the colony – they are especially intolerant of other adult males that may compete for females. This may contribute to the tendency for males to disperse from the colony of their birth in their second summer. Females form groups of closely related individuals: sisters, or a mother and her daughters. Most of their interactions are amicable, but they are less tolerant of unrelated females.

Mating occurs within two weeks of emergence from hibernation. Females produce a single litter of 4 young (range: 3 to 8) after a gestation period of about 30 days. Although females may produce litters in consecutive years, in severe environments they often breed in alternate years. Newborn Yellow-bellied Marmots weigh about 34 grams; by 30 days they emerge above ground, all litter-mates emerging on the same day. After they are weaned at about two months, the young grow rapidly, tripling their body weight before they enter hibernation. The only breeding data for British Columbia is a pregnant female with six well-developed embryos that was taken on April 20 at Okanagan Landing near Vernon. This fits closely with Leo Couch's observation that young Yellow-bellied Marmots are born in mid April in the Okanogan region of northern Washington. The precise timing of emergence in the spring and mating is closely linked to elevation and spring snowmelt. In some alpine areas in the Rocky Mountains of the United States where there is late snowmelt, young-of-the-year Yellow-bellied Marmots appear above ground as late as July or early August. This species reaches sexual maturity at two years of age, but the age of first reproduction can be delayed until six years of age. Females stop breeding after 10 years. In the wild, males live up to 9 years; females survive longer, a few reaching 14 or 15 years. Most mortality in this species is from predation. Mammalian predators include the Badger, Black Bear, Bobcat, Cougar, Coyote, Marten and Long-tailed Weasel. The Golden Eagle is the most important avian predator.

Range
The Yellow-bellied Marmot is found in the western Cordillera from southern British Columbia and Alberta to California, Nevada and New Mexico. In B.C., it inhabits the Fraser and Thompson plateaus, and the southern mountains, including the Cascades, Monashees and Selkirks. The western limit of the range is on the east side of the Fraser River; the northern limit is the Williams Lake area, although there is a historical museum specimen taken in the 1950s at Prince George.

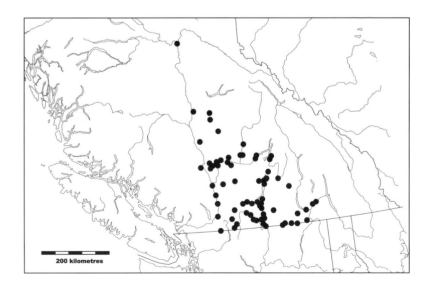

200 kilometres

Taxonomy
Eleven subspecies are recognized; one occurs in British Columbia:
Marmota flaviventris avara (Bangs) – a small pale race found in
Nevada, Oregon, Idaho, Washington and B.C.

Conservation Status
British Columbian populations are not of conservation concern.

Remarks
In the 1980s a colony of several dozen Yellow-bellied Marmots was
discovered in North Vancouver just north of the Second Narrows
Bridge in an industrial area. Far outside the known range in British
Columbia, it is suspected that this colony was derived from animals
that were transported in an industrial shipment. Although surrounded by major highways and industrial development, this colony persisted for several years until it was transplanted to a site near
Manning Provincial Park in 1989.

Selected References: Armitage 1991, Couch 1930, Frase and
Hoffmann 1980, Johns and Armitage 1979, Schwartz et al. 1998.

Woodchuck   *Marmota monax*

Other Common Names: Groundhog.

Description

The Woodchuck is our smallest marmot. It has brown dorsal fur grizzled with silver-grey, and reddish-brown ventral pelage. The head is dark brown with no white markings on the face; the fur on the sides of the neck does not contrast sharply in colour with the fur on the flanks and back. The front legs are covered with reddish-brown hairs; the dorsal surface of the hind feet is dark brown to nearly black. The short, nearly flat, bushy tail varies from dark brown to blackish. Melanism is common in British Columbian Woodchucks: in some populations in northern and southeastern British Columbia, black animals are more prevalent than the normal brown colour morph. The posterior pad on the sole of the hind foot is oval in shape (figure 40). The skull is straight in dorsal profile, with a distinct depression in the interorbital region, and the postorbital processes are at right angles to the main axis of the skull. The posterior border of the nasal bones form a V-shaped notch. The upper toothrows are parallel.

Measurements:
  total length: 496 (388-597) n=45
  tail vertebrae: 126 (98-171) n=44
  hind foot: 72 (58-81) n=49
  ear: 28 (25-31) n=5
  weight: 1.59 kg (1.06-2.40) n=7

Dental Formula:
  incisors: 1/1
  canines: 0/0
  premolars: 2/1
  molars: 3/3

Identification:
The Woodchuck could be confused with the Hoary Marmot (*Marmota caligata*) or the Yellow-bellied Marmot (*M. flaviventris*) – see those accounts for distinguishing

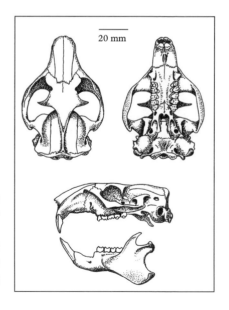

20 mm

traits (pages 124 and 130). Curiously, young Woodchucks have been misidentified as the Mountain Beaver (*Aplodontia rufa*), a southern coastal mammal not found within the range of the Woodchuck, and readily distinguished by long claws on its front and hind feet and a rudimentary stubby tail.

Natural History

Associated with valley bottoms, lowlands and the lower slopes of mountains, the Woodchuck inhabits open forests, recently cleared forests, agricultural fields, meadows, ravines associated with streams and rivers, road embankments, and campgrounds. In agricultural areas, Woodchuck burrows are most concentrated in edge habitats bordering fields and cleared areas. Its elevational range in British Columbia is from 350 to 1250 metres.

Because no studies have been done on the Woodchuck in western Canada, virtually all information on its basic biology is derived from studies done in eastern North America. Estimates of population densities range from 0.1 to 3.3 animals per hectare. The home-range size varies from 0.3 to 10.6 ha. A consistent pattern reported in a number of studies is that males have larger home ranges than females. Evidently, yearlings and older males travel long distances from their home site in the early spring.

Although the Woodchuck is regarded as the most solitary of the North American marmots, its social structure appears to be somewhat variable. Animals living in high population densities will occasionally develop rudimentary family groups that resemble those of other marmots, but most adult Woodchucks are solitary, occupying a single burrow system. Generally, this species is territorial, with sexually mature females especially intolerant of one another. Adult males also have mutually exclusive home ranges, although Paul Meier described a population in Ohio where the home ranges of males overlapped with the home ranges of several females. Evidently, some males are nomadic and wander over large areas throughout the breeding season. Young-of-the-year Woodchucks usually disperse from their birth site shortly after they are weaned at about two months of age. But in habitats supporting high population densities, some juvenile females may remain within their birth site until the following spring.

Woodchuck burrows have one to eleven entrances and several tunnels. A simple burrow has a single entrance that terminates in a nest chamber. More complex systems can have several entrances into a maze of tunnels up to 12 metres long and 1.5 metres deep leading to blind chambers and several nest chambers.

The Woodchuck hibernates for three to five months of the year. In the eastern United States it may emerge from hibernation as early as February, but in eastern Canada this species emerges between early March and the first week of April, often when there is still snow cover. Adult males (two years or older) appear first above ground, and adult females and yearlings emerge up to a month later. The Woodchuck steadily increases in weight throughout the summer, accumulating heavy fat deposits that reach a thickness of 14 mm in some parts of the body. In southern Ontario, the Woodchuck disappears from above ground in September and most animals begin hibernation by the first of October. Jean Ferron found that animals in northern Quebec entered hibernation between September 18 and November 16. Unlike our other marmot species, the Woodchuck hibernates alone or, rarely, in pairs. The hibernaculum is located in a site with well-drained soils and a southern exposure. In northern Canada, this species faces severe winter conditions, comparable to those of alpine marmots such as the Hoary Marmot, with burrow temperatures falling as low as 2°C.

The Woodchuck is mainly diurnal. In early spring most activity is concentrated around midday, and in summer its activity peaks in early morning and mid afternoon. Vocalizations include a sharp alarm-call whistle and a two-part whistle emitted in the burrow, consisting of an intense shriek followed by a warble or chuckling sound. The Woodchuck scent-marks its territory by rubbing its muzzle on various objects around its burrow. It feeds on a variety of plants, mostly clovers, Alfalfa and dandelions, and opportunistically on invertebrates. This species will climb into trees and shrubs to feed on leaves.

Mating takes place above ground shortly after animals emerge from hibernation in late February to April. The young are usually born from early April to mid May after a gestation period of 31 to 32 days. Other than a nursing female taken May 21 in Mount Revelstoke National Park, no data exits on the breeding season in British Columbia. Females produce one litter per year. Litter sizes ranges from 1 to 9, with estimates for average litter size from 3.1 to 4.0 young. Weighing about 27 grams, the newborn young are helpless, blind, and naked except for short hairs on the muzzle, chin and head. By four weeks their eyes open; by four to six weeks they are weaned and begin to emerge above ground. By the time they enter hibernation in autumn the young will weigh about three kilograms; they reach full adult size in their second or third year. Although some Woodchucks reach sexual maturity the spring after their birth, most breed when they are two years old. This species can survive 4 to 6 years in the wild, and one animal in captivity lived for 10 years. Predators include the Coyote, Red Fox, Bobcat and various raptors.

Range
The Woodchuck has the broadest distributional area of any North American marmot, ranging across eastern North America and central Canada through western Canada to central Alaska. In British Columbia, it inhabits the Columbia Mountains, southern Rocky Mountains and most of the central and northern interior. It is generally absent from coastal regions, but there are records from the Skeena and Stikine river valleys in the Coast Mountains. The Woodchuck also appears to be absent from most of the dry grasslands of the central and southern interior, where it is replaced by the Yellow-bellied Marmot.

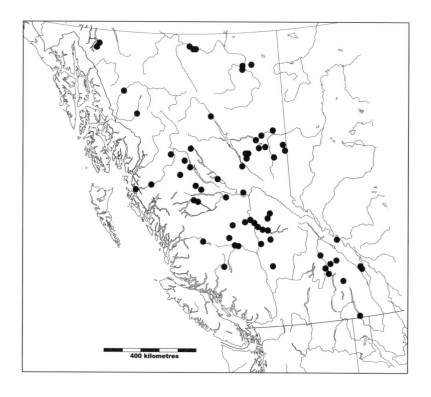

Taxonomy

Nine subspecies are recognized, defined mostly from minor differences in fur colour; three are found in British Columbia:

Marmota monax canadensis (Erxleben) – an extensive distribution across North America from the Maritimes and New England to Alberta, the southern Northwest Territories and the Peace River and Fort Nelson lowlands in eastern B.C.

Marmota monax ochracea Swarth – a pale race found in central Alaska, the southern Yukon Territory, and extreme north-central (Liard River) and northwestern (Haines Triangle to Atlin) regions of B.C.

Marmota monax petrensis Howell – from northern Idaho and extreme northeastern Washington across southeastern and central B.C. as far north as the Stikine River.

Conservation Status

British Columbian populations are not of conservation concern.

Remarks

Most of our knowledge on the basic biology of this species is based on studies done in the agricultural areas of eastern North America. Comparative studies on Woodchucks living in the mountainous regions of British Columbia are needed to determine if the montane populations show any major difference in their life history. Another intriguing research area would be the relationships between the distributions of the Woodchuck, Hoary Marmot and Yellow-bellied Marmot in B.C., and the possible effects of competition between them. Although the Hoary Marmot and Woodchuck overlap broadly in their general distributions, they appear to be separated by elevation, with the Hoary Marmot restricted to alpine areas. Similarly, the Woodchuck is absent from areas of the dry interior inhabited by the Yellow-bellied Marmot other than in a narrow zone around the Williams Lake area where both marmots evidently come into contact.

Selected References: Cowan 1933, Ferron 1997, Grizzell 1955, Kwiecinski 1998, Meier 1992.

Vancouver Island Marmot
Marmota vancouverensis

Other Common Names: Vancouver Marmot.

Description

The Vancouver Island Marmot is a large marmot with grizzled brown to black dorsal pelage, a white nose, and attractive white markings on the forehead, chin and belly. Fresh pelage tends to be nearly black, but as the fur ages it fades to tan or light brown. Old animals often have variegated dark-and-pale fur. The posterior pad on the sole of the hind foot is round (figure 41). The skull is straight in dorsal profile, the interorbital region has a distinct depression and the postorbital processes are at right angles to main axis of skull. The posterior border of the nasal bones form a V-shaped notch.

Measurements:
 total length: 668 (580-750) n=23
 tail vertebrae: 200 (162-300) n=23
 hind foot: 97 (80-105) n=22
 ear: 29 (24-35) n=13
 weight: 3.76 kg (3.20-4.40) n=3

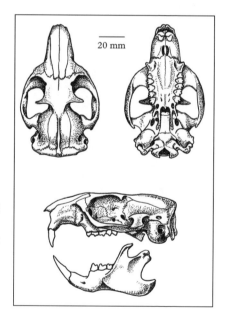

20 mm

Dental Formula:
 incisors: 1/1
 canines: 0/0
 premolars: 2/1
 molars: 3/3

Identification:

This is the only marmot found on Vancouver Island. The fur colour resembles that of melanic Hoary Marmots (*Marmota caligata*) but the two species do not overlap in their distributions.

Natural History

Typically this species inhabits open subalpine meadows (1000 to 1400 metres elevation) on south to west facing slopes where avalanches and snow creep inhibit tree growth. These natural habitats, relatively uncommon on Vancouver Island, usually have scattered Mountain Hemlock, Subalpine Fir or Yellow-cedar trees, Bracken Fern, grasses, sedges, and various forbs such as Spreading Flox, Arctic Lupine, Subalpine Daisy, Woolly Eriophyllum, Pearly Everlasting, Leafy Aster and paintbrush. Large rocks or tree stumps that provide lookouts are an essential component of the habitat. Vancouver Island Marmots also live in talus slides or cliffs, dense willow thickets bordering open meadows, ski hills, road banks, mine tailings and heather meadows. They also exploit logged habitats – colonies have been found as low as 730 metres elevation in clear-cuts. Research by Andrew Bryant has shown that marmots may occupy clear-cuts up to 20 years after logging, but these are temporary habitats that eventually become unsuitable when trees and shrubs regenerate.

The Vancouver Island Marmot is Canada's most endangered mammal. Because of this status, researchers and government biologists have been conducting comprehensive population inventories on this marmot for several decades. The current population (estimated in 2004) is fewer than 50 animals distributed in six separate locations. In the mid 1980s, 300 to 350 animals lived in more than 30 colonies. The population may have increased at this time due to new habitats being created by logging in subalpine areas, but the numbers began declining in the late 1980s and they continue to decline.

The Vancouver Island Marmot lives in colonies of one or two family groups. A typical family group has a breeding-age male, one or two breeding-age females, and immature animals two years old and younger. The largest colonies, which tend to be concentrated in the centre of the range, can have as many as 16 young-of-the-year and a dozen animals one year or older. But most colonies, especially at the periphery of the range, are much smaller with a few young-of-the-year and two to five animals one year or older. After two years of age some animals disperse from the colony of their birth. Tagged animals studied by Andrew Bryant moved about 7.4 km between sites, but dispersing animals are probably capable of moving much greater distances. A number of solitary Vancouver Island Marmots have appeared at low-elevation locations on Vancouver Island far from any known colony.

Art Martell and Robert Milko identified the remains of ferns, lichens, fungi, moss, juniper, grasses, sedges, forbs and blueberry in marmot feces collected from natural habitats. But they found that only a few plant species form the bulk of the diet. In spring, Vancouver Island Marmots feed mostly on grasses such as Timber Oat-grass, sedges and phlox. In summer, they shift to forbs, mostly lupines and Woolly Eriophyllum. Although their food habits in natural meadows are relatively well known, their diet in logged habitats has yet to be studied.

Vancouver Island Marmots use their burrows for raising young, hibernating, and escaping predators or temperature extremes. Research by Andrew Bryant has revealed that marmots use the same burrows in different years. The most elaborate burrows are those used for giving birth – constructed under large rocks or tree roots, they tend to have several entrances. Escape burrows are usually shallower and provide temporary shelter from predators. Hibernacula are situated in sites below steep slopes where the snow cover persists throughout winter; presumably, a deep cover of snow insulates the hibernacula from temperature variations. An entire family group will share a hibernaculum. The marmots enter it in November and plug the entrances with vegetation before settling down to hibernate.

Vancouver Island Marmots emit piercing whistle-like alarm calls when they see a predator or intruder – to people studying them, their distinctive call is often the first indication that a colony is nearby. These marmots can also hiss, growl, scream and call "kee-aw". This is a highly social rodent. Douglas Heard described 13 distinct behavioural interactions among Vancouver Island Marmots. The most common were greetings where two or more marmots touch cheeks or noses, and play-fighting where two standing marmots push against each other's chests with their forelimbs. The breeding structure of the Vancouver Island Marmot has been debated at length by biologists. It appears that most family groups are monogamous. But Andrew Bryant observed three colonies where a single male mated with two or three females in the same year.

The Vancouver Island Marmot is most active above ground in mornings and evenings, when it is either feeding or resting. In summer, it curtails its above-ground activity when the temperature exceeds about 20°C. As the summer progresses, above-ground activity steadily declines until animals enter hibernation in early October. The hibernation period lasts seven to eight months, until animals emerge in early May.

Mating presumably occurs above ground within the first few weeks after the marmots emerge from hibernation. The young are born in the underground burrow following a gestation period of 28 to 33 days. The litter size ranges from two to five but is usually three. The length of the nursing period is unknown, but the first young appear above ground in late June or early July. Although a few females produce litters in consecutive years, they generally breed in alternate years. Most females do not reach sexual maturity until they are four years of age, but they are still capable of breeding at nine years.

Based on ear-tagged individuals, Andrew Bryant documented that Vancouver Island Marmots can survive 10 years in the wild, but

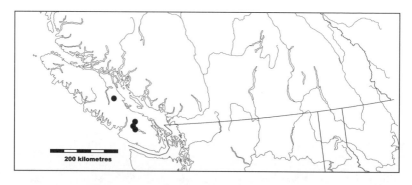

few reach this age. Mortality is highest for young-of-the-year. Survival varies from year to year and with habitat. Survival rates of all ages appear to be poorer in clear-cut habitats created by logging. Known predators include the Golden Eagle, Bald Eagle, Black Bear, Cougar and Grey Wolf.

Range

The Vancouver Island Marmot is confined to south-central Vancouver Island. Although it was found on 30 sites on 13 mountains during the inventories conducted in the 1970s and 1980s, by 2004 the population comprised only six colonies. Five are in the Nanaimo Lakes area northwest of Lake Cowichan. An isolated colony also occurs at Mount Washington, about 74 km from the others. Curiously no active colonies have been found in the mountains of Strathcona Provincial Park, although old abandoned burrows occur there. Both historic and prehistoric records suggest that this species was more widespread on Vancouver Island, having ranged as far north as the Hankin Range east of Nimpkish Lake and west to the mountains above Tahsis.

Taxonomy

No subspecies are recognized. More research is needed to verify the species status of the Vancouver Island Marmot. It is most closely related to the Hoary Marmot (*Marmota caligata*), differing only in pelage and some skull features, traits that could be attributed to rapid divergence in a small island population. Two recent genetic studies with DNA have demonstrated that *Marmota vancouverensis* shows less genetic divergence from *Marmota caligata* than expected for a rodent species, raising some doubts about its species status.

Conservation Status

Canada's most endangered mammal, the Vancouver Island Marmot appears on provincial, national and international lists of endangered species. This rodent is rare because its habitat occurs in small scattered patches. Reasons for its decline and disappearance from large areas on Vancouver Island are still unclear. Vegetation changes associated with a warm period that occurred some 10,000 to 7,000 years ago may account for its disappearance from low elevation sites in the mountains east of Nimpkish Lake. The recent warming trend may have resulted in trees invading some alpine areas and reducing the amount of marmot habitat. But high-elevation logging (above 800 metres) within the past few decades has had the most dramatic impact on recent populations. Andrew Bryant has shown that marmots living in clear-cut habitats have lower survival and poorer reproduction rates. More significantly, with Vancouver Island Marmots dispersing into nearby clear-cuts, rather than outlying colonies, many of the peripheral colonies have gone extinct. A major recovery program is attempting to restore this species through captive breeding and introductions.

Remarks

This is the only mammalian species endemic to British Columbia. Fossils recently found in a sea cave at Port Eliza dated from about 16,000 years ago, indicating that marmots (presumably the Vancouver Island Marmot) occupied low elevations on Vancouver Island during the last glacial advance. Some biologist have argued that the Vancouver Island Marmot survived on the ice-free mountain tops of Vancouver Island throughout the last ice age, 15,500 to 14,000 years ago, but there are no fossil remains dating from the last ice age to support this. Prehistoric bones dating from 800 to 2500 years ago found in five archaeological sites demonstrate that aboriginal peoples once hunted the Vancouver Island Marmot for pelts, which they used to make robes.

Selected References: Bryant 1996, 1997; Bryant and Janz 1996; Heard 1977; Martell and Milko 1986; Nagorsen 1987; Steppan et al. 1999.

Eastern Grey Squirrel
Sciurus carolinensis

Other Common Names: Grey Squirrel.

Description

The Eastern Grey Squirrel is a large tree squirrel with a long bushy tail. It exhibits three distinct colour phases: grey, intermediate and black. The grey morph has a grey dorsal pelage with a red or cinnamon tinge; the ventral pelage is white. Patches of white hairs occur behind the ears, and the tail is edged with long white hairs. The intermediate and black morphs are forms of melanism – the black morph is pure black, the intermediate morph is a brownish colour that results from reddish-brown or blonde bands in the hairs. All three colour morphs occur in southwestern British Columbia – the black and intermediate morphs dominate the lower Fraser River valley population, whereas the population on southeastern Vancouver Island consists almost entirely of the grey morph. The skull has parallel zygomatic arches that are not flat in a horizontal plane; the ventral border of the orbit is opposite the first upper molar. Most individuals have two upper premolars, but the first tiny peg-like premolar may be missing in some.

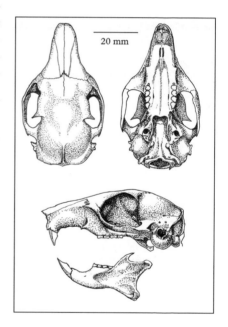

Measurements:

total length: 476 (425-526) n=21
tail vertebrae: 216 (186-250) n=21
hind foot: 67 (60-72) n=22
ear: 28 (23-32) n=19
weight: 534.8 (309.3-700.6) n=21

Dental Formula:
  incisors: 1/1
  canines: 0/0
  premolars: 2/1
  molars: 3/3

Identification:

The Eastern Grey Squirrel is easily distinguished from the native Douglas' Squirrel (*Tamiasciurus douglasii*) and Red Squirrel (*T. hudsonicus*) by its pelage colour and larger size (tail vertebrae length > 180 mm, hind foot length > 60 mm, skull length > 55 mm, and maxillary toothrow > 10 mm). The skull also differs from those of the native squirrels in having the anterior border of the orbit opposite the first upper molar. Although the Eastern Grey Squirrel superficially resembles the Eastern Fox Squirrel (*Sciurus niger*) – another recent introduction to the province – the distributions of these two squirrels are widely separated in British Columbia. The Eastern Fox Squirrel can be distinguished by its yellow-orange ventral pelage, a reddish-orange patch behind the ear and a skull with only one pair of upper premolars.

Natural History

Throughout its native range in eastern North America, the Eastern Grey Squirrel inhabits deciduous forests and wooded habitats in urban areas. In British Columbia, it is restricted to low elevations in the Coastal Douglas-fir biogeoclimatic zone. Based on an analysis of sightings in six broad habitat types, Emily Gonzales concluded that Eastern Grey Squirrels in southwestern B.C. are most common in residential areas and avoid agricultural and undeveloped habitats. Nevertheless, the concentration of observations from residential areas may simply mean that there are more observers there or that squirrels are easier to detect in these areas. Don Robinson and Ian McTaggart Cowan's study in Stanley Park, Vancouver, is the only detailed field study done on habitat use in B.C. They found that the Eastern Grey Squirrel preferred deciduous and mixed forests of Vine Maple, Beaked Hazelnut, Western Hemlock and Douglas-fir, but avoided stands of coniferous trees. It used stands of Vine Maple mostly for feeding, and made dens in habitats with a mix of deciduous and coniferous trees. In other parts of the Fraser River valley, this squirrel inhabits deciduous and mixed forests in urban and suburban landscapes. The Vancouver Island population is mainly associated with Garry Oak and Bigleaf Maple stands in urban and rural areas of Victoria. In the Sooke and Malahat regions, Eastern Grey Squirrels are also established in mixed and coniferous forests of Douglas-fir, Grand Fir and Western Hemlock.

In eastern North America, population densities are usually less than three animals per hectare, although much higher populations can occur in urban parks. The only data available for B.C., from Robinson and Cowan's study in Stanley Park, estimates 1.7 to 2.2 animals per hectare. In an urban park setting in Toronto in eastern Canada, Donald Thompson reported that average home-range size ranged from 0.9 to 2.0 ha outside the breeding season, but expanded to 1.5 to 8.8 ha during the breeding season with males occupying larger areas than females. No estimates of home-range size are available for the B.C. populations.

This species is strictly diurnal. In spring and summer it has morning and afternoon activity peaks, but in winter only one, around midday. Unlike Red and Douglas' squirrels, the Eastern Grey Squirrel is not highly territorial. The home ranges of neighbouring animals overlap extensively with no segregation of the sexes or different age groups; but residents show a clear dominance hierarchy, with males dominant over females. Except in the breeding season, when the young animals have to establish a home range, aggressive behaviour among resident Eastern Grey Squirrels is mainly directed toward unfamiliar intruders. Although some young animals disperse to new areas, most tend to remain within the general area of their birth. Males, however, occasionally disperse several kilometres from their birth site. Adult Eastern Grey Squirrels emit an assortment of alarm calls consisting of various combinations of four distinct sounds: a buzz, "kuk", "quaa" and a moan. Females in heat advertise their breeding condition to males by emitting a distinctive "quaa"/moan call.

The Eastern Grey Squirrel constructs nests either in tree cavities or on branches. In B.C., this species usually nests in tree dens located in the cavities of large hollow Western Redcedar, Bigleaf Maple or Garry Oak. Leaf nests located on tree branches are less common. Of 26 leaf nests found in Stanley Park by Robinson and Cowan, 24 were in large Western Hemlock trees and 4 were in large Western Redcedars. The nests (about 40 cm long, 36 cm wide and 24 cm high) consisted of a thick outer layer of twigs and cedar bark lined with maple leaves, shredded cedar bark or hemlock twigs. They sat on branches 11 to 14 metres above ground either in the crown of the tree or near the main tree trunk. In eastern North America, this species nests communally, though sexes usually nest separately. Unrelated males will nest together, but female nesting groups usually consist of related individuals. The largest nesting groups (up to nine animals per den) occur in autumn and winter, suggesting that communal nesting may be a strategy to conserve heat. In urban areas this species will occupy the attics of buildings.

In the deciduous forests of eastern North America, nuts are the major food source of this species. In southwestern B.C., where coniferous and mixed forests are dominant, nut-bearing trees are rare. Robinson and Cowan observed that Vine Maple and Bigleaf Maple provided about 75% of the annual food for the Eastern Grey Squirrels in Stanley Park. In spring, they ate mainly the buds of maple and oak, and throughout the rest of the year fed mostly on the fruits of maple and Beaked Hazelnut; and they supplemented their diet with the fruits from other plants, fungi and food provided by humans. In other parts of Greater Vancouver, this squirrel's diet includes nuts from introduced exotic trees such as walnut, chestnut and filbert. The diet of the southeastern Vancouver Island population is presumably similar, although unlike the population in the Greater Vancouver area, the Vancouver Island population has access to acorns of the native Garry Oak tree. In late summer, Eastern Grey Squirrels collect nuts to store for winter. They carry the nuts in the mouth, then bury them shallowly (< 5 mm) in the soil. Throughout winter, they eat most of the nuts they have cached, locating them largely by smell. In urban areas, many Eastern Grey Squirrels become habituated to bird feeders, which are a major food resource, particularly in winter.

Throughout its range in North America, the Eastern Grey Squirrel demonstrates two breeding periods: one in late winter and a second in mid summer. Robinson and Cowan observed breeding males as early as January in the Stanley Park population, but based on the dates they first observed young they estimated that successful matings occurred from early March to early April. A second breeding period occurs from mid June to mid July. The gestation period is 44 days. The average size of 11 litters for the Stanley Park population was 1.5, which is lower than estimates of 1.8 to 3.7 for populations living in eastern North America. At birth, Eastern Grey Squirrels are naked and weigh 13 to 18 grams. They first appear at the entrance of the den around 57 days, by 65 days they can climb in tree branches, and by 73 days they make their first descents to the ground. They are weaned at 7 to 10 weeks. In B.C., young from the first breeding period appear from early June to early July; young from the summer litter emerge from mid July to the end of August. In some regions of eastern North America, during years of abundant food resources, as many as 36% of females in a population will produce two litters in a year, but the percentage of females that produce two litters a year in southwestern B.C. is unknown. In eastern Canada, females breed at 10.5 to 11 months; males breed at about 15 months. In the wild, this species can live up to 12 years; a captive female survived more than 20 years. Major predators in southwestern B.C. are the Barred Owl, Northern Goshawk, Great Horned Owl, Western Screech-Owl and Domestic Cat.

Range

Native to eastern Canada and the United States, the Eastern Grey Squirrel has been introduced into a number of areas in western North America including two separate introductions to British Columbia. The population in the Lower Mainland in the Fraser River valley originated from six to eight animals that were obtained from the New York Parks Department and introduced to Stanley

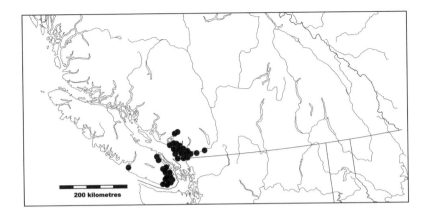

200 kilometres

Park, Vancouver, about 1914. For many decades this population remained confined to the park, but in the 1970s animals appeared outside the park and, by the 1980s, this species began to spread throughout the greater Vancouver area. It has now spread to North Vancouver, east to Chilliwack and south to Boundary Bay. Emily Gonzales also reported sightings on Bowen Island in Howe Sound, the Squamish area, and at 100 Mile House and Quesnel in B.C.'s interior.

A second population, restricted to southeastern Vancouver Island, was introduced in the late 1960s when three captive Eastern Grey Squirrels from Ontario escaped from a farm in the Metchosin area near Victoria. It has now spread as far north as Duncan and west to the Sooke area. There are also recent sightings from the Ladysmith area and from Bamfield on the west coast.

The recent spread of this species in southwestern B.C. has been accelerated by humans capturing and releasing squirrels into new areas. Several operators of wildlife rehabilitation centres have admitted to me that they released Eastern Grey Squirrels into new areas after the animals recovered from injuries.

Taxonomy

Five subspecies are recognized; both populations in British Columbia probably originated from animals brought from Ontario:

Sciurus carolinensis pennsylvanicus Ord – an eastern race found throughout eastern Canada and the northeastern United States.

Conservation Status

Except for its potential impact on our native tree squirrels (*Tamiasciurus douglasii* and *T. hudsonicus*) and Garry Oak trees, this species is not of conservation concern.

Remarks

Populations of Eastern Grey Squirrel now found on southeastern Vancouver Island and the lower Fraser River valley are derived from very small founder populations. Evidently the original animals brought to Stanley Park were a mix of colour morphs. This may account for the greater range of colours in that population. Don Robinson and Ian McTaggart Cowan estimated that in 1954 about 84% of the Eastern Grey Squirrels living in Stanley Park were melanic; Emily Gonzales recently estimated that about 70% of the Eastern Grey Squirrels in the greater Vancouver region were melanic. Colour morphs of the three original animals brought to Vancouver Island are unknown; only the grey morph was known from the island until 1998, when several brown (intermediate morph) animals appeared in Beacon Hill Park, Victoria.

There has been considerable speculation about the possible impact of this species on our native tree squirrels: the Red Squirrel and Douglas' Squirrel. Reports of declining populations of Douglas' Squirrel in the greater Vancouver region and Red Squirrel in the Victoria region are largely anecdotal. Moreover, rapid urban growth in these regions has destroyed many stands of coniferous trees, the critical habitat of our native tree squirrels. Of more concern is the effect of the Eastern Grey Squirrel on acorns of the native Garry Oak, an endangered tree on southeastern Vancouver Island. In eastern North American, this squirrel kills White Oak acorns by gnawing out the seed embryos to prevent the acorns from germinating after they are buried in caches. Given the high interest and publicity associated with declining Garry Oak habitat, it is surprising that no research has been done on the impact of Eastern Grey Squirrels on Garry Oak acorns.

Selected References: Gonzales 2000, Guiguet 1975, Koprowski 1994a, Robinson and Cowan 1954, Thompson 1978.

Eastern Fox Squirrel   *Sciurus niger*

Other Common Names: Fox Squirrel.

Description
The Eastern Fox Squirrel is a large, attractive tree squirrel with a long, bushy tail. In eastern North America, it has a number of distinct colour morphs, including melanism; but only the most common colour morph – a grey dorsal pelage washed with light orange or tan – is found in British Columbia. The ventral pelage is whitish with a pale orange wash. There are patches of orange or tan hairs behind the ears. The dorsal surface of the tail has tricoloured hairs that are tan at the base, black in the middle and orange at the distal end. The underside of the tail is deep orange. The skull has parallel zygomatic arches that are not flat in a horizontal plane; the ventral border of the orbit is opposite the first upper molar.

Measurements:
 total length: 454 (417-480)
  n=3
 tail vertebrae: 203 (201-
  204) n=3
 hind foot: 65 (62-67) n=3
 ear: 22 (18-27) n=3
 weight: 529.8 (529.6-
  530.0) n=2

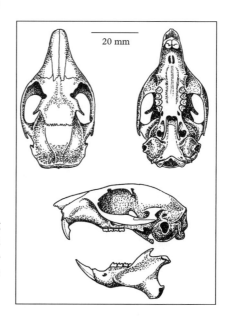

20 mm

Dental Formula:
 incisors: 1/1
 canines: 0/0
 premolars: 1/1
 molars: 3/3

Identification:
The only species resembling this squirrel is the Eastern Grey Squirrel (*Sciurus carolinensis*); see that account (page 148) for distinguishing traits.

Natural History

In its native range in the eastern and mid-western United States, the Eastern Fox Squirrel inhabits deciduous woodlots of oak, hickory, or walnut and cottonwood along streams and rivers. It also thrives in parklands associated with urban settings. Unlike the Eastern Grey Squirrel, which inhabits dense forests, the Eastern Fox Squirrel is adapted to open forest and edge habitats with sparse tree cover. In British Columbia, this squirrel lives in urban areas, open forests and orchards in the southern Okanagan Valley. Nothing is known about its biology in the province and most of the information in this account is based on the native populations in eastern North America. In its native range, population densities can reach as high as 12 animals per hectare but most population estimates range from 0.10 to 3.5 animals per hectare. Population densities seem to be closely linked to the annual production of nuts. Home-range estimates are 0.9 to 17.2 ha for females and 1.55 to 42.8 ha for males.

Eastern Fox Squirrels are strictly diurnal and active throughout the year. In winter they show a single peak of activity around midday; in other seasons they are most active after sunrise and a few hours before sunset. They exhibit limited territorial behaviour and their home ranges overlap extensively. But there is a clear hierarchy within a population, based on sex and age. Males dominate females, and juveniles tend to be subordinate to adults. Juveniles usually disperse from their birth site in late summer or autumn. Unlike the Eastern Grey Squirrel, this species rarely nests communally. Eastern Fox Squirrels use two types of nests: leaf nests on tree branches, used mostly for rearing young in summer, and nests in tree cavities. They also occupy nest boxes, especially in winter; one of the available museum specimens from B.C. was an animal found dead in an owl nest box. A highly arboreal species, the Eastern Fox Squirrel can move swiftly through the tree tops. Its repertoire of vocalizations includes barks, whines and screams.

The Eastern Fox Squirrel eats mainly seeds and nuts, but also flowers, tree buds, fruits, corn and invertebrates. It often caches nuts by burying them in shallow holes in the soil. The only native nut-bearing tree in the Okanagan Valley is the Beaked Hazelnut, but non-native trees such as elm, maple and oak have been planted in southern towns. Cherries, plums and peaches from the many orchards in this region may also be seasonal foods, and bird feeders provide a ready supply in winter.

In its native range, the Eastern Fox Squirrel has two distinct breeding periods: one from November to February and the other from April to July. Females produce two to three young after a gestation period of 44 or 45 days. Although a few females produce two litters a year, most have only one. Naked and blind, newborns weigh about 13 to 18 grams. At about 5 weeks their eyes open and they are weaned by about 12 weeks. Although females can breed at eight months of age, most do not breed until they are more than a year old. Males may live up to 8 years and females up to 13 years in the wild. Predators include the Ferruginous Hawk, Great Horned Owl, Northern Goshawk, Red-tailed Hawk, Coyote, Long-tailed Weasel, Domestic Cat and Domestic Dog.

Range

Natural populations of the Eastern Fox Squirrel range throughout the central and eastern United States; in Canada they are restricted to southern Manitoba and Pelee Island, Ontario. There have been a number of introductions in the western United States including Washington counties adjacent to British Columbia. This squirrel

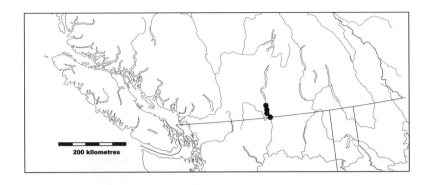

first appeared in B.C. about the mid 1980s around Osoyoos in the extreme southern Okanagan Valley. My first observation of this squirrel in B.C. was in Gyro Park, Osoyoos, in the summer of 1985. The B.C. population now ranges as far north as Okanagan Falls. The source of these Eastern Fox Squirrels is unknown, but biologists suspect that they originated from the introduced population in Okanogan County, Washington.

Taxonomy
The origin of this introduced population is obscure and its racial affinities are unknown.

Conservation Status
An alien species, the Eastern Fox Squirrel is not of conservation concern in British Columbia except for its possible impact on native species.

Remarks
As an alien species recently introduced to the province, the Eastern Fox Squirrel should be closely monitored to document its range expansion. With potential habitat in the larger urban centres of Penticton and Kelowna in the central Okanagan Valley and Vernon in the northern Okanagan Valley, this squirrel may spread throughout the entire region, especially if people capture and release animals into new areas. To help monitor this species, report sightings to local wildlife agencies.

Selected References: Koprowski 1994b, Wright and Weber 1979.

Columbian Ground Squirrel
Spermophilus columbianus

Other Common Names: None.

Description

The Columbian Ground Squirrel has grey dorsal fur washed with brown, and a spotted or mottled pattern with yellow-brown spots. The side of the neck is greyish, and the nose and front of the head are orange-brown. The underside of the tail is a mix of black and white. The hind legs and dorsal surface of the feet are reddish-brown. The skull is convex in dorsal profile, with the zygomatic arches converging anteriorly in a horizontal plane.

Measurements:
   total length: 341 (280-395) n=164
   tail vertebrae: 97 (78-120) n=161
   hind foot: 51 (42-59) n=163
   ear: 16 (10-24) n=52
   weight: 421.4 g (195.0-625.0) n=35

Dental Formula:
   incisors: 1/1
   canines: 0/0
   premolars: 2/1
   molars: 3/3

Identification:

The only other ground squirrel with similar pelage colour and markings and a similar sized skull is the Arctic Ground Squirrel (*Spermophilus parryii*). With their ranges separated by about 300 km in British Columbia, identification is not a problem. Nevertheless, the Arctic Ground Squirrel is distinguished by paler fur on the sides of its neck and the underside of the tail, and white rather than brownish

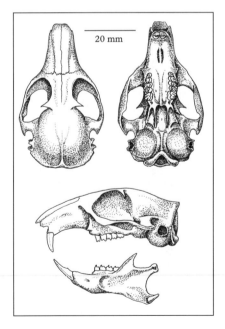

20 mm

spots, particularly in the mid dorsal region. Skulls of the two species overlap considerably in size and cannot be distinguished by any single measurement.

Natural History

The Columbian Ground Squirrel lives in alpine meadows, open forests, clearings, rangeland and agricultural land in valley bottoms. It seems to use moist meadows and grassy areas more than talus or rocky habitats, and it readily exploits open grassy meadows such as roadsides, campgrounds and logged areas. The main habitat requirements for this species are soils suitable for burrowing and an abundant supply of forbs for food. Its elevational range in the province is from about 350 to 2500 metres.

This species lives in large colonies that may contain as many as 1000 individuals. Population estimates for several colonies in the foothills of the Rocky Mountains in southwestern Alberta range from 11.6 to 70.3 animals per hectare. These estimates were based on adult and yearling animals; after the young appear above ground in early summer, the population can double. Home-range size for males averages about 4200 m² (range: 3000 to 6000 m²), decreasing from early spring to summer. Females occupy much smaller home ranges of about 200 to 700 m². Columbian Ground Squirrels make occasional excursions up to 500 metres beyond their normal home-range boundaries. Yearling animals, particularly males, usually disperse from their birth site to new colonies as far as 4 km away – a few tagged males have been recovered up to 8.5 km from their birth site.

A colonial species, the Columbian Ground Squirrel lives in groups of territorial males and females. Within their home range, males defend smaller territories against other adult males. But their territory may include the ranges of several females and their associated young. Fighting and territorial displays associated with boundary disputes are most prevalent among males early in the breeding season when they compete for females. Females are also territorial, defending a portion of their home range from other adult females. Females become most aggressive and territorial when their young are born.

Strictly diurnal, the Columbian Ground Squirrel is most active in the early morning and late afternoon. As the season progresses, it remains active later in the day, but spends increasingly less time above ground. Most of its activity is centred around the burrow. In

established colonies, well-worn pathways extend between the burrow entrances and nearby patches of vegetation where feeding is concentrated. A typical burrow system consists of a number of interconnected tunnels up to 60 cm below ground, with several nest chambers. The system has numerous surface entrances that provide ready escape from predators. When alarmed, a Columbian Ground Squirrel stands upright facing the direction of the intruder or predator and emits shrill chirps that alert other members of the colony. Distinctive flicking movements of the tail from side-to-side or vertically acts as visual signals of alarm or uneasiness. Other alarm calls include a soft chirp and "churr" calls emitted from the burrow. After retreating underground, this squirrel often returns quickly to its burrow entrance to investigate. Females use small inconspicuous burrows located at the edge of the main burrow systems for raising their young. They often plug the entrances to their brood burrows with soil, presumably to protect their young from predators and extreme weather. Columbian Ground Squirrels can climb trees and shrubs.

The Columbian Ground Squirrel hibernates for seven to eight months of the year. Depending on elevation and the timing of spring snowmelt, it emerges from hibernation from mid April to early May.

Adult males are the first to appear above ground, followed a week or so later by adult females. Yearling animals are the last to emerge from hibernation appearing above ground in May. This species depends on its stored fat reserves for winter survival. Small amounts of food have been reported in the hibernation burrows of colonies living in agricultural regions, but no food caches have been found in the hibernacula of those living in the foothills of the Rocky Mountains. Animals begin to go underground for hibernation in late July; in most regions all above-ground activity stops by the end of August. They follow the same sequence for entering hibernation as for emerging: adult males first, followed by adult females, then yearlings; young-of-the-year are the last to hibernate. Young Columbian Ground Squirrels hibernate near their mothers and siblings, but each has a separate hibernation burrow. The hibernaculum consists of a spherical nesting chamber about 30 to 80 cm below the surface lined with dried plant material. A short tunnel (5 to 15 cm) connects the chamber to the main burrow tunnels. The small exit tunnel (3 to 6 cm in diameter) that leads to the surface is plugged with loose soil from within the burrow. Evidently, Columbian Ground Squirrels occasionally arouse from hibernation during winter and are active above ground for short periods – tracks have been observed in the snow a few metres from their hibernaculum.

The Columbian Ground Squirrel feeds on plant material. In grazed pasture land, the diet is predominately clover. Other plants eaten include Yarrow, Alpine Timothy, fleabane, Arrow-leaved Balsamroot, Silky Lupine and grasses, particularly Bluebunch Wheatgrass and Creeping Bentgrass.

The breeding season begins in late April and lasts until early June. At high elevations where spring emergence from hibernation and breeding is closely linked with snow depth, timing of the breeding season can vary considerably from year-to-year, but it is often three to four weeks later than the breeding schedule for populations living at lower elevations. Within a week after they emerge from hibernation, females mate underground in their burrow. The young are born in May or June after a gestation period of 24 days. The average litter size ranges from 2.7 to 5.4, with populations at high elevations producing fewer young. Typically, only two or three young will survive to weaning stage. At birth, Columbian Ground Squirrels are naked, blind and weigh 7 to 8 grams. At 21 to 23 days their eyes open; by 23 to 24 days they eat solid food. Young first appear above ground in late June or early July. By the time they enter hibernation in late summer, young-of-the-year have tripled in body weight. Columbian Ground Squirrels continue to increase in weight during their second and third summers only attaining their full adult body size by their fourth summer.

Females have only one litter per year. Most Columbian Ground Squirrels do not reach sexual maturity until they are at least two years old. But in productive habitats, particularly at low elevations or in years of mild weather, a few females breed in the spring after their birth. The young remain in the mother's territory during their first summer, then disperse to a new area the following spring when they are a year old.

The Columbian Ground Squirrel is the principal food of the endangered Badger. Other mammalian predators include the Bobcat, Coyote, Grizzly Bear, Long-tailed Weasel and Marten. Avian predators include the Golden Eagle, Northern Goshawk and Red-tailed Hawk. A study done in the foothills of western Alberta found that Columbian Ground Squirrels accounted for about 80% of the prey eaten by Golden Eagles.

Range
Associated with the western cordillera of North America, the Columbian Ground Squirrel ranges from northeastern Oregon and Washington, northern Idaho and northwestern Montana to south-ern British Columbia and Alberta. In B.C., it occurs as far north as

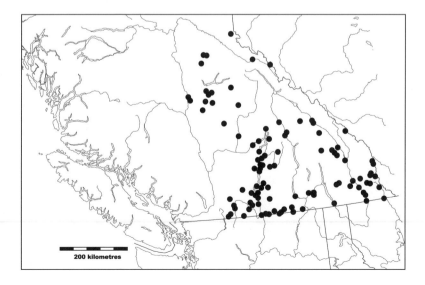

200 kilometres

Intersection Mountain north of Yellowhead Pass in the Rocky Mountains and Bowron Lake Provincial Park in the Cariboo Mountains. Western limits of its range are in the Cascade Mountains and the east side of the Fraser River in the Interior Plateau.

Taxonomy
Two subspecies are recognized; one occurs in British Columbia:
Spermophilus columbianus columbianus (Ord) – a widespread race that ranges across the northern United States and western Canada.

Conservation Status
British Columbian populations are not of conservation concern.

Remarks
There is some evidence that the Columbian Ground Squirrel has expanded its range west into the Cascade Mountains of British Columbia. In the mid 1950s, it was confined to the eastern slopes of the Cascades with historical records limited to Apex Mountain, the Ashnola River and Princeton area. Populations are now established as far west as Allison Pass and the Three Brothers area in Manning Provincial Park, and at Tulameen, west of Princeton. One of the most accessible areas to see this rodent is at Manning Park Lodge by Highway 5. Aboriginal people used pelts of the Columbian Ground Squirrel for robes.

Selected References: Elliot and Flinders 1991, Festa-Bianchet and Boag 1982, Harestad 1986, Murie and Harris 1978.

Golden-mantled Ground Squirrel
Spermophilus lateralis

Other Common Names: Mantled Ground Squirrel.

Description

The Golden-mantled Ground Squirrel is our most distinctive ground squirrel. Its dorsal pelage has a longitudinal white stripe on each side bordered by a pair of prominent black stripes. The dark, solid stripes contrast sharply with the white stripes. None of the stripes extend onto the head or face. A conspicuous rusty-red mantle is usually present on the head and shoulders, although it may be indistinct in some individuals. The ventral pelage is creamy white. The underside of the tail is pale yellow; the dorsal surface of the hind feet is pale yellow to nearly white. The skull is convex in dorsal profile; the zygomatic arches converge anteriorly in a horizontal plane.

Measurements:
total length: 268 (171-360) n=114
tail vertebrae: 97 (70-132) n=113
hind foot: 43 (35-55) n=114
ear: 17 (12-23) n=30
weight: 196.8 g (165.0-216.7) n=10

Dental Formula:
incisors: 1/1
canines: 0/0
premolars: 2/1
molars: 3/3

Identification:
Chipmunks (*Tamias*) are smaller (skull length < 40 mm, maxillary toothrow length < 7 mm) and they have five dark stripes and four light stripes that extend onto the head. The Columbian Ground Squirrel (*Spermophilus columbianus*) and Arctic Ground Squirrel (*S. parryii*) are mottled with no stripes

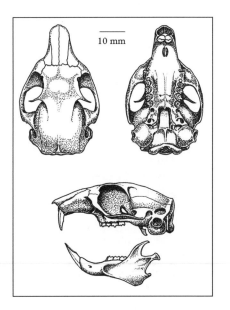

10 mm

and have larger skulls (skull length > 49 mm, maxillary toothrow length > 10 mm). The Golden-mantled Ground Squirrel is similar to the Cascade Mantled Ground Squirrel (*S. saturatus*), but their distributions do not overlap in British Columbia. Externally, the Cascade Mantled Ground Squirrel can be discriminated by its darker and duller pelage, a dull indistinct mantle, a poorly defined innermost dark stripe, and dark stripes that do not contrast sharply from its whitish stripe. The skull of the Cascade Mantled Ground Squirrel averages slightly larger than that of the Golden-mantled Ground Squirrel, but all skull measurements overlap between the two species.

Natural History
The Golden-mantled Ground Squirrel is most common in rock slides, boulder fields and talus in open subalpine and alpine areas. It also inhabits open forests of Subalpine Fir, Engelmann Spruce, Western Redcedar and Lodgepole Pine. A low-elevation population in Okanagan Mountain Provincial Park lives in open Ponderosa Pine forest along the edge of Okanagan Lake. This species readily exploits habitats disturbed by fire or logging, and human-made habitats such as rock debris found at the sides of railroad beds. Necessary features of Golden-mantled Ground Squirrel habitat are rocks, stumps, fallen trees or logging debris that provide cover from predators. Its elevational range in the province is from 397 metres in the Okanagan Valley to 2600 metres in the Rocky Mountains.

The Golden-mantled Ground Squirrel is essentially a solitary animal that does not share its home range with other adults. Females tolerate their young only during the brief nursing period. After weaning, the young disperse from their birth site to establish their own home ranges. Although neighbouring animals commonly chase each other, fight and engage in threat displays, the extent to which they maintain a defended territory is unknown. There have been no studies of movements or population densities.

Depending on elevation, this species hibernates for six to seven months. In the Sierra Nevada Mountains of California, Michael Bronson found that this squirrel emerges from hibernation between early April and late May, with males tending to emerge several weeks before the females. Animals living at lower elevations emerge earlier than those living at higher elevations. Throughout summer, Golden-mantled Ground Squirrels steadily gain weight for hibernation. This rodent does not store food in its burrow, but relies solely on fat

reserves. At high elevations, adults begin to go underground as early as late August – by late October most individuals are in hibernation.

The Golden-mantled Ground Squirrel is diurnal, displaying two peaks in feeding activity: in early morning and in late afternoon. It has several types of high-pitched "tick" and "tsp" alarm calls. Rocks and logs provide lookout sites where the squirrel watches for predators or intruders. This species is a capable climber, up to 10 metres high, in shrubs and trees. It excavates burrows under rocks or fallen trees, using them for shelter, protection from predators and winter hibernation. In subalpine or alpine areas, it locates the burrow entrance under rocks. A typical burrow system has one or more entrances, about 7 to 8 cm in diameter, leading to several horizontal tunnels and side passages about 20 to 90 cm below ground. Each burrow system has several nest chambers located deep in the main or side tunnels. The nest chamber is lined with dried leaves, grass and shredded bark.

The Golden-mantled Ground Squirrel eats seeds, forbs, fungi, invertebrates and carrion. Its major food items are the flowers, leaves, stems and seed heads of plants such as asters, butterweeds, lupines, vetches and dandelions. In late summer it consumes large amounts of truffles. Occasionally, this squirrel eats animals, such as invertebrates, carrion, bird eggs, and nestling birds, but these are minor items in the diet.

Mating occurs shortly after animals emerge from hibernation in spring. Remarkably there are no reproductive data for British

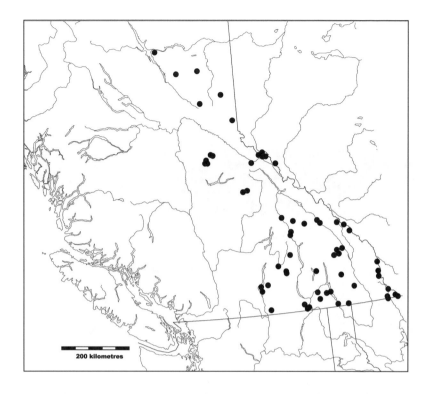

200 kilometres

Columbian populations. In other regions the breeding season begins in March or April. Females have only one litter a year averaging about five young (range: three to eight). After a gestation period of 26 to 33 days, young are born in May or June, depending on elevation. Newborns are naked with closed eyes and weigh about six grams. At 22 to 31 days their eyes open, and by 28 to 32 days they begin to eat solid food. Weaned young leave their natal nest at 35 to 40 days, and shortly after, disperse from their birth site to establish their own home range. Both sexes reach sexual maturity in their second spring, but in some high-elevation populations many animals one year old – and even two years old – fail to breed. This species survives up to seven years in the wild and eight years in captivity. Mammalian predators include the Coyote, Ermine and Bobcat; avian predators include the Golden Eagle and various species of large hawks.

Range
The Golden-mantled Ground Squirrel is found throughout the western Cordillera from California, New Mexico and Arizona north to eastern British Columbia and western Alberta. In B.C., it occupies the Monashee, Selkirk, Purcell, Cariboo and Rocky mountains ranging as far north as Mount Selwyn in the Rocky Mountains and Bowron Lakes Provincial Park in the Cariboo Mountains. The western limit of its range is the Okanagan Valley on the east side of Okanagan Lake. West of the Okanagan Valley, it is replaced by the ecologically similar Cascade Mantled Ground Squirrel.

Taxonomy
Thirteen subspecies are recognized; one occurs in British Columbia:
Spermophilus lateralis tescorum (Hollister) – a large richly coloured race that occupies western Canada, Idaho, and northwestern Montana.

Conservation Status
British Columbian populations are not of conservation concern.

Remarks
Although it tends to be quiet and somewhat secretive, this is one of the more common small mammals of alpine meadows and talus slopes in the high country. Look for it living in association with the American Pika, Columbian Ground Squirrel and Hoary Marmot. Despite being conspicuous and easily observed, particularly in open habitats at the treeline, no ecological or behavioural studies have been done on the British Columbian populations.

Selected References: Bartels and Thompson 1993, Bihr and Smith 1998, Bronson 1979, Ferron 1985, McKeever 1964.

Arctic Ground Squirrel
Spermophilus parryii

Other Common Names: Parry Ground Squirrel, Yukon Ground Squirrel.

Description

The Arctic Ground Squirrel has grey dorsal fur washed with brown and spotted or mottled with whitish or tan spots. In some individuals the spots are indistinct. The sides of the neck are tawny brown; the nose and front of the head are orange-brown. The underside of the tail is reddish with a black tip; the hind legs and dorsal surface of the feet are grey to yellowish-brown. In arctic regions, some populations may show a high incidence of melanism, but black individuals are rare in British Columbia. The skull is convex in dorsal profile; the zygomatic arches converge anteriorly in a horizontal plane.

Measurements:
  total length: 332 (300-375) n=84
  tail vertebrae: 87 (72-111)
    n=94
  hind foot: 52 (45-57)n=96
  ear: 15 (9-18) n=8
  weight: 462.8 g (350.0-
    557.0) n=14

Dental Formula:
  incisors: 1/1
  canines: 0/0
  premolars: 2/1
  molars: 3/3

Identification:
The only similar British Columbian mammal is the Columbian Ground Squirrel, but its range does not overlap with the Arctic Ground Squirrel. See its account for identification traits.

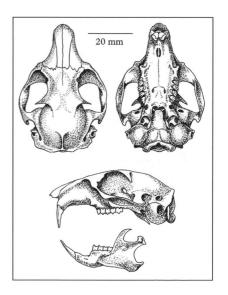

20 mm

Natural History

The Arctic Ground Squirrel lives in alpine meadows, clearings adjacent to lakes, river banks, hillsides, open shrub meadows and patchy spruce forests with openings. Essential habitat features are well drained soils suitable for burrowing, a close supply of food plants and open areas where the squirrel has a non-restricted view of potential predators. Elevational records in British Columbia range from 400 to 2000 metres.

Estimates of population densities for the southern Yukon Territory range from about two to six animals per hectare. Home-range size, which is determined by the availability of food, is from 3.1 to 20.2 ha for males and 2.3 to 3.1 ha for females. This mobile species travels up to a kilometre during its daily activities.

Dispersing young-of-the-year travel considerable distances. Andrea Byom and Charles Krebs found that yearling males dispersed an average distance of 500 metres from their birth site – a few animals moved as far as 3.8 km. Females remained closer to home, within about 120 metres, although a few moved as far as 1800 metres.

This is a social rodent living in groups of related females and territorial males. Shortly after males emerge from hibernation in early spring they establish and defend territories that contain the home ranges of several females. Highly aggressive, males chase or fight with any male intruder. Females do not appear to select a male but simply mate with the dominant male where its burrow is located. Home ranges of females may overlap, but each occupies a separate burrow system. Although females are more tolerant of other individuals than males, they become more aggressive after the young are born. By late summer, the breeding territories break down, but the Arctic Ground Squirrel still defends small areas around its hibernation burrow.

Each Arctic Ground Squirrel has several resident burrows in its home range, using them for shelter, raising young and hibernation. A typical burrow has 6 to 15 active surface openings, a number of tunnels 10 to 15 cm wide and more than a metre deep and an expanded cavity with a nest chamber; the spherical nest is lined with dry grass, fur and leaves. The home range also contains a number of small, shallow transient burrow systems that provide temporary shelter from predators when the squirrel is foraging or moving between resident burrows. Although it is heavily dependant on burrows for shelter in open meadows and tundra, some Arctic Ground Squirrels living in forested habitats nest in cavities of dead trees 1.5 to 2 metres above ground.

In arctic regions with severe winters, the hibernation period is 8 to 10 months. In the subarctic forest of the southern Yukon Territory and British Columbia where the winters are less severe, the Arctic Ground Squirrel hibernates for seven to eight months. It emerges from hibernation in early to mid April, the males appearing above ground one to two weeks before females. Mature animals tend to emerge before yearlings of the same sex. Although males will initially lose weight for several weeks after they emerge in the spring because of the stress of mating and defending a territory, Arctic Ground Squirrels steadily accumulate fat reserves throughout the summer in preparation for hibernation. The first to enter hibernation are adult females. They disappear as early as late July and are last seen above ground in early September. Young-of-the-year females begin to hibernate in late August, and by early October they have disappeared. Males are the last group to hibernate – some remain active until the first snowfalls in October. Because they store food reserves in their burrow, males do not lose much weight during their hibernation. In contrast, the females do not store any food reserves and they lose considerable weight over the winter. Arctic Ground Squirrels face the lowest winter temperatures of any hibernating rodent – the temperature near their nests in the burrow can fall to -27°C. Laboratory experiments have shown that hibernating Arctic Ground Squirrels can allow their body temperatures to as fall as low as -2.9°C without freezing.

Recent observations of populations living in forested habitats have shown that this species frequently climbs trees, deadfalls and stumps up to six metres in height. These structures may provide temporary refuge and a clear view for sighting predators. Arctic Ground Squirrels produce six distinct types of vocalizations including two types of alarm calls. Terrestrial predators provoke a three-note chatter call, audible for several metres, alerting other members of the colony. When a predator approaches, the squirrel emits five or more fading chatter calls as it retreats to its burrow. Aerial predators, such as hawks, owls and jaegers, elicit a bird-like whistle call. Other squirrels respond to this call by freezing and watching the sky or running to their burrow.

Arctic Ground Squirrels eat the shoots, flowers and seeds of forbs, grasses, horsetails and shrubs. Legume plants, especially lupines, hedysarums, locoweeds and vetches, seem to be the most common plants in its diet. This ground squirrel may occasionally feed on arthropods, berries, fungi and carrion. Males spend considerable time in July and August carrying seeds to their burrows to store for winter; they also cache grasses, sedges and willow leaves.

Mating occurs above ground as soon as females emerge from hibernation in late April. After a gestation period of 25 days, they give birth to litters of four to seven young. The young remain underground in their natal burrow for about four weeks where they are nursed. The young are weaned within one or two weeks after emerging from their natal burrow; by two to three weeks they disperse from their birth site to establish a home range. Both sexes reach maturity by their second spring; females produce one litter per year.

This is an important prey species for a number of mammalian and avian predators in the north. Young animals dispersing from their birth sites are particularly vulnerable. Dominant mammalian predators are the Ermine, Red Fox, Wolverine and Grizzly Bear; bears excavate burrows after the spring thaw to find Arctic Ground Squirrels. Avian predators include the Northern Goshawk, Northern Harrier, Red-tailed Hawk and Great Horned Owl.

Range

The Arctic Ground Squirrel has a vast range across northeastern Siberia, Alaska, the Yukon Territory and Northwest Territories, east to Hudson Bay and as far south as northern British Columbia. Its distributional area in the province ranges from the Haines Triangle through the Cassiar Mountains and the Skeena Mountains as far south as Tatlatui Lake near the headwaters of the Finlay River. Curiously, it is absent from the northern Rocky Mountains.

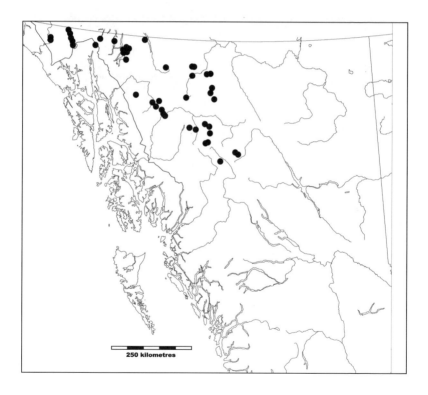

250 kilometres

Taxonomy

In the original *Mammals of British Columbia*, Cowan and Guiguet classified this species as *Spermophilus undulatus*, but this name now applies to a species that inhabits southern Siberia, Mongolia and northeastern China. Northeastern Siberian and North American populations are treated as a separate species, *S. parryii*. Depending on the authority, six or seven subspecies are recognized in North America; one occurs in British Columbia:

Spermophilus parryii plesius Osgood – the smallest and palest North American race, found in eastern Alaska, southern Yukon Territory, west of the Mackenzie River in the Northwest Territories and northwestern B.C.

Conservation Status

British Columbian populations are not of conservation concern.

Remarks

Aboriginal groups hunted the Arctic Ground Squirrel for meat and clothing. In 1999, the 550-year-old frozen remains of a man were found at the edge of a melting glacier in Tatshenshini-Alsek Wilderness Park in northern British Columbia. With these remains, known as *Kwäday Dän Ts'inchi* (Long Ago Person Found), was an elaborate robe made from 95 Arctic Ground Squirrel pelts.

Selected References: Byrom and Krebs 1999, Carl 1971, Green 1977, McLean and Towns 1981, Melchior 1971.

Cascade Mantled Ground Squirrel
Spermophilus saturatus

Other Common Names: Cascade Golden-mantled Ground Squirrel.

Description

The dorsal pelage of the Cascade Mantled Ground Squirrel has a longitudinal white or tan stripe on each side, bordered by a pair of indistinct black stripes. The body stripes do not extend onto the head. It also has a dull, poorly defined mantle of brown or reddish fur on the head and shoulders, though some individuals lack the mantle. The ventral pelage is tan or pale brown. The dorsal surface of the hind feet are tan; the underside of the tail is yellow to light orange. The skull is convex in dorsal profile; the zygomatic arches converge anteriorly in a horizontal plane.

Measurements:
total length: 300 (253-320) n=30
tail vertebrae: 103 (89-115) n=29
hind foot: 46 (40-53) n=30
ear: 22 (21-23) n=3
weight: 245.0 g (162.0-314.0) n=3

Dental Formula:
incisors: 1/1
canines: 0/0
premolars: 2/1
molars: 3/3

Identification:
The only similar mammal in British Columbia is the Golden-mantled Ground Squirrel; see that account (page 165) for identification traits.

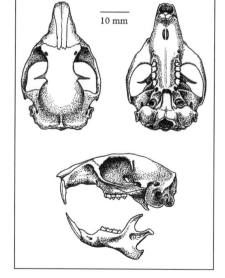

10 mm

Natural History

In British Columbia, the Cascade Mantled Ground Squirrel ranges from 700 to 2500 metres elevation with most occurrences between 1000 and 2000 metres. It lives in closed forests of Douglas-fir, Engelmann Spruce and Amabilis Fir, in subalpine stands of Subalpine Fir and Subalpine Larch, in recent clear-cut areas, on talus slopes, and in open meadows. There are no estimates of population density for B.C. populations, but Stephen Trombulak estimated populations in the eastern Cascade Mountains of Washington to be 4.5 animals per hectare in open meadows and 2 per hectare in coniferous forests.

The Cascade Mantled Ground Squirrel is strictly diurnal, spending about seven hours a day outside its burrow. Feeding activities reach a peak during the breeding season. Although the social system of this squirrel has not been studied, it is probably a solitary animal. There are no estimates of home-range size. Young animals will disperse 55 metres in forested habitat and 230 metres in open meadows. Cascade Mantled Ground Squirrels can climb three to five metres into bushes and coniferous trees, and they have been observed sitting on fence posts. Their burrows, located under rocks, stumps, logs or roots of bushes, are critical for hibernation, raising young and escaping predators. A typical burrow has several entrances and tunnels leading to a nest chamber. The nest is made from dried grass formed into a cup and lined with fresh vegetation.

This squirrel's diet includes fungi, vetch leaves, grasses, forbs, dandelion heads, bark, corms, Salal berries, huckleberries, and the seeds of conifers, grasses and lupines. In spring, green plant material dominates the diet. In late summer as plant material dries up, the squirrel switches its diet to fungi, especially truffles.

Because the Cascade Mantled Ground Squirrel does not store food, it depends solely on its fat reserves and hibernation to survive winter. After breeding, these squirrels steadily gain body weight as they accumulate fat. Animals disappear from mid August to late September to hibernate for six to eight months. Because adults and yearlings are able to accumulate sufficient fat reserves before the young-of-the-year, they enter hibernation well before the young (up to 45 days). The Cascade Mantled Ground Squirrel appears above ground from early April to mid May – the timing of spring emergence varies from year-to-year and with elevation. Reproductively active males are the first to emerge, appearing about a week before the first adult females, which emerge before the yearlings.

Reproduction studies in the eastern Cascade Mountains of Washington revealed that the mating period there lasts about two weeks in late April. After a gestation period of 28 days the young are born in underground nests. Birth dates range from May 13 to 28, with most births in the last two weeks of May. In B.C., three pregnant females were captured from May 8 to 18, suggesting a similar breeding season. The litter size averages four young (range: one to five). Females produce only one litter per year. The newborn young weigh about six grams. After nursing in the underground nest for 36 days, young appear above ground in late June or early July. Most females are capable of breeding the spring after their birth, but a few males (10%) breed as yearlings. The Cascade Mantled Ground Squirrel can survive four years in the wild. Predators include Great Horned Owl, Long-eared Owl, Northern Goshawk, Red-tailed Hawk, Coyote, Marten, Long-tailed Weasel and Bobcat.

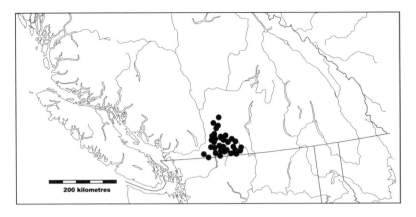

200 kilometres

Range

This species has a small distributional area confined to the Cascade Mountains of Washington and British Columbia. In B.C., it has been found as far north as Iron Mountain near Merritt, as far west as William's Peak in the Chilliwack River valley and as far east as Hedley and the mouth of the Ashnola River.

Taxonomy

Although now treated as a distinct species, a few authorities have considered the Cascade Mantled Ground Squirrel to be a subspecies of the Golden-mantled Ground Squirrel (*Spermophilus lateralis*). No subspecies are recognized for *Spermophilus saturatus*.

Conservation Status

Nationally, the Cascade Mantled Ground Squirrel was ranked not at risk by COSEWIC. Maria Leung speculated that it disappeared from the eastern edge of its historical range near the Similkameen River because of competition with the Columbian Ground Squirrel, which recently expanded its range westward in the Cascade Mountains of British Columbia. Nonetheless, the supposed range contraction of this species may largely reflect inadequate surveys at historical sites. Since Leung's surveys in the early 1990s, Cascade Mantled Ground Squirrels have been observed at several sites within in the eastern part of the historical range. Curiously, the impact of forest harvesting on this mammal has not been assessed.

Remarks
One of the best areas to observe this animal is at the lodge in Cathedral Lakes Provincial Park. More research is needed to assess the impacts of forest harvesting on this rodent and its possible competition with the Columbian Ground Squirrel.

Selected References: Leung 1991; Leung and Cheng 1997; Trombulak 1987, 1988.

Yellow-pine Chipmunk

Yellow-pine Chipmunk   *Tamias amoenus*

Other Common Names: Northwestern Chipmunk.

Description

The dorsal pelage of the Yellow-pine Chipmunk has five dark and four light stripes, with the middle dark stripe extending from the back of the head to the rump and the other dark stripes extending from the side of head to the rump. Most populations are brightly coloured with yellow-brown sides, a buff or tawny belly, and yellow to orange on the underside of the tail. But this is the most variable of any chipmunk in British Columbia: the five subspecies in the province demonstrate considerable colour variation, from rich-brown coastal forms (*Tamias amoenus felix*) to pale grey forms in the dry interior (*T. a. affinis*). The skull has a simple, rounded infra-orbital opening piercing the zygomatic plate. The baculum is small (shaft length 2.2 to 2.7 mm) with a tip length more than 30% of the shaft length. The baubellum is also small (greatest length 1.0 to 1.8 mm) with a short, indistinct base that may be notched (figure 119).

Measurements:
   total length: 213 (198-230) n=58
   tail vertebrae: 94 (79-110) n=58
   hind foot: 32 (28-34) n=61
   ear: 16 (11-18) n=47
   weight: 55.0 (42.0-77.0)
      n=50

Dental Formula:
   incisors: 1/1
   canines: 0/0
   premolars: 2/1
   molars: 3/3

Identification:
The distribution of this chipmunk overlaps with the ranges of three others in B.C.: Least Chipmunk (*Tamias minimus*), Red-tailed Chipmunk (*T. ruficaudus*) and Townsend's Chipmunk (*T. townsendii*). Townsend's Chipmunk is easily distinguished

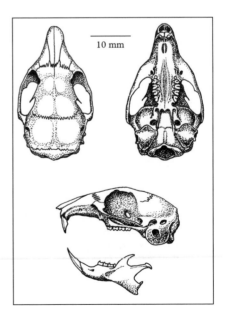

10 mm

by its dull, indistinct light stripes, frosted tail and larger size (body length > 130 mm, skull length > 37 mm, maxillary toothrow length > 6 mm).

In the Purcell and southern Rocky mountains of southeastern B.C., the Least Chipmunk is smaller (total length < 195 mm, skull length < 32.5 mm, lower jaw length < 17.5 mm) and it has a white or grey belly. But in central B.C. where it approaches the Yellow-pine Chipmunk in size and some have a similarly white belly, positive identification can only be made from genital-bone morphology.

Although the Rocky Mountain population of the Red-tailed Chipmunk has more reddish-brown on the neck, sides and underside of the tail than the Yellow-pine Chipmunk, its size and pelage traits are not completely reliable, particularly for identifying immature animals. In the Selkirk Mountains, where Yellow-pine and Red-tailed chipmunks converge in size and colour, positive identification can only be made from their genital bones.

Natural History
The Yellow-pine Chipmunk occupies the widest range of habitats of any chipmunk in British Columbia from lowland forest and arid steppe-grasslands to subalpine forests and open alpine. It ranges from sea level in coastal regions to 2130 metres in the Coast Mountains and 2300 metres in the Rocky Mountains. It lives in open forests of Western Hemlock and Douglas-fir in southwestern B.C., and forests of Ponderosa Pine, Douglas-fir, Engelmann Spruce, Lodgepole Pine, White Spruce and Subalpine Fir in the interior. This chipmunk also inhabits rocky talus and tree islands above the treeline. Open forests with extensive ground cover of woody debris seem be the optimum habitat. Thomas Sullivan and Walt Klenner's study in Lodgepole Pine forests in B.C.'s interior clearly showed this species preference for open forests. Young forests supported significantly higher populations than mature forests, and young forests commercially thinned to promote tree growth had 1.3 to 3.8 times as many chipmunks as those unaltered by thinning.

Harold Broadbrooks reported population densities of 0.7 to 1.3 animals per hectare for the Cascades of Washington, but densities as high as 12.8 per hectare were found in the open Ponderosa Pine forests of Oregon. In Lodgepole Pine forests of B.C.'s interior, Sullivan and Klenner reported average summer populations of 0.1 to 1.0 animals per hectare in mature forests, 0.5 to 3.1 per hectare in young forests, and up to 4.8 per hectare in commercially thinned young forests. The home-range size is from 0.1 to 3.2 ha, with males having the largest and nursing females tending to have smaller home ranges than nonbreeding females. Yellow-pine Chipmunks are solitary. Home ranges overlap among individuals in an area, but breeding females are territorial and will chase off other females.

The Yellow-pine Chipmunk uses both ground and tree nests. Each of the 13 burrows in eastern Washington studied by Broadbrooks had a single entrance 38 to 100 cm in length. Some burrows were simple, consisting of a single tunnel, but others had several side passages. This chipmunk builds a spherical nest, about 15 cm in diameter, from dried grass, lichen, feathers or fur. The only description of a tree nest for this species was found 2.6 metres above ground in a clump of willow trees; it was 30 cm wide and 15 cm deep, constructed from dried grass.

No information is available on the timing or duration of the Yellow-pine Chipmunk's hibernation period in British Columbia. In other regions, this chipmunk appears above ground in early April, males emerging before females. In the Rocky Mountains of Alberta, David Sheppard observed females as early as April 15. In most regions the Yellow-pine Chipmunk remains active above ground until late autumn. Broadbrooks observed it above ground until mid November in the Cascade Mountains of Washington. As is typical of most chipmunks, this species depends on stored food to survive winter. It uses brief periods of torpor, relying on reduced winter activity and a well insulated nest to conserve its winter energy demands. There are reports of this chipmunk appearing above ground in mid winter during warm spells.

Yellow-pine Chipmunks eat fungi, plants, invertebrates and bird eggs. Seeds are the most important food source for most populations. In spring and summer they eat the soft, immature seeds of ripening flowers and in late summer harvest mature seeds for winter food stores. The major plants found in food stores are Arrow-leaved Balsamroot, dandelion, arnica, penstemons, Silverleaf Phacelia, Saskatoon, Common Snowberry, knotweed, thistles, Great Mullein, Yarrow, Oceanspray, Ponderosa Pine, grasses and sedges. This chipmunk also eats conifer seeds after the cones are sufficiently mature for their scales to open. An excellent climber, Yellow-pine Chipmunks have been observed 15 metres above ground in Ponderosa Pines. In humid coastal regions, truffles or subterranean fungi are a major food. In autumn, this species makes large caches of seeds and corms in its nest for winter. Three food caches analysed

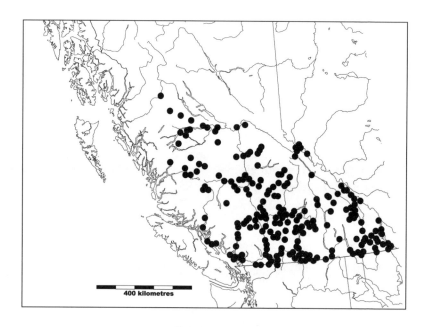

400 kilometres

by Harold Broadbrooks weighed 70 to 190 grams – they contained as many as 67,970 seeds and pieces of corm.

Yellow-pine Chipmunks mate in late March and April; females bear young in April, May or June following a gestation period of about 30 days. They produce one litter a year of four or five young. After nursing for 35 to 40 days in the nest, the young emerge above ground. In the Cascades of Washington, births occur from late April to early June, with the first young emerging above ground in early June. In the Rocky Mountains of Alberta, however, David Sheppard observed pregnant females from May 5 to June 24, and saw the first young above ground on July 5. Yellow-pine Chipmunks are capable of breeding in the year following their birth, but the proportion of yearling animals breeding in a population seems to vary considerably among populations and from year-to-year within a population. At birth, the young are naked with closed eyes and weigh about 2.6 grams. Their eyes open at about 30 days; by 6 weeks they are able to consume solid food. In the wild, Yellow-pine Chipmunks survive up to 5 years. Predators include the Badger, Long-tailed Weasel, Bobcat and Northern Goshawk.

Range

The Yellow-pine Chipmunk is distributed across western North America from California and Nevada to British Columbia and western Alberta. It is widespread across southern and central B.C., ranging as far north as Mount Robson Provincial Park in the Rocky Mountains and Hazelton and Prince George in the interior. The only population not on the mainland is on Savary Island, a small island off the Malaspina Peninsula near Powell River; it is not known if this population is native or introduced.

Conservation Status

British Columbian populations are not of conservation concern.

Taxonomy

Fourteen subspecies are recognized; five occur in British Columbia. But results from recent DNA research by John Demboski and Jack Sullivan are inconsistent with this taxonomy. Some of the named subspecies show no genetic differences. Moreover, some geographic populations of the Yellow-pine Chipmunk show strong genetic divergence. For example, populations from the Rocky and Columbia mountains and the Okanagan have sufficiently distinct DNA from that of coastal populations to suggest that the two groups represent distinct species. Jeff Good and colleagues found that a number of *Tamias amoenus* in B.C.'s Rocky Mountains carried the mitochondrial DNA of *T. ruficaudus* (Red-tailed Chipmunk), suggesting that there has been some interbreeding between the two species.

Tamias amoenus affinis Allen – a pale form associated with the eastern slopes of the Cascades in Washington and B.C., and the southern dry interior of the province from the Okanagan and Similkameen valleys north to Sorenson Lake.

Tamias amoenus felix Rhoads – a dark coastal race that ranges from the Cascade Mountains in extreme northern Washington to Bute Inlet in B.C. It has also been recorded from Savary Island off the coastal mainland.

Tamias amoenus ludibundus (Hollister) – the central Rocky Mountains of Alberta and B.C., and west to the Bowron Lakes region.

Tamias amoenus luteiventris Allen – Montana, Idaho, Wyoming, northeastern Washington, the Rocky Mountains of south-central Alberta, and B.C., where it inhabits the Columbia and Rocky mountain ranges as far north as Glacier National Park.

Tamias amoenus septentrionalis (Cowan) – restricted to the province where it inhabits the Interior Plateau area west of the Fraser River as far north as the Skeena River valley.

Remarks

Distribution of the Yellow-pine Chipmunk in British Columbia is influenced in part by the presence or absence of other chipmunk species. In the Cascade Mountains, where its range overlaps with Townsend's Chipmunk, it is usually confined to high elevations; in the Kootenay region it is replaced at high elevations by the Least Chipmunk or the Red-tailed Chipmunk. In areas where it is the sole chipmunk species (Coast, Monashee, northern Selkirk and Purcell mountains), the Yellow-pine Chipmunk ranges from the valley bottoms to alpine habitats above the treeline.

Selected References: Broadbrooks 1958, 1974; Demboski and Sullivan 2003; Good et al. 2003; Sheppard 1965, 1971; Sullivan and Klenner 2000; Sutton 1992.

Least Chipmunk   *Tamias minimus*

Other Common Names: Western Chipmunk.

Description

Our smallest chipmunk, the Least Chipmunk has the typical striping pattern of five dark and four light stripes. Generally, it has greyish-brown dorsal pelage with a grey rump, a white or grey belly, and pale yellow to light orange fur on the neck, sides, dorsal surface of the hind feet and the underside of the tail. Populations in southeastern British Columbia tend to be smaller and paler than northern populations. Melanism occurs in some northern populations. Black specimens have been taken at Talatui Lake, Bennett Lake, Bear Lake and the headwaters of the Stikine River. The skull has a simple rounded infraorbital opening piercing the zygomatic plate; it is the smallest skull of any chipmunk in the province. The baculum is small (shaft length 2.6 to 2.9 mm) with a tip that is less than 30% of the shaft length. The baubellum is small (greatest length 0.9 to 1.1 mm) with a distinct U-shaped base that tapers sharply at its proximal end (figure 118).

Measurements:
 total length: 198 (172-222) n=78
 tail vertebrae: 87 (69-101)
  n=79
 hind foot: 31 (27-34) n=85
 ear: 13 (10-16) n=75
 weight: 45.5 (36.0-62.0)
  n=76

Dental Formula:
 incisors: 1/1
 canines: 0/0
 premolars: 2/1
 molars: 3/3

Identification:
This is the only species of chipmunk throughout most of northern British Columbia, but its range overlaps with the Yellow-pine Chipmunk (*Tamias amoenus*) in the southern Rocky Mountains, Purcell

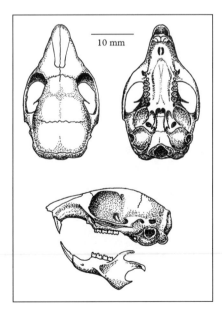

10 mm

Mountains and possibly a narrow zone around Babine Lake in the north. See the species account for the Yellow-pine Chipmunk (page 181) for diagnostic traits for discriminating the two species. The Least Chipmunk also co-occurs with the Red-tailed Chipmunk (*T. ruficaudus*) in the Rocky Mountains of extreme southeastern B.C. The Red-tailed Chipmunk is easily distinguished by its larger size (total length > 205 mm; skull length > 34 mm) and darker pelage with reddish-brown on the sides and underside of the tail.

Natural History
In northern British Columbia the Least Chipmunk ranges from 300 to 1700 metres elevation where it lives in open forests of Trembling Aspen, White Spruce, Black Spruce or Engelmann Spruce, clearings, rock slides, and open ridges or alpine areas above the treeline. In contrast, in the southern Rocky Mountains and Purcell Mountains, it is confined to elevations generally above 2000 metres, in open alpine habitats characterized by a few shrubs, stunted trees such as Subalpine Fir, Subalpine Larch, White Bark Pine, and dense mats of alpine plants such as such as Mountain Avens. In this region, the Least Chipmunk is associated with talus, rockslides or tree islands presumably because they provide cover from predators. In the northern boreal forests, this chipmunk colonizes recent burns and clear-cuts.

The Least Chipmunk's restriction to habitats above the treeline in the Columbia and southern Rocky mountains of Canada reflects its preference for rocky talus habitats, its physiological adaptations for living in harsh conditions and the effects of competition with the Yellow-pine and Red-tailed chipmunks. Better adapted to forests, these larger species exclude the Least Chipmunk from forested slopes and valleys. Yet, because it is smaller with lower food requirements and is able to tolerate arid conditions, the Least Chipmunk appears to have a competitive edge over Yellow-pine and Red-tailed chipmunks in the harsh conditions of the alpine zone.

No population estimates exist for B.C. populations. In other parts of North America, population estimates range from 0.3 to 22.2 animals per hectare. David Sheppard estimated the average home size at 1.22 ha (range: 0.39 to 3.35 ha) for males and 0.66 ha (range: 0.22 to 1.51 ha) for females for his study area in the southern Rocky Mountains of Alberta. He reported maximum distances between captures ranging from 100 to 530 metres. Least Chipmunks are most active one to three hours after sunrise and from mid to late

afternoon until sunset. They use both ground and tree nests. The shallow underground burrows are typically situated 14 to 50 cm below the surface. The burrow entrance, 3 to 5 cm in diameter, is located under a rock or the base of a shrub. The burrow may be a simple structure consisting of a single passageway (0.2 to 1.5 metres in length) with several openings or complex structures reaching 3.5 metres in length with multiple passages or openings. A burrow system terminates at a single nest chamber containing a nest made of hair, bird feathers, dried shredded grass, bark, or downy fibres from willow or poplar catkins. This chipmunk is an agile climber that often feeds in shrubs or the lower branches of trees, but researchers have reported only one tree nest, 3.7 metres above ground in the cavity of a cottonwood tree. The Least Chipmunk produces an assortment of vocalizations including a low "cluk" call and a strong "tsk" call.

Least Chipmunks eat flower heads, leaves, seeds, fruits, fungi and invertebrates such as grasshoppers. Seeds are the most critical food, because they are stored as the main winter food source. In the Rocky Mountains of Alberta, David Sheppard found that the proportion of Least Chipmunks carrying seeds in their cheek pouches reached a peak in late summer as animals accumulated their winter food stores. The most common seeds identified in the cheek pouches of chipmunks there were Kinnikinnick, cinquefoil, gooseberry, Lodgepole Pine, rose, sedge, Small-flowered Blue-eyed Mary, Soapberry and timothy. In the northern boreal forests this chipmunk eats raspberries, sedges, Small-flowered Wood-rush and Soapberry. It caches large amounts of winter food in its burrow, near the nest. Stuart Criddle described a 465 gram food cache found in Manitoba that contained 6369 cherry kernels, 5116 grains of wheat, 63 shelled acorns, 40 rye seeds and a small amount of millet seed.

The Least Chipmunk is a shallow hibernator, entering short bouts of torpor that last 80 to 110 hours. There are few data on the hibernation period in British Columbia. Anecdotal observations and studies done in the high alpine of the Rocky Mountains of Alberta, suggest that Least Chipmunks emerge from mid to late April through early May and remain above ground until mid October. In northern B.C., they have been observed above ground until the end of September.

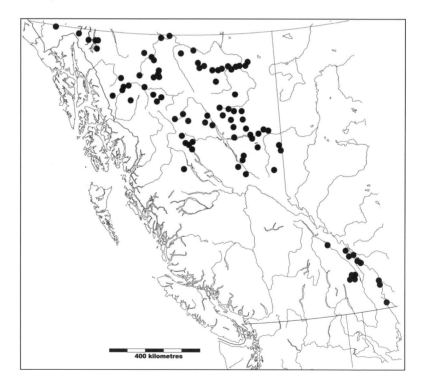

400 kilometres

In the southern Rocky Mountains of Alberta, Sheppard found that mating occurred from late April through May with young born in June after a gestation period of about 30 days; the latest birth was June 24. The young first appear above ground in mid July after nursing in their underground burrow for 35 to 40 days. Females produce only one litter each year, which averages about four young. Newborn young are naked, blind and weigh about two grams. Least Chipmunks can breed in the spring after their birth, but in western Alberta as many as one third of yearling chipmunks fail to breed. Reproductive data for northern populations are scanty, but it appears that young are born before mid June and first appear above ground by the first week of July. Embryo counts from museum specimens taken near Fort Nelson in northeastern B.C. ranged from five to seven among pregnant females taken from May 16 to 30. A female captured on May 26 was nursing and young-of-the-year were captured from June 25 to 30. This chipmunk may live up to six years in captivity. Known predators include Ermine, Long-tailed Weasel, Marten, Red-tailed Hawk, Cooper's Hawk and the Northern Goshawk.

Range

The Least Chipmunk has the largest range of any chipmunk in North America inhabiting Canada from the Yukon Territory to western Quebec and the north-central and western United States. In northern British Columbia it occurs as far south as the Omineca and Babine mountains, and the Rocky Mountains south of the Peace River. South of 55°N latitude in B.C., this species is restricted to areas above the treeline in the Rocky Mountains. In southeastern B.C., an isolated population inhabits alpine areas of the southern Purcell Mountains.

Taxonomy

Eighteen subspecies are recognized; four occur in British Columbia:

Tamias minimus borealis J.A. Allen – a large northern form that ranges from northern Montana and across Canada from the Yukon Territory and B.C. to northern Ontario; it inhabits the northeastern and north-central parts of B.C. as far north as Babine Lake and the Peace River and the Rocky Mountains as far south as Kicking Horse Pass.

Tamias minimus caniceps (Osgood) – a weakly defined race that inhabits the Yukon Territory, the western Northwest Territories, and northwestern and north-central B.C. as far south as Spatzizi Plateau and east to the western slopes of the northern Rocky Mountains; this subspecies may be indistinguishable from *T. m. borealis.*

Tamias minimus oreocetes (Merriam) – a small, pale form restricted to the Rocky Mountains of northern Montana, extreme southwestern Alberta north to the Bow River and extreme southeastern B.C. as far north as Kicking Horse Pass; its taxonomic distinctness from *T. m. borealis* is not clear.

Tamias minimus selkirki (Cowan) – a small, isolated form confined to B.C.'s southern Purcell Mountains. Although originally described by Ian McTaggart Cowan from only a few specimens, recent research has confirmed that this form is distinct.

Conservation Status

Two subspecies are listed provincially as of conservation concern: *Tamias minimus selkirki* appears on the Red List and *T. m. oreocetes* on the Blue List, primarily because of their few known occurrences in British Columbia. Genetic studies are required to confirm the taxonomic validity of *T. m. selkirki*, but its distinct morphology and isolated range endemic to the Purcell Mountains suggest that it is distinct from populations of *T. minimus* in the Rocky Mountains. There are no clear threats to this population. Because it is localized in only two areas in the Purcell Mountains (Mount Brewer and Spring Creek Basin) and isolated there, it is appropriate to consider it as potentially at risk. This subspecies is also listed as Vulnerable in the IUCN Red Book.

More taxonomic research is needed to confirm the validity of the subspecies *T. m. oreocetes*. It is known from only a few locations in B.C.'s southern Rocky Mountains, but the few occurrences reflect inadequate sampling from inaccessible alpine areas, rather than rarity. There is no evidence that this population is at risk. Much of this subspecies' range in Canada is protected, falling within the boundaries of national and provincial parks.

Remarks

The Least Chipmunk is wary and secretive, rarely venturing far beyond the protective cover of rock slides, woody debris or shrub cover. Its ventriloquist-like calls make it extremely difficult to locate even when one is nearby. Except for some taxonomic research, the biology of the British Columbian populations has not been studied.

Much remains to be learned about the ecology, behaviour and distribution of this poorly known chipmunk in the province.

Chipmunk Peak, south of Spatzizi Plateau near the headwaters of the Skeena River, was presumably named for this animal. The adjacent mountain, Melanistic Peak, was likely named for the predominately melanic population of Least Chipmunks that inhabit the area.

Selected References: Cowan 1946; Criddle 1943; Nagorsen et al. 2002; Meredith 1977; Sheppard 1965, 1971; Verts and Carraway 2001.

Red-tailed Chipmunk   *Tamias ruficaudus*

Other Common Names: Rufous-tailed Chipmunk.

Description

The Red-tailed Chipmunk is a large chipmunk with the typical striping pattern of five dark and four light stripes. The Rocky Mountain subspecies (*Tamias ruficaudus ruficaudus*) has bright reddish fur on the underside of the tail; its sides and shoulders and the back of its head are washed with reddish-brown or deep orange. The fur on the dorsal surface of its hind feet is rufous or cinnamon coloured. The Selkirk Mountains subspecies (*T. r. simulans*) is paler, with tan or yellow fur on the underside of the tail; its sides and shoulders and the back of its head are washed with tan or yellow. The fur on the dorsal surface of the hind feet is tawny. In both forms, the winter pelage is paler. The skull has a simple rounded infraorbital opening piercing the zygomatic plate. The baculum and baubellum are large and robust (baculum shaft length 3.4 to 4.7 mm, greatest length of the baubellum 2.0 to 3.2 mm).

Measurements:
total length: 223 (207-237) n=47
tail vertebrae: 99 (89-115) n=46
hind foot: 33 (30-35) n=46
ear: 17 (13-19) n=37
weight: 59.5 (44.2-78.7) n=37

Dental Formula:
incisors: 1/1
canines: 0/0
premolars: 2/1
molars: 3/3

Identification:
See the species accounts for the Yellow-pine Chipmunk (page 181) and Least Chipmunk for identification problems (page 187). The large robust genital bones of this species are diagnostic and cannot be confused with those of any other British Columbian chipmunk.

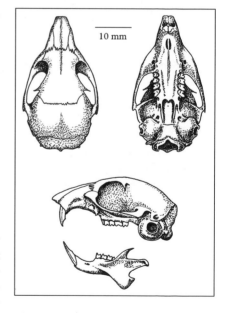

10 mm

Natural History

In the Selkirk Mountains of southern British Columbia, Red-tailed Chipmunks inhabit a broad elevational range, from 560 metres in the floodplain of the Creston Valley to 1829 metres at the Salmo-Creston Summit. It lives in forests of Western Hemlock, Engelmann Spruce and Western Redcedar of any age, from mature to recent clear-cuts. In B.C.'s Rocky Mountains, the Red-tailed Chipmunk is restricted to a narrow elevational belt, from about 1785 to 1950 metres, where it lives in subalpine coniferous forests. In Akamina-Kishinena Provincial Park, Mark Fraker found this species in a mature forest of Engelmann Spruce and Subalpine Fir with an understorey of Thimbleberry and Black Gooseberry. In Middle Kootenay Pass, I found it in the same type of habitat as well as disturbed open areas created by a recent forest fire.

The Red-tailed Chipmunk seems to be most common in disturbed habitats, such as clearings, roadsides and forest edges, with abundant shrubs and woody debris. A study in Idaho revealed that this chipmunk occupies forests of various ages, but it is most abundant in early and mid successional stages (less than 80 years old) that follow timber harvesting. Even recent clear-cut habitats (1 to 10 years old) can support large populations of this chipmunk.

The Red-tailed Chipmunk co-occurs with Least and Yellow-pine chipmunks in the southern Canadian Rocky Mountains. Generally, they are separated by elevation: the Least Chipmunk in alpine areas above 2000 metres, the Yellow-pine Chipmunk in forests below 1850 metres and the Red-tailed Chipmunk inhabiting the mid elevations. Nonetheless, there are narrow zones where they overlap in their distributions. At Middle Kootenay Pass for example, I found all three chipmunks coexisting in an elevational transect of a few hundred metres. In B.C.'s southern Selkirk Mountains, where the Least Chipmunk is absent and the Yellow-pine Chipmunk is rare, the Red-tailed Chipmunk inhabits lowland valleys, mid-elevation forests and the alpine.

The only information on population density is Mirza Beg's estimate of 4 to 13 animals per hectare in Montana. The Red-tailed Chipmunk typically moves between 90 and 150 metres a day, with males having larger home ranges than females. Most of the maximum distances between captures in Beg's study were between 70 and 155 metres – the longest movement was 458 metres. Both sexes move greater distances in the spring before the young emerge. An excellent climber, the Red-tailed Chipmunk does much of its foraging above ground in shrubs and trees, and it often nests in live or dead standing trees. This chipmunk is diurnal. Its vocalizations include a scolding note, a trill and a bell-like call.

The Red-tailed Chipmunk locates its den underground in burrows or above ground in trees. It uses the burrows mostly in winter and spring, and the tree nests in summer. The burrows are about 90 cm long and 28 cm deep with a nest of dried plant material, such as grass or lichen, at one end. Three tree nests found in eastern Washington were all in Engelmann Spruces 12 to 24 metres tall. These grass nests were situated near the trunks in dense growths of branches and small twigs 5.8 to 18.3 metres above ground. In Idaho, cavities in a dead fir 5.4 metres high supported six Red-tailed Chipmunks. The young are born in nests or burrows. A tree nest in Washington studied by Harold Broadbrooks was occupied by a female and her nursing young; he suspected that the young were born in a burrow, then moved to the tree nest when they were capable of climbing. In Idaho, a nest (17.8 x 25.4 cm) made of dried grass located in the forks of a bush, contained four young that were only about one-third grown and had unopened eyes.

The Red-tailed Chipmunk's diet includes green plant material, fruits, seeds and invertebrates; seeds are the most common food. In spring, this chipmunk forages mostly on the forest floor, eating flowers and leaves; flower heads of Common Dandelion and Heart-leaved Arnica are favourite food items at this time of year. In summer and especially in autumn, the diet shifts to fruits and seeds, with much of the foraging activity above ground in shrubs or trees. Mirza Beg identified 51 types of seeds in the cheek pouches of this chipmunk. Its staple foods are the seeds of Bull Thistle, Ponderosa Pine, Common Snowberry, Saskatoon and Mallow Ninebark, which form the bulk of seeds cached for winter.

The Red-tailed Chipmunk gains little weight in the autumn, relying on its store of seeds for winter survival. It usually remains inactive in its underground burrow from late October until early April. In Montana, animals have been seen above ground as late as December 4 and as early as March 29. Males emerge earlier than females in spring.

Reproduction in British Columbian populations has not been studied. In Montana, Mirza Beg found that mating occurs from April 28 to May 24, shortly after animals emerge from hibernation. The timing of the breeding season correlated with elevation: populations in high elevations lagged 7 to 10 days behind those in lower areas. Red-tailed Chipmunks breed in the spring after their birth, although in the Montana populations only 11 to 15% of the yearling females were pregnant. They produce only one litter each year, averaging 4.9 young (range: 4 to 6). Females more than two years old produce larger litters than younger animals. Young are born in late May through June. They nurse for about 31 days, and first appear above ground in mid July when they are 39 to 45 days old. They obtain their permanent teeth at about 79 days of age. Red-tailed Chipmunks may live six to seven years in the wild; a captive animal survived eight years.

Range

The Red-tailed Chipmunk has a small distributional area confined to northern Montana and Idaho, northeastern Washington, southeastern British Columbia and southwestern Alberta. There are two disjunct populations in B.C. One is restricted to the southern Selkirk Mountains where its range is delimited by the Kootenay and Columbia rivers on the west, and Kootenay Lake and the Creston Valley on the east. Another population inhabits the southern Rocky Mountains from the Akamina Pass - Wall Lake area north to Middle Kootenay Pass. In the original *Mammals of British Columbia*, Cowan

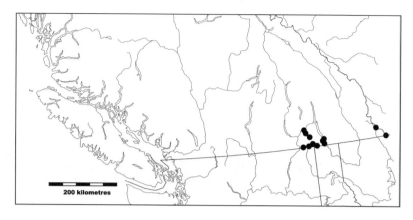

and Guiguet showed this chipmunk inhabiting the Purcell Mountains and the Rocky Mountain trench, but my research has shown that Red-tailed Chipmunk is absent from those areas. Historical museum specimens from the Purcell Mountains originally identified as Red-tailed Chipmunks are actually Yellow-pine Chipmunks.

Taxonomy
Two distinct subspecies are recognized; both occur in British Columbia. My studies have demonstrated that B.C. forms of the two subspecies differ in a number of traits – size, pelage colour and genital bone morphology – and are geographically separated by about 100 km. Recent DNA studies done by Jeff Good and Jack Sullivan revealed that the two subspecies differ genetically. Because they found genetic intergrades in a narrow contact zone in central Idaho, they concluded that the two forms were best treated as well-defined subspecies.

Tamias ruficaudus ruficaudus (Howell) – a large reddish form found in western Montana and the Rocky Mountains of extreme southwestern Alberta and the southeastern corner of B.C.

Tamias ruficaudus simulans (Howell) – a smaller paler form inhabiting northwestern Montana, northern Idaho, northeastern Washington and the southern Selkirk Mountains of B.C.

Conservation Status
Both subspecies are listed by the province as being of conservation concern, although there are no clear threats to this species' habitat.

Tamias ruficaudus ruficaudus is on the Red List because it has a limited range (three locations) in British Columbia restricted to a narrow elevational belt in the Rocky Mountains. Nonetheless, in Canada much of this subspecies' distributional area falls within the boundaries of protected areas such as Akamina-Kishinena Provincial Park and Waterton Lakes National Park.

Although *T. r. simulans* is on the province's Blue List, largely on the basis of its small distributional area, this population does not appear to be at risk. In contrast to *T. r. ruficaudus, T. r. simulans* occupies a wide elevational range and a variety of habitats including flood-plain valley bottoms, mid elevation forests (mature and logged) and subalpine forests.

Remarks

The natural history of Red-tailed Chipmunk is based almost entirely on research done in Idaho and Montana. Comparable ecological studies in British Columbia should be done to determine its habitat requirements and reproductive biology, and the behaviour of populations at the northern edge of its range. The precise distributional limits of this species in the province is also poorly documented. It seems likely that the Rocky Mountain subspecies may range as far north as Crowsnest Pass. Potential habitat also exists in mountains on the west side of the Flathead River.

Selected References: Beg 1969, 1971; Best 1993; Broadbrooks 1974; Good and Sullivan 2001; Nagorsen et al. 2002; Patterson and Heaney 1987.

Townsend's Chipmunk   *Tamias townsendii*

Other Common Names: None.

Description

Our largest chipmunk, Townsend's Chipmunk has a rich brown pelage and the typical chipmunk striping pattern, although the four dorsal light stripes tend to be indistinct. It has conspicuous white or grey patches behind the ears. The fur on the dorsal surface of the tail has a frosted appearance, because the black hairs are tipped with silver-grey. The undersides are creamy white to grey. Coastal populations (*Tamias townsendii townsendii*) are dark and dull with indistinct stripes; populations in the Cascade Mountains (*T. t. cooperi*) are paler with brighter stripes. The skull is the largest of any British Columbian chipmunk; it has a simple rounded infraorbital opening piercing the zygomatic plate. The baculum is small (shaft length 2.2 to 2.6 mm) with the tip length more than 30% of the shaft length. The baubellum is small (greatest length about 1 mm) with a U-shaped base that tapers sharply at its proximal end (figure 118).

Measurements:
total length: 260 (234-286) n=43
tail vertebrae: 114 (103-130) n=43
hind foot: 36 (32-40) n=42
ear: 19 (17-23) n=12
weight: 77.8 (65.2-89.0)
  n=11

Dental Formula:
incisors: 1/1
canines: 0/0
premolars: 2/1
molars: 3/3

Identification:
The Yellow-pine Chipmunk (*Tamias amoenus*) is the only chipmunk species that overlaps the range of Townsend's Chipmunk; see that account (page 181) for diagnostic traits.

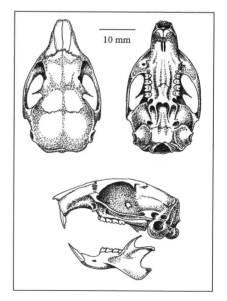

10 mm

Natural History

Townsend's Chipmunk lives in coniferous and deciduous coastal forests in upland and riparian areas. Typical habitats are forests of Western Hemlock, Douglas-fir, Western Redcedar, Bigleaf Maple and Red Alder. Although found in open clearings and roadsides, Townsend's Chipmunk seems to prefer habitats with a thick understorey of Vine Maple or Salmonberry. At higher elevations in the Cascade Mountains, Townsend's Chipmunk is found in forests or tree islands of Mountain Hemlock, Subalpine Fir and Engelmann Spruce. Its elevational range in British Columbia is from sea level in the lower Fraser River valley to 1800 metres in the Cascade Mountains. The elevational range in the southern Coast Mountains is unknown. Townsend's Chipmunk readily exploits sites disturbed by forest harvesting or fire, but the impact of forestry practices on its populations is not clear. In one study in coastal Oregon the largest populations were in young (25 to 50 years old) and old-growth forests, and the smallest populations were in mature forests 130 to 200 years old. Another study in coastal Oregon revealed that populations were greatest in old-growth forests. Some researchers have concluded that forest structure such as the abundance of shrub cover is more important than forest age for determining the abundance of this chipmunk.

In the United States, population densities have been estimated at 0.4 to 10.3 animals per hectare, although densities as high as 35 animals per hectare have been reported. The only estimate available for B.C. is for a population living in logged habitats at Maple Ridge in the Fraser River valley that was studied by Thomas Sullivan and colleagues. They found that Townsend's Chipmunk exhibited an annual cycle of abundance with low spring densities increasing to a peak of 2.0 animals per hectare in late summer or early autumn. The home-range size for populations in the Cascade Mountains of Washington was estimated at 0.6 to 0.8 ha. Males move greater distances than females, travelling up to 400 metres.

In the Cascade Mountains, this species hibernates for about 4.5 months, with animals disappearing by early November and emerging above ground in late March. It also seems to hibernate in mild coastal regions – the population at Maple Ridge appears to hibernate from November to early March. Little is known about the burrows and dens of this chipmunk. Although a capable climber, no tree nests have been described. An underground burrow containing newborn young found by William Shaw in the Olympic Mountains was 15 to 20 cm below the surface. It had a single tunnel about 1.3 metres long that terminated in a nest made of dried sedge and Methuselah's Beard, a long lichen that hangs from tree branches.

Townsend's Chipmunk is active throughout the day but is most active in late morning and early afternoon. This species often forages above ground in shrub thickets and small trees. It produces a number of vocalizations including a chuck-like call and a musical bird-like alarm call. Various underground fungi or truffles are a major part of this species' diet and Townsend's Chipmunk plays a major role in spreading the spores of these fungi in coastal forests. Other foods include conifer seeds, Bigleaf Maple seeds, huckleberries, Thimbleberries, Salmonberries, Wild Strawberries, Kinnikinnick berries, fireweed tubers, lichens and invertebrates.

In the Cascade Mountains of Washington, mating occurs in the last week of April and first week of May, with young born in late May or early June. In coastal lowland areas, the breeding season may begin earlier. A museum specimen noted as pregnant was taken at Huntingdon near Abbotsford on April 11. For the Maple Ridge population, Thomas Sullivan and colleagues estimated that mating occurred in the second and third week of April, and females began nursing their young in early to mid May. Females produce one litter a year consisting of three to four young. At birth, Townsend's

Chipmunks weigh about 3.5 grams and are naked with closed eyes. By 39 days they begin to eat solid food. The young emerge above ground in early July when they are about half the size of adults. Although this chipmunk is capable of breeding in the spring after its birth, 30 to 40% of yearling animals do not breed in most populations.

This species can live seven years in the wild. Predators include the Long-tailed Weasel, Western Spotted Skunk, Bobcat, Cooper's Hawk and Great Horned Owl.

Range

Townsend's Chipmunk inhabits coastal lowlands and coastal mountain ranges from Oregon to extreme southwestern British Columbia where it occupies the Fraser River valley and the Cascade Mountains east to Allison Pass and Treasure Mountain on the eastern slopes. Most occurrences in the province are from the south side of the Fraser River. The northern limits of its range on the north side of the Fraser River are unknown, but existing records suggest that it may range into the southern Coast Mountains as far north as Garibaldi Provincial Park. Although range maps in several publications show this species inhabiting southern Vancouver Island, chipmunks are not native to the island nor are there any established populations despite several releases in the 1900s. The Vancouver Island population shown in these publications is presumably based on a specimen housed at the Museum of Natural History in London, England, reputedly collected at Esquimalt in 1858 by the British naturalist John Keast Lord. Thirty-six chipmunks from

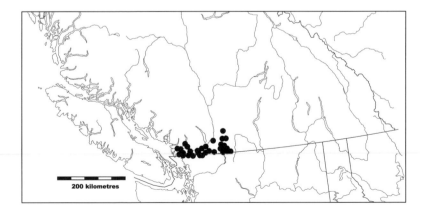

200 kilometres

Oregon released on Sidney Island in 1961 were probably Townsend's Chipmunk; the status of this population is unknown.

Taxonomy
Originally, five subspecies were recognized, but three of those, inhabiting southern Oregon and northern California, are now regarded as full species. Both of the currently recognized subspecies of *Tamias townsendii* occur in British Columbia:

Tamias townsendii cooperi Baird – the Cascade Mountains of Oregon, Washington and B.C.

Tamias townsendii townsendii Bachman – a dark coastal race associated with the lowlands and lower western slopes of the Cascades in Oregon and Washington, the Olympic Mountains in Washington and the lower Fraser Valley in B.C.

Conservation Status
British Columbian populations are not of conservation concern.

Remarks
Remarkably, little research has been done in British Columbia on this secretive rodent, and its distributional limits and habitat requirements in the province are poorly known. A quiet, wary chipmunk that is well camouflaged in dense dark forests with its dull fur colour, Townsend's Chipmunk is rarely noticed by most naturalists. Its soft calls can be easily mistaken for bird calls.

Selected References: Kenagy and Barnes 1988, Shaw 1944, Sullivan et al. 1983, Sutton 1993.

Douglas' Squirrel   *Tamiasciurus douglasii*

Other Common Names: Chickaree.

Description

Douglas' Squirrel is a small tree squirrel with olive-brown dorsal pelage, a flat bushy tail, tufted ears and a prominent white eye ring. A prominent black band along each side separates the dorsal and ventral pelage, which ranges from pale yellow to orange. The tail hairs are tipped with white or pale yellow. The winter pelage tends to be paler with a dull, reddish-brown dorsal stripe. The skull has the infraorbital opening modified as a canal that passes between the zygomatic plate and the side of the rostrum, the zygomatic arches are parallel and not flat in a horizontal plane. The anterior border of the orbit is opposite the second upper premolar.

Measurements:
  total length: 308 (271-351) n=144
  tail vertebrae: 122 (90-150) n=144
  hind foot: 49 (44-57) n=125
  ear: 26 (20-31) n=151
  weight: 203.5 g (157.4-270.6) n=154

Dental Formula:
  incisors: 1/1
  canines: 0/0
  premolars: 1/1
  molars: 3/3

Identification:

The only similar species is the Red Squirrel (*Tamiasciurus hudsonicus*), whose range overlaps that of Douglas' Squirrel along the crest and eastern slopes of the Cascade and southern Coast mountains. The Red Squirrel is discriminated by its grey to olive-brown dorsal pelage, white to grey belly, and tail hairs tipped with red. Although the Red Squirrel tends to be larger, the two

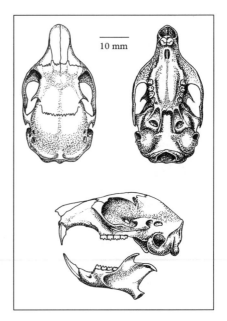

10 mm

species overlap extensively in their skull and body measurements and cannot be reliably identified from size. With experience one can learn to recognize the two species from their vocalizations. Their chirp alarm calls are particularly distinctive, with the calls of Douglas' Squirrel having longer notes; and the territorial rattle call of Douglas' Squirrel also has longer notes, spaced at longer intervals. Nonetheless, identification of some individuals is problematic in parts of the Cascade Mountains (e.g., Manning Provincial Park, Fraser River canyon) where some squirrels appear intermediate in pelage, vocalizations and skull morphology.

Natural History

Douglas' Squirrel lives in the coniferous forests of coastal British Columbia from sea level to 1830 metres. Thomas and Druscilla Sullivan estimated population densities of about 0.2 to 0.9 animals per hectare for a 40- to 50-year-old forest of Western Hemlock, Western Redcedar and Douglas-fir at Maple Ridge. Similar population densities have been reported for the coastal forests of Washington. In most seasons abundance is generally similar in various aged forests, but a study done in Washington revealed that old-growth forests supported much higher populations in winter, presumably because these older forests yield more conifer cones. Douglas' Squirrel populations also vary dramatically from year to year in response to changes in conifer cone crops. The Sullivans demonstrated the contribution of food supply to abundance when they increased the population density in their study area by 5 to 10 times simply by supplying extra food, such as sunflower seeds and whole oats. In Washington and Oregon, the Douglas' Squirrel is most abundant in areas where populations of Northern Flying Squirrel and Townsend's Chipmunk are low. These species may compete with Douglas' Squirrel for fungi and seed resources.

Douglas' Squirrel is diurnal with activity peaks in morning and late afternoon, though during winter it is most active around midday. This squirrel places stick nests in tree branches (lining the nest with moss, lichen or shredded bark), in tree cavities, or in holes in the ground.

This is a highly territorial species. In his study area in B.C.'s Cascade Mountains, Christopher Smith found that Douglas' Squirrels maintained non-overlapping territories from 0.2 to 1.2 ha. They defended their territory from individuals of either sex – females tolerated males only during brief periods in the breeding

season. Evidently, they also defend their territories against Townsend's Chipmunks. Douglas' Squirrel defends by vocalizing, chasing and, rarely, fighting. It often uses a rattle call to advertise its territory or respond to an intruder – this "chirr"-like call lasts less than 10 seconds and carries up to 130 metres. Most intruding squirrels respond by running away from the caller. The screech call – a series of high-pitched "tsew" sounds – is another important signal of aggression for deterring intruders. The squirrel often emits a screech call immediately after a rattle call. Other vocalizations used in maintaining territories are growls and buzz calls. Douglas' Squirrel also produces bark calls that are short, loud "chip" or "chuck" sounds. While it emits these calls, the squirrel stamps its feet and flicks its tail with sharp jerking motions above its back. Often emitted in response to predators, bark calls have been interpreted as alarm calls by some researchers, but others have suggested that these calls are a response to a potential conflict.

Seeds from the cones of coniferous trees are the most important food of Douglas' Squirrel. Christopher Smith documented its use of 10 conifer species in British Columbia: Amabilis Fir, Douglas-fir, Grand Fir, Engelmann Spruce, Lodgepole Pine, Mountain Hemlock, Ponderosa Pine, Subalpine Fir, Western White Pine and Western Hemlock. Douglas' Squirrel harvests conifer cones in late summer and autumn before they shed their seeds. It clips the cones from their branches and carries them to caches or middens in damp sites where they are stored for a winter food supply. To extract seeds from a cone, the squirrel holds the ends of the cone in its paws and turns it in its mouth, biting off the outer protective scales. Douglas' Squirrel seems to prefer the cones of Amabilis Fir, Douglas-fir and Engelmann Spruce, probably because they are soft, have high-energy seeds and are easy to detach from trees. If a cone is too large to be carried easily (e.g., those of Douglas-fir), the squirrel chews off the outer scales to reduce it to a manageable size. Although Smith speculated that the Douglas' Squirrel was less efficient than the Red Squirrel at using hard cones, such as Lodgepole Pine, because it is smaller and has weaker jaw muscles, he could find no differences in diets among the two species in the Cascade Mountains. Because Western Hemlock cones retain their seeds throughout winter until March, they provide a supplement to cones stored in caches. Truffles and mushrooms are a major food source in years of cone crop failures. Other foods eaten include conifer pollen, berries, fruits, seeds of various deciduous trees and shrubs, invertebrates, and birds.

The breeding season is four to five months in British Columbia. Males are in reproductive condition from March to July; nursing females have been found from May to September. According to Christopher Smith there are two peaks in the breeding season, in early spring and early summer. Females produce four to eight young after a gestation period of about 40 days. Estimates of average litter size in B.C. range from 4.0 to 5.7. This species may produce two litters a year especially in low elevation habitats, but successful breeding is closely linked to food supply. In years with poor cone production, breeding is delayed, litter sizes are small and many females fail to reproduce. Young Douglas' Squirrels are weaned about 64 to 78 days after birth. Little is known about their growth and development. Survival of young-of-the-year is determined by the abundance of conifer cones and their ability to find and defend a territory after weaning. Predators include raptors such as Cooper's Hawk, Northern Goshawk, Great Horned Owl and Spotted Owl, and mammalian carnivores such as Ermine and Marten.

Range

Douglas' Squirrel inhabits the coastal lowlands and coastal mountain ranges from California to southwestern British Columbia where it occurs as far north as the south side of Owikeno Lake in Rivers

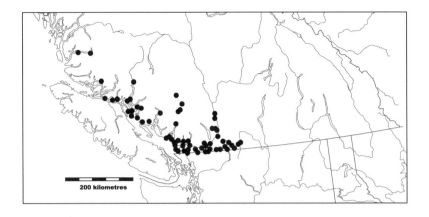

200 kilometres

Inlet. In the Cascade Mountains, it ranges east to Boston Bar in the Fraser River canyon and Copper Creek on the eastern edge of Manning Provincial Park. Eastern limits of the range in the Coast Mountains are poorly documented, but I have observed it as far east as the Pemberton Valley. Douglas' Squirrel is known to inhabit six coastal islands in the province: Bowen and Gambier islands in Howe Sound, and Cortes, East Redonda, Stuart and Subtle islands near Desolation Sound.

Taxonomy
In parts of the Cascade Mountains, where distributions of Douglas' Squirrel (*Tamiasciurus douglasii*) and the Red Squirrel (*T. hudsonicus*) overlap in a narrow zone about 30 km wide, some squirrels appear intermediate in pelage colour, skull size and vocalizations. Christopher Smith interpreted these intermediate animals as evidence for hybridization among the two species. But Stephen Lindsay, who found that eight specimens from Manning Provincial Park with intermediate pelage traits could be confidently assigned to either species on the basis of their skull morphology, concluded that there was no evidence of hybridization in British Columbia. Nevertheless, he did not assess hybridization in the Boston Bar area of the Fraser River canyon, where Smith reported that nearly half the squirrels appeared to be hybrids. A DNA study by Brian Arbogast and colleagues found little genetic differentiation among the two species, raising more doubts about their taxonomy. Ranges of the two species presumably also come into contact along the eastern edge of the Coast Mountains in British Columbia, but the

rugged inaccessible terrain has prohibited any taxonomic studies in that region.

Four subspecies are recognized; one occurs in the province:

Tamiasciurus douglasii mollipilosus (Audubon and Bachman) – a small coastal form ranging from northern California to B.C.

Conservation Status
British Columbian populations are not of conservation concern.

Remarks
Field studies in the Coast Mountains are needed to determine the eastern limits of this species range and its boundary with that of the Red Squirrel. Because Douglas' Squirrel and the Red Squirrel have a similar diet of conifer cones and both are highly territorial, Christopher Smith suggested that competition prevents them from coexisting over most of their ranges. This idea is supported by the distributional pattern of the two species on B.C.'s southern coastal islands, where the two species live close to each other but never on the same island. For example, Cortes Island supports only Douglas' Squirrel, whereas nearby Quadra Island supports only the Red Squirrel.

Selected References: Arbogast et al. 2001; Lindsay 1982; Smith 1968, 1981; Steele 1999; Sullivan and Sullivan 1982b.

Red Squirrel   *Tamiasciurus hudsonicus*

Other Common Names: Pine Squirrel.

Description

The Red Squirrel is a small tree squirrel with olive-brown dorsal pelage, a flat bushy tail, tufted ears and a prominent white eye ring. The ventral pelage is whitish, sometimes with a pale yellow wash; a prominent black band along the sides separates the dorsal and ventral pelage. The tail hairs are tipped with yellow or reddish-brown. The winter pelage is paler with a bright reddish-brown dorsal stripe. The skull has the infraorbital opening modified as a canal that passes between the zygomatic plate and the side of the rostrum; the zygomatic arches are parallel and not flat in a horizontal plane. The anterior border of the orbit is opposite the second upper premolar.

Measurements:
total length: 318 (192-363) n=346
tail vertebrae: 125 (99-150) n=342
hind foot: 49 (32-57) n=359
ear: 25 (14-32) n=328
weight: 224.5 (147.8-295.2) n=365

Dental Formula:
incisors: 1/1
canines: 0/0
premolars: 1/1
molars: 3/3

Identification:
In some parts of the Cascade and Coast mountains this species can be confused with Douglas' Squirrel (*Tamiasciurus douglasii*); see that account (page 205) for diagnostic traits.

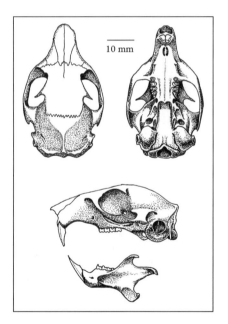

10 mm

Natural History

Associated with coniferous and mixed forests, the Red Squirrel ranges from sea level to 2330 metres elevation occurring in all of the province's biogeoclimatic zones. Estimates of average population densities for western North America range from about 0.2 to 2.0 animals per hectare. The lowest populations are in Black Spruce forests; higher densities occur in forests dominated by pines. Results from studies done in northern spruce forests suggest that the Red Squirrel is more abundant in mature forests. But Thomas Sullivan and colleagues found that managed second-growth and mature Lodgepole Pine forests in British Columbia's interior supported similar population densities; they also observed that Red Squirrels were 2 to 5.5 times less abundant in stands of Lodgepole Pine that had been thinned. Population densities of the Red Squirrel are closely tied to food supply – year-to-year fluctuations in abundance correspond to annual variations in conifer cone crops. Researchers have confirmed the link between food supply and population density when they increased the population in an area by adding supplemental food such as sunflower seeds.

This highly territorial squirrel maintains non-overlapping territories centred around a food midden. It advertizes and defends its territory against other squirrels by threatening and chasing them away. The Red Squirrel's vocalizations are similar to Douglas' Squirrels (see page 207). The "chirr"-like rattle call lasts less than 10 seconds and carries up to 130 metres, scaring away most intruders. The screech call – a series of high-pitched "tsew" sounds – is another aggressive signal for deterring intruders, often emitted immediately after a rattle call. Its bark is a short, loud "chip" or "chuck", made while stamping its feet and flicking its tail sharply above its back.

Territorial boundaries remain remarkably stable from year to year even when their occupants die and are replaced by new squirrels. Although long distance dispersal movements up to 5.6 km have been recorded, this is a sedentary mammal with estimates of average territory size ranging from about 0.24 to 0.98 ha. Young-of-the-year Red Squirrels remain in the territory where they were born making short distance exploratory trips until they can acquire their own territory. Most young animals will establish territories within a few hundred metres of the site where they were born.

The Red Squirrel constructs nests in tree cavities, on branches and in underground burrows. A branch nest, against the tree trunk or in witches' broom, consists of an outer layer of grass, moss or lichen and an inner lining of shredded bark, leaves and fur. The Red Squirrel usually nests in coniferous trees 2 to 14 metres tall, but more important than the tree's height is the its proximity to the

squirrel's food middens. A common feature of a nest tree is its crown touching those of adjacent trees – interconnecting branches presumably provide the squirrel with an escape route from predators. The Red Squirrel also constructs burrows in and under cone middens, and it may occupy old, abandoned ground-squirrel burrows.

The Red Squirrel is diurnal, exhibiting morning and late-afternoon activity peaks in summer and a single midday peak in winter. Even in northern regions with a severe climate, this squirrel remains active throughout the winter. Some of its strategies for surviving winter are: storing food in middens close to its nest, using underground burrow systems insulated by a deep snow cover, restricting activity above ground on the snow to a brief midday period when temperatures are warmest, and remaining in its nest or burrow when temperatures fall below about -32°C.

The Red Squirrel's major food is conifer seeds: Black Spruce, Douglas-fir, Grand Fir, Engelmann Spruce, Lodgepole Pine, Mountain Hemlock, Amabilis-fir, Ponderosa Pine, Subalpine Fir,

Western White Pine, Western Hemlock and White Spruce. It harvests cones in late summer or autumn, before they shed their seeds, and stores them in large middens near the centre of its territory. These central middens, which may cover 30 to 40 m² and be nearly half a metre deep, contain thousands of fresh-cut cones; some large middens reach 1.2 metres in depth and hold the remains of cones accumulated over decades by several generations of squirrels. Smaller hoards of fewer than 20 cones are also throughout the squirrel's territory. The Red Squirrel eats truffles and mushrooms too, storing dried mushrooms in tree cavities, woodpecker holes or old bird nests. Some of these caches contain a remarkable variety of fungi: a cache recovered from a rotten Douglas-fir stump on Mount McLean near Lillooet had 59 specimens of fungi representing 14 species. Other foods eaten by the Red Squirrel include conifer flowers, catkins of deciduous trees and shrubs, berries, fruits, tree shoots, cambium, sapwood, bark and resinous secretions associated with pine galls, bird eggs, young birds, and invertebrates.

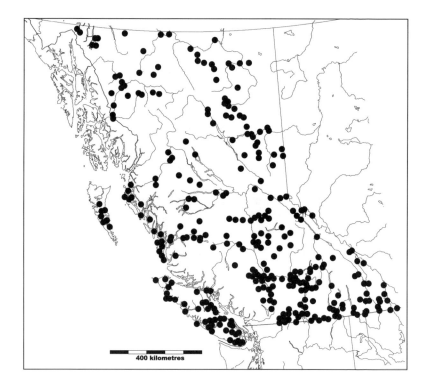

400 kilometres

The breeding season is two to six months in British Columbia. Throughout most of the province females have a single litter each year, but on Vancouver Island, John Millar found that females may produce a second litter from July to August in some years. Thomas Sullivan suspected that several females in his study area in B.C.'s central interior produced second litters when he supplemented their food supply with sunflower seeds. Most males are in reproductive condition from January to April, although the male breeding season may extend until August on Vancouver Island. Nursing females have been found from April to September. Females produce two to five young after a gestation period of 31 to 35 days. Estimates of average litter size in B.C. range from 2.8 to 3.8. In years with poor cone production, many females fail to reproduce and those that do breed tend to have small litters. Newborn young are naked, pink, blind and weigh six to eight grams. Their eyes open at about 27 to 35 days. They are first observed out of the nest at four to six weeks. The young are weaned at seven or eight weeks, and within a week later become independent of the mother. By 13 to 18 weeks of age, they shed and replace their deciduous milk premolars. Red Squirrels do not breed as young-of-the-year; although they are capable of breeding in the spring after their birth, many females will not breed until their second year. There are records of Red Squirrels living up to 10 years in the wild or captivity, but few animals survive beyond 5 years. The highest mortality is in the first autumn and winter when young-of-the-year have to establish a territory and obtain a sufficient supply of cones to ensure their winter survival. Many avian and mammalian carnivores prey on the Red Squirrel – major predators include Cooper's Hawk, Northern Goshawk, Ermine, Long-tailed Weasel and Marten.

Range

This squirrel has a vast range across Alaska, Canada and the United States. It occupies the entire mainland of British Columbia except for the southwestern coast region south of Rivers Inlet and west of the crest of the Cascade and Coast mountains, where it is replaced by Douglas' Squirrel. The Red Squirrel occurs on more than 50 islands in the province including Vancouver Island and its associated smaller islands, islands in Johnstone Strait, and numerous islands along the central and north coast. The presence of the Red Squirrel on Vancouver Island is curious as the adjacent mainland and its associated islands are inhabited by the Douglas' Squirrel. A boreal species well-adapted to cold environments, the Red Squirrel may have reached Vancouver Island shortly after the last glaciation before the Douglas' Squirrel colonized the province. An introduced population transplanted from Vancouver Island is now established on at least 11 of the Queen Charlotte Islands.

Taxonomy

Taxonomic relationships with Douglas' Squirrel (*Tamiasciurus douglasii*) are discussed in the previous account (see page 209). Twenty-five subspecies of the Red Squirrel (*T. hudsonicus*) are recognized; seven occur in British Columbia:

Tamiasciurus hudsonicus columbiensis Howell – western Alberta, the southern Yukon Territory, and northern and central B.C., where it ranges east of the coastal mountain ranges from Lillooet east to Mount Robson and north to the Liard River; may be indistinguishable from *T. h. preblei*.

Tamiasciurus hudsonicus lanuginosus Bachman – the coastal mainland from Rivers Inlet north to the Skeena River and various islands off the coastal mainland (Calvert, Campbell, Chatfield, Dufferin, Horsefall, Hunter, McCauley, Pitt, Porcher, Swindle, Townsend, Yeo); it also inhabits Vancouver Island and numerous adjacent islands. This small, dark form shows some similarities in pelage and skull morphology to *T. douglasii*.

Tamiasciurus hudsonicus petulans (Osgood) – southeastern Alaska, the southwestern Yukon and extreme northwestern B.C. in the Haines Triangle area.

Tamiasciurus hudsonicus picatus (Swarth) – mainland and islands of southeastern Alaska and the coastal mainland of northwestern B.C. from Telegraph Creek south to the Skeena River.

Tamiasciurus hudsonicus Howell – Alaska, the Yukon, the Northwest Territories, Alberta, Saskatchewan and the Peace River - Fort Nelson region in B.C.

Tamiasciurus hudsonicus richardsoni (Bachman) – Oregon, Idaho, Montana, extreme southwestern Alberta and southeastern B.C., where it occupies the Rocky, Purcell and Selkirk mountain ranges as far north as Field.

Tamiasciurus hudsonicus streatori (Allen) – northern Idaho and Washington, and south-central B.C., where its eastern limits extend from Midway to Glacier and its western limits are on the east slopes of the Cascade and southern Coast mountains.

Conservation Status

The British Columbian populations are not of conservation concern. But a local population in the Victoria region on Vancouver island could be affected by the introduced Eastern Grey Squirrel, which is expanding its range on southeastern Vancouver Island.

Remarks

Using six animals captured on Vancouver Island, the British Columbia Game Commission introduced Red Squirrels to the Queen Charlotte Islands in 1950, presumably as an ill-conceived attempt to provide a prey source for the native Marten or to assist with the dispersal of conifer cones. The Red Squirrel is now established on 11 islands there: Burnaby, Bischof, East Limestone, Graham, Huxley, Kat, Louise, Lyell, Moresby, Talunkwan and Wanderer. Little research has been done on the impact of this introduced mammal to the islands' native flora and fauna, but there is concern about its impact on nesting song birds.

Because it harvests cones and damages trees by clipping young shoots and removing bark, this species may have a considerable impact on reforestation. Classified as fur-bearers under the provincial Wildlife Act, squirrels (Douglas' Squirrel, Red Squirrel) are a surprisingly important species in the province's fur harvest. From 1919 to 1993, nearly 10 million squirrels were harvested for their fur in British Columbia. Most were Red Squirrels taken on trap lines in northern regions.

Selected References: Millar 1970; Smith 1968, 1981; Steele 1998; Sullivan 1990; Sullivan and Moses 1986.

Family Geomyidae: Pocket Gophers

Pocket Gophers are fossorial. Thirty-five species are recognized; one occurs in British Columbia.

Northern Pocket Gopher
Thomomys talpoides

Other Common Names: None.

Description

The Northern Pocket Gopher is a chipmunk-sized rodent with a thickset, muscular body, short legs, a short, nearly naked tail, and small eyes and ears. The front feet have five long claws. Two external, fur-lined cheek pouches extend from the face to shoulder area (figure 6). The dorsal pelage is generally brown, although some populations show a high incidence of melanism; most of the Northern Pocket Gophers living at Lavington, in the Coldstream Creek valley, are melanic. Albinos have also been reported in British Columbia. Some populations have prominent white markings on their chin, forearms and abdomen. The skull is flat with wide zygomatic arches and an infraorbital opening positioned on the side of the rostrum in front of the zygomatic plate. The upper and lower premolars are indented on the side to form an "8" shape (figure 101).

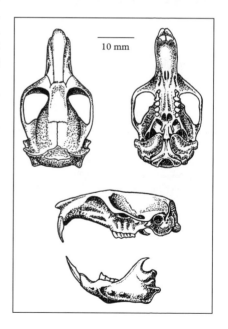

10 mm

Measurements:
  total length: 199 (154-230) n=413
  tail vertebrae: 65 (48-82) n=410
  hind foot: 26 (22-36) n=228
  ear: 5 (4-7) n=17
  weight: males 90.7 g (69.1-113.0) n=24
          females 83.2 g (60.8-116.4) n=31
Dental Formula:
incisors: 1/1
canines: 0/0
premolars: 1/1
molars: 3/3
Identification:
As the only pocket gopher in British Columbia, the Northern Pocket
Gopher cannot be confused with any other rodent.

Natural History
In British Columbia the Northern Pocket Gopher lives in dry grasslands, open subalpine forests, and subalpine or alpine meadows. It
also thrives in disturbed human-made habitats, such as logging
roads, railroad embankments, clearings under power lines, hay or
Alfalfa fields, orchards and gardens. It tends to avoid heavily cultivated fields, because farm machinery damages its burrows. The elevational range extends from 300 metres in the valley bottoms to
2100 metres in the Cascade Mountains and 2130 metres in the
Monashee Mountains. In the Rocky Mountains, the highest known
occurrence is at 1350 metres in the Crowsnest Pass. Distribution
and abundance is closely linked to soil type. Although the Northern

Pocket Gopher lives in many soil types, even dense, compacted clay soils with large stones, it is most common in soils that are deep and loamy or light and friable, where it can dig easily.

Habitat and elevation affect the Northern Pocket Gopher's abundance. The most productive habitats are alpine meadows, and hay or Alfalfa fields. Disturbances such as fire, logging or road construction often create openings with rich grasses and forbs resulting in a population increase. In the Okanagan Valley, Thomas Sullivan and colleagues estimated population densities of 15.4 to 31.5 animals per hectare in old field habitats, but only 4.5 to 17.7 per hectare in orchards. Populations follow a distinct annual cycle, lowest in spring and highest in autumn.

Although the Northern Pocket Gopher spends most of its life in underground tunnels, it does venture above ground. Trapping studies of marked animals have revealed that most adults remain in the same area for their entire lives with distances between successive captures usually less than 30 metres. In summer, the adults spend much of their time in their burrows, but they will emerge above ground to collect food within a metre of their burrow openings. Young animals dispersing from their natal nest travel up to 300 metres above ground. In an experimental study using released animals, Terry Vaughan found that most moved about 240 metres, although one animal in his study travelled 790 metres. Northern Pocket Gophers are active in winter under the snow pack at the surface of the ground, and there are even a few observations of it moving on top of the snow in mid winter. This rodent can swim – one was seen swimming 90 metres across the Assiniboine River in Manitoba. Northern Pocket Gophers are active all hours of the day.

The Northern Pocket Gopher feeds almost entirely on plant matter. It seems to change its diet with season and habitat. In summer, it feeds on succulent stems and leaves of forbs and grasses; in subalpine meadows the leaves of lupines, Yarrow and agoseris dominate its diet; in cultivated agricultural lands, Alfalfa is the main food. The pocket gopher collects most of its food above ground at night close to its burrow entrance. It clips off plants at the base, then drags them into the burrow and cuts them into short lengths that can be carried in the cheek pouches. The leaves may be taken to the nest for immediate consumption or cached in feeding tunnels for future use. During autumn and winter, the Northern Pocket Gopher uses shallow underground feeding tunnels to gain access to roots, tubers and corms. It stores this food in underground caches located in lateral chambers of the shallow tunnel system, and in caches beneath the snow on the surface of the ground or in the snow itself. Underground stores may house anywhere from 160 to 600 grams of food. The pocket gopher's winter food is diverse, but in alpine habitats it seems to prefer dandelion roots, tubers of starwort, and the bulbs of Western Spring-beauty and Yellow Glacier-lily. During winter, it will also girdle the bark of small fruit trees and conifer seedlings.

The Northern Pocket Gopher is highly territorial excluding other individuals from its burrow system. A typical burrow system consists of two types of tunnels: shallow feeding tunnels in the top 10 to 30 cm of soil and deep tunnels that can be more than 3 metres below the surface. Deep tunnels, which house the nest chamber and food storage chambers, are separated from the surface tunnels by one or more perpendicular passageways. The feeding runs form a complex network of branching tunnels. Many simply end blindly where the animal stopped digging; others are blocked by excavated soil and refuse. Some of the lateral branches end at the surface as fan-shaped mounds of freshly excavated earth brought to the surface – these are the characteristic pocket-gopher diggings visible in fields or alpine meadows. Inactive surface openings are plugged from beneath, leaving concentric rings of soils. The burrow's occupant plugs any holes or damage to the burrow system within 24 hours. The nest chamber is lined with dried vegetation. In winter, Northern Pocket Gophers dig shallow burrows in the surface of the soil under the snow. The spring snowmelt often reveals the cylindrical casts of soil formed from these winter diggings (figure 13). As soon as the ground thaws in spring, this rodent begins to excavate its tunnel systems. Burrowing activity declines in late spring and early summer as the soil dries out, but another peak of burrow activity occurs in late summer.

In most areas of western North America the breeding season extends from March to July with births from early May to the end of June. The gestation period is about 18 days. The scanty breeding data for British Columbian populations are consistent with this pattern. Fourteen museum specimens noted as pregnant were taken from April 17 to June 10, and five nursing females were taken between May 4 and June 15. Some authorities have reported that females produce only one litter a year, while others report two litters – there is evidence for two litters in B.C.'s Okanagan Valley. The average litter size in B.C. (based on embryo counts for the 14 female museum specimens) is 4.6 (2 to 9). At birth, the young are naked, blind and weigh 2.5 to 3.6 grams. By 17 days they can consume solid food; by 5 to 7 weeks they disperse from the nest. They reach adult size by about 100 days, but will not breed until the year following their birth. The maximum life span is five years, but most animals survive no more than two years.

The Northern Pocket Gopher's fossorial lifestyle does not protect it totally from predators, especially owls such as the Great Horned Owl and Long-eared Owl. Young pocket gophers dispersing above ground are especially vulnerable to owls, although adult remains also show up in owl pellets. Another avian predator is the Ferruginous Hawk. Mammalian predators include the Coyote, Long-tailed Weasel and Badger.

Foresters and farmers regard the Northern Pocket Gopher as a pest – they use traps, poisons and various repellents to control gopher populations. By pruning roots and stems and girdling the bark of seedlings, this rodent may seriously hinder the regeneration of coniferous trees in managed forests. It can also cause extensive damage in fruit orchards. Pocket gopher burrows provide shelter for a number of reptiles, amphibians and small mammals. In grasslands, the local distribution of adult Tiger Salamanders appears closely linked to the presence of pocket-gopher burrows.

Range
Found throughout the Great Plains and western Cordillera of North America, the Northern Pocket Gopher has the largest range of any pocket gopher. In British Columbia, it inhabits the southern interior from the east Kootenays to the west slopes of the Cascade

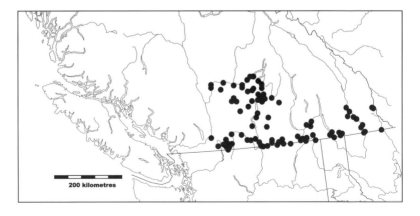

200 kilometres

Mountains. The western limits of the range are on the west slopes of the Cascade Mountains near Hope and the south side of the Thompson River. In the Kootenay region, it ranges north in the Rocky Mountain trench to Wasa; in the Rocky Mountains it is found in Crowsnest Pass and the Flathead River valley.

Taxonomy
Populations demonstrate considerable variation in size and fur colour; this has led to the description of 58 subspecies. Walter Johnstone recognized six races in British Columbia. He considered the Columbia and Kootenay rivers to be major barriers to pocket gopher movements, delimiting subspecies boundaries in the Rocky Mountain trench and Creston valley.

Thomomys talpoides cognatus Johnstone – extreme southeastern B.C. from the Crowsnest Pass to the Montana border and as far west as the east side of the Kootenay River in the Rocky Mountain trench.

Thomomys talpoides fuscus Merriam – Idaho, northern Washington and the southern Monashee Mountains of B.C.

Thomomys talpoides incensus Goldman – a weakly defined subspecies in the southern dry interior of B.C. from the Thompson River valley to the Washington border. The eastern limit is the Monashee Mountains; the western limit is the eastern slopes of the Cascade Mountains.

Thomomys talpoides medius Goldman – restricted to the southern Selkirk Mountains of B.C. where the Kootenay River, Kootenay Lake and Columbia River delimit its range.

Thomomys talpoides saturatus Bailey – northeastern Idaho, northwestern Montana and the southern Kootenays in B.C. where it ranges from the east side of the Creston valley to the western side of the Kootenay River in the Rocky Mountain trench.

Thomomys talpoides segregatus Johnstone – an isolated race with a high incidence of white markings on its chest and abdomen; restricted to a small area at Wynndel on the east side of the Kootenay River north of Creston.

Conservation Status
The Wynndel subspecies (*T. t. segregatus*) is an isolated race confined to 10 km^2 on benchlands on the east side of the Kootenay River north of Creston. Mark Fraker and colleagues found that the range of this subspecies has not changed from when it was first discovered

50 years ago. Because of its restricted range and small population, it appears on the provincial Red List and is designated Lower Risk (near threatened) by the IUCN. Genetic studies are needed to verify the taxonomic validity of this race.

Remarks

People living in the interior of the province often confuse the Northern Pocket Gopher with a mole, but moles live on the coast and none are found in B.C.'s interior. The Coast Mole co-occurs with the Northern Pocket Gopher at the eastern edge of its range in a small area around Hope and the Skagit River valley.

Selected References: Fraker et al. 1997, Johnstone 1954, Sullivan et al. 2001, Teipner et al. 1983, Vaughan 1974, Verts and Carraway 1999.

Great Basin Pocket Mouse

Family Heteromyidae: Heteromyids

Heteromyids show a number of adaptations for living in arid regions. This family has 59 species, including kangaroo mice, kangaroo rats and pocket mice; one species occurs in British Columbia.

Great Basin Pocket Mouse
Perognathus parvus

Other Common Names: None.

Description

The Great Basin Pocket Mouse is a mouse-like rodent with a long tail (longer than the head and body combined), soft, sleek pelage, small ears, and external fur-lined cheek pouches. The dorsal pelage is dull grey-brown and the undersides are pure white. A prominent band along the sides demarcates the dorsal and ventral pelages. The tail is bicoloured. The skull has a large circular infraorbital opening on the side of the rostrum, slender zygomatic arches and thin compressed incisors that have a distinct groove on their anterior face. Males tend to be larger and significantly heavier than females.

Measurements:
  total length: 175 (151-202) n=241
  tail vertebrae: 93 (77-112) n=238
  hind foot: 23 (20-27) n=237
  ear: 6 (4-10) n=55
  weight: males 21.2 (16.5-28.0) n=15
          females 18.5 (16.0-27.0) n=16

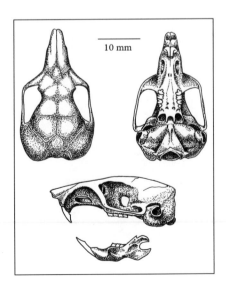

10 mm

Dental Formula:
incisors: 1/1
canines: 0/0
premolars: 1/1
molars: 3/3

Identification:
The Deer Mouse (*Peromyscus maniculatus*) is the only British Columbian rodent that could be confused with the Great Basin Pocket Mouse. The Deer Mouse has larger ears, a shorter tail and no cheek pouches; its skull has a V-shaped infraorbital opening at the base of the zygomatic arch and only three pairs of cheek teeth (no premolars).

Natural History

Restricted to the dry grasslands of the province, the Great Basin Pocket Mouse inhabits arid valley bottoms (300 to 400 metres elevation) and open slopes on hillsides. The maximum elevation inhabited by this rodent in British Columbia is 1370 metres on Mount Kobau in the southern Okanagan Valley. The most productive habitats are grassland-steppe with Bluebunch Wheatgrass, Antelope Bush, Big Sagebrush, Prairie Sagewort and Common Rabbit-Brush. The Great Basin Pocket Mouse exploits habitats disturbed by fire or grazing and will live in grasslands with alien plants such as Cheatgrass, Crested Wheatgrass and Russian Thistle. Rare in open Ponderosa Pine forest, this species also avoids irrigated agricultural land such as Alfalfa fields and vineyards. Optimum habitats for this rodent have loose, sandy soils, which are ideal for excavating burrows. The sandy bench land covered with Antelope Brush bordering the east side of Osoyoos Lake is a good example of ideal pocketmouse habitat. This species also inhabits clay-till soils with some rocks, and I have found it in disturbed sites with loose soil, such as road or railway embankments.

Population densities may reach 40 to 74 animals per hectare in some shrub-steppe habitats in the south Okanagan. Old fields support 10 to 20 animals per hectare; low densities occur in orchards and Ponderosa Pine forests (1 to 5 animals per hectare). Populations undergo a distinct annual cycle, reaching a peak in early summer; but unlike those of grassland voles (such as the Montane Vole) that undergo pronounced year-to-year cycles, Great Basin Pocket Mouse populations remain relatively stable from one year to the next. Thomas O' Farrell and colleagues reported only minor year-to-year variations in numbers during their six-year study, although their populations did increase fivefold in a year with heavy precipitation and high plant productivity.

Home-range size varies from year-to-year depending on food resources and among animals of different sex and age. Males occupy larger home ranges than females, and adults have larger home ranges than young animals. For the southern Okanagan Valley, Stuart Iverson estimated home-range sizes of 898 m^2 for males and 656 m^2 for females. Considerably larger home-range estimates (3125 to 3796 m^2 for adult males; 1941 to 2151 m^2 for females) were reported for populations in the Columbia Plateau of Washington. Despite the variations in home-range size, the core activity areas of Great Basin Pocket Mice are quite small (about 6 m^2 for females and 16 m^2 for males) and stable. Although individual home ranges overlap to some degree, this is a solitary mammal that does not share its burrow with other adults. In B.C., the Great Basin Pocket Mouse co-occurs with the Deer Mouse, Western Harvest Mouse and Montane Vole.

Strictly nocturnal, the Great Basin Pocket Mouse is most active above ground about an hour after dark. Stuart Iverson described two types of burrows: escape and permanent. Escape burrows are simple, shallow (20 to 30 cm deep) structures with several entrances that lack a nest chamber or food stores. Most pocket mice maintain several escape burrows within their home range for temporary refuge from predators. They also construct complex permanent burrow systems, with a nest and food chambers, that they defend from other pocket mice. The nest chamber, located below the frost line, houses a nest made of fine twigs, dried grass or seed husks. Permanent burrows have several entrances and some tunnels may extend more than a metre below ground. Their distinct circular entrances, about 2 cm in diameter, are located at the base of shrubs (such as Antelope-Bush or Big Sage) or under small rocks. In mid summer, small heaps of fine soil reveal the locations of burrow systems.

The Great Basin Pocket Mouse is one of our few small rodents that hibernates during winter, though not deeply. It enters bouts of torpor 3 to 240 hours in duration. In mid winter, 60 to 80% of its time may be spent in torpor. Captive pocket mice will spontaneously enter torpor at 6 to 19°C. In the wild, they probably enter torpor in any season. Because he captured few animals in his study area in August, Stuart Iverson suggested that the pocket mouse may enter torpor to avoid the stress of extreme summer temperatures.

Above 730 metres elevation in the Okanagan Valley, the Great Basin Pocket Mouse ceases above-ground activity by the last week of October, but in the valley bottoms it remains active until the end of November. Thomas O' Farrell and colleagues estimated that it requires 60 to 90 days of activity outside of its burrow to gather sufficient food stores for winter. Therefore, animals active in late autumn probably are young-of-the-year that were born in late-summer litters. Although the pocket mouse increases in weight throughout late summer, it does not accumulate heavy fat deposits for winter. Instead it relies on torpor, an insulated nest and seed stores in its burrow to meet its winter energy requirements. A Great Basin Pocket Mouse requires the equivalent of 46 to 60 grams of Cheatgrass seeds to survive the winter. Temperatures in its burrow seldom fall below 5°C, even in mid winter.

In the southern Okanagan Valley it emerges from hibernation in late March or early April. Males emerge first, the young males emerging before older males, possibly because they were not able to cache enough seeds to last them through to April. Females emerge up to a month after the males. Evidently, soil temperature provides the cue for spring emergence. This is our only rodent with special adaptations for surviving in arid desert-like conditions. Most of its water comes from metabolic water, a by-product generated from eating dry seeds or moisture in succulent plant food. Other strategies to avoid water deprivation include: the construction of burrows that provide a cool humid environment, the ability to produce a highly concentrated urine and foraging at night to avoid extreme daytime temperatures.

Although usually described as a seed-eater, many of the early diet studies on this rodent were based on the analysis of cheek pouch contents, which tend to be biased for seeds. Using stomach contents, Stuart Iverson demonstrated that the Great Basin Pocket Mouse also eats green plant material and invertebrates. Nevertheless, seeds are, by far, its most common food, particularly in late summer when they are abundant. Though it eats the seeds of many plants, it seems to prefer grass seeds. In the Columbia Plateau of Washington, one of the dominant foods is the seed of Cheatgrass, an alien grass introduced to western North America in the 1800s. Captive Great Basin Pocket Mice readily ate Antelope-Bush seeds, even though they have a thick hull and contain toxic compounds. This rodent collects seeds from the ground and carries them in its cheek pouches back to its burrow to store them for later consumption. These seed caches can be large. Burrows excavated by Iverson in July housed seed caches from 25 to 300 cm³. In summer, vegetative plant material, such as succulent grass leaves and stems, form 25 to 40% of the diet and may be a source of water. Invertebrates (beetles, insect larvae and pupae) are also eaten throughout the summer. The Great Basin Pocket Mouse rarely transports vegetative plant material and invertebrates in its cheek pouches.

The precise timing for the onset of breeding in British Columbia is not clear, but Iverson observed males with enlarged testes in April and the first pregnant females in late May. Males remain in breeding condition until mid August; females may breed until late August or early September. The gestation period is 21 to 25 days. Embryo counts for 21 British Columbian females averaged 4.9 (range 4 to 6). Females generally produce one or two litters, although Iverson found that those living in the valley bottoms occasionally produce three litters. High-elevation populations produce fewer litters and have a shorter breeding season than low-elevation populations. Young males do not breed until the following spring, but at low elevations females born in the early spring litters may breed in their first summer. Three-day-old young weigh 2.2 grams; however, little else is known about the growth and development of this species.

The Great Basin Pocket Mouse is surprisingly long lived for a small rodent. Captive animals have survived 4.5 years; wild animals can live to their fourth year, although few survive beyond their second. This species is the dominant vertebrate prey of the endangered Burrowing Owl. Other predators include the Western Screech-Owl, Long-eared Owl, Short-eared Owl, Badger, Striped Skunk and Western Rattlesnake.

Range

The Great Basin Pocket Mouse inhabits the arid intermontane regions of the United States, from California and northern New Mexico to Washington. In Canada, where it is at the northern periphery of its range, it is restricted to the southern dry interior of British Columbia. It occupies three isolated regions in the province: the Thompson River valley from Ashcroft to Kamloops, the north Okanagan Valley at Okanagan Landing - Vernon, and the southern Similkameen and Okanagan valleys east to Midway and Grand Forks in the Kettle River valley. The Thompson and north

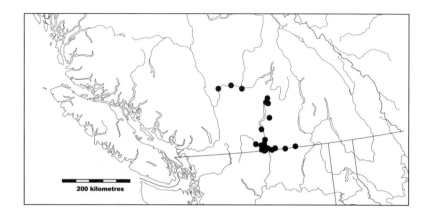

200 kilometres

Okanagan populations are isolated from each other and the south Okanagan population by unsuitable habitat.

Taxonomy

A study of DNA by Carolyn Ferrell revealed two strongly divergent genetic lineages (Columbia Plateau, Great Basin) that may be distinct species. Eleven subspecies are traditionally recognized; two occur in British Columbia: *Perognathus parvus laingi* and *P. p. lordi*. According to Cowan and Guiguet in the original *Mammals of British Columbia*, *P. p. laingi* and *P. p. lordi* represent high- and low-altitude forms. Although these forms differ in their life histories and pelage colour, Stuart Iverson found no major morphological differences that would support their recognition as distinct subspecies. Moreover, there are no barriers isolating them. Valley-dwelling and high-elevation populations are connected by mid-elevation hillside populations. Genetic studies are needed to assess the taxonomy of British Columbian populations.

Perognathus parvus laingi Anderson – a small dark race restricted to B.C.; originally described from specimens taken at Anarchist Summit, Cowan and Guiguet in the original *Mammals of British Columbia* described its range as the Thompson River valley, the Vernon area in the northern Okanagan Valley, and Anarchist Mountain east of the Okanagan Valley.

Perognathus parvus lordi (Gray) – a pale race found in northern Washington, northwestern Idaho and low elevations in the Okanagan, Similkameen River and Kettle River valleys of B.C.

Conservation Status

The Great Basin Pocket Mouse appears on the province's Blue List. Although this species survives in grassland habitats modified from grazing or the introduction of alien plants, it is a species of concern because agriculture and urban development have eliminated much of its native grassland habitat, resulting in fragmented populations. The last known record from the Thompson River valley was a specimen taken in 1949 near Kamloops – it is not known if the Great Basin Pocket Mouse still persists in this valley. The population in the northern Okanagan Valley appears to be isolated from the populations in the southern areas of the valley by unsuitable forested habitats and habitat loss from urban development at Kelowna. Although a number of records exist for the Vernon region, they are all historical specimens taken by early museum collectors. This area has undergone recent urban growth and irrigation of the Vernon Commonage. Given that the last Great Basin Pocket Mouse record from the Vernon area is a museum specimen taken in 1951, it is not clear if a population still persists in the north Okanagan.

Remarks

At the northern limits of its range, the distribution of the Great Basin Pocket Mouse in British Columbia is probably limited by winter temperature and the length of the summer growing season. Potential grassland-steppe habitat exists in the Nicola, Hat Creek and Fraser River valleys, but this rodent appears to be absent from these regions. Short summers combined with the cool winters in these valleys may allow insufficient time for the young-of-the-year to reach an adequate body weight and store enough seeds to survive over winter.

The Great Basin Pocket Mouse readily enters baited live traps, and its conspicuous burrow entrances are easily identified. Yet its conservation status in the province has received little attention. A comprehensive survey in the Thompson River valley and north Okanagan is needed to determine its precise distributional area and the extent that it has disappeared from its historic range.

Selected References: Iverson 1967, O' Farrell et al. 1975, Ferrell 1995, Schreiber 1978, Verts and Kirkland 1988.

Family Castoridae: Beavers

This family has two species; one occurs in North America.

Beaver   *Castor canadensis*

Other Common Names: None.

Description

Our largest rodent, the Beaver is a heavy muscular animal with a dorsally flat, scaly tail. Its front and hind legs are relatively short; each has five toes. The front feet have well-developed claws; the hind feet are webbed. The claw on the second digit of the hind foot is split, forming a comb-like structure used for grooming fur. The eyes and ears are small and inconspicuous. The pelage has long, rich brown glossy guard hairs and a soft fine underfur. The skull is massive with high-crowned cheek teeth, large orange incisors and a prominent depression in the basioccipital region.

Measurements:
  total length: 1041 (900-1212) n=58
  tail vertebrae: 297 (146-500) n=54
  hind foot: 175 (146-200) n=57
  ear: 35 (30-44) n=41
  weight: 19.2 kg (11.6-35.3) n=43

Dental Formula:
  incisors: 1/1
  canines: 0/0
  premolars: 1/1
  molars: 3/3

Identification:
This rodent cannot be confused with any other British Columbian mammal.

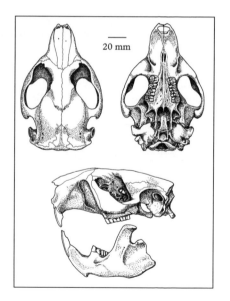

20 mm

Natural History

An aquatic rodent, the Beaver must have a stable year-round source of water – a lake, pond, marshy area or slow-flowing stream. Because Beavers construct dams to control water levels in their habitat, they prefer narrow streams or lakes with outlets. The water level must be of sufficient depth to avoid freezing solid in winter and to accommodate the Beaver's lodges or burrows, dens, and food caches. The other essential habitat requirement is abundant woody plants, such as Trembling Aspen, poplar, willow and alder, that it uses for food and lodge-building material. Disturbed habitats regenerating from fire, forest harvesting or insect kills tend to be highly productive for Beavers, because they support stands of young aspen or poplar. Trembling Aspen stands of 20 to 30 years old seem to be the most productive. In British Columbia, most of these habitats are in the Cariboo and northern regions. Although widespread throughout the coast, the Beaver tends to be less successful in coastal forests than interior wetlands. This rodent is limited to lowlands and valleys – subalpine or alpine lakes and ponds are too shallow and poor in plant productivity.

Biologists usually measure Beaver populations as the number of active lodges or colonies (i.e., families) per length of a stream or unit of area. Across the boreal forest region, population densities range from 0.15 to 3.90 families per hectare or 0.04 to 0.44 colonies per kilometre. Regional variation in population densities can be attributed to variation in water quality and food resources. The only population-density estimates for B.C. are Dave Hatler's of 0.55 to 0.71 colonies per kilometre along the Nechako River. They represent some of the highest densities recorded for North America. Beaver populations do not experience cyclic fluctuations. Because this rodent's daily movements follow the shoreline of a lake, pond or stream, its home-range size is determined in part by the size and shape of water bodies. Home-range size is larger in summer than in autumn. Along streams, daily movements can range from 0.6 to 1.6 km. Estimates of home-range size for Beavers living in ponds or lakes range from 2.5 to 20.5 ha. Major dispersals occur either when a colony abandons a pond or lake and moves to another, or when the two-year-olds leave their birth site to establish a new colony. Dispersing young can move up to 5.6 km and transplanted Beavers more than 100 km.

A Beaver colony is a family group consisting of a breeding pair, newborn young and yearling animals born the previous year; a few non-breeding animals older than a year may be found in some colonies. This species is monogamous – the breeding pair stay together for years, though because of age difference, they rarely remain together for life. Two-year-old Beavers disperse from the family lodge or den site after the breeding female gives birth to new young. Families are highly territorial, excluding other Beavers by fights and aggressive behaviours. Presumably to mark their territory, they construct mounds of mud and vegetation up to 0.6 metres high near the edge of a water body. They scent mark these mounds using castoreum, a chemically complex, yellowish substance produced in their paired castor sacs. To signal danger or threaten an intruder, a Beaver slaps its tail loudly against the water's surface. Beavers also communicate with an assortment of vocalizations and body postures.

The Beaver's construction skills are renowned. Populations living in fast-flowing or deep streams will excavate burrows in banks or stream-sides, but the most familiar structure is the dome-shaped lodge or house (see figure 11) constructed from peeled sticks and mud. A typical lodge may be up to 10 or 12 metres in diameter and sit several metres above the water level. It contains a nest chamber located above water with several exit holes in the floor leading directly into the water. A hole at the peak of the lodge's roof provides ventilation. The lodge is essential for survival, protecting the colony from predators and extreme environments. Several studies monitoring winter temperatures within a lodge have shown that the nest chamber is significantly warmer and less variable in temperature than the outside environment. Beavers dredge canals and construct dams to control the water level of their habitat. They build dams in narrow sections of a stream or outflow, using mud and stones as a foundation, then placing sticks across the current until the dam is higher than the water level; they quickly repair any damage to a dam.

In winter, Beavers confine their activity to the lodge, except when swimming under the ice to reach their food caches. This rodent's insulating fur protects it from the cold water – the body temperature may decrease by 1.0° to 1.5°C while swimming in winter, but it quickly recovers to normal after the Beaver returns to the lodge. Northern populations show seasonal changes in body weight and activity that are linked to winter energy demands. The tail, which contains a high amount of fat, particularly in winter, acts as an energy reserve.

The Beaver is most active at dawn or dusk. It is our most specialized rodent for an aquatic lifestyle, and swims well on the surface and underwater. It propels itself with the webbed hind feet, holding the front feet against the body to reduce drag. The tail provides stability and may move from side to side when the Beaver swims quickly. A Beaver can swim up to five kilometres per hour. Other adaptations for an aquatic lifestyle include a membrane that protects the eyes, lips that seal the mouth behind the incisor teeth, and nostrils and ears that close when under water. It also has a number of specializations for diving – reducing its heart rate, shunting blood to the brain, tolerating high levels of carbon dioxide and the capacity to exchange most of the air in its lungs. The average dive lasts 4 to 5 minutes, but a Beaver can remain submerged for 15 minutes. The Beaver can swim in sea water; it probably reached most of B.C.'s coastal islands by swimming from the mainland or other islands. On land, a Beaver moves with a clumsy, waddling gait, though it can walk upright on the hind legs, and stand upright while gnawing a tree trunk.

Strictly herbivorous, Beavers eat the leaves, twigs and bark of a wide range of woody and herbaceous plants – in summer, mainly aquatic plants and the leaves of trees or shrubs; in autumn and winter, mostly the bark and twigs of woody plants. Water lilies are a major summer food. Deciduous trees and shrubs are the most common woody plants in the diet, though Beavers also eat coniferous trees, such as Lodgepole Pine and spruce. In northern regions, they prefer Trembling Aspen, Cottonwood, alders and willows. In coastal areas, where aspen is rare, they feed on Red Alder, Western Hemlock, Western Redcedar and Lodgepole Pine.

Beavers cut down trees mostly in autumn when they construct their winter food caches. They fell live trees by gnawing through the trunk with their incisors while standing on hind legs. Most of the trees felled by Beavers are less than 10 cm in diameter, though some have trunks up to a metre thick. Beavers usually leave the larger trees standing, even if they gnaw off the bark at their base. They generally take trees near the water's edge, but when food is scarce they harvest trees several hundred metres from the lodge. After falling a tree, the Beavers cut it into smaller lengths for transport. A colony constructs a single winter food cache on the water near their lodge. The cache consists of a large pile of woody branches and twigs capped with alder or spruce. As the pile becomes waterlogged it sinks and its cap freezes in the ice. In northern British Columbia, Brian Slough found that the winter caches consisted of Trembling Aspen, willows and alders.

A Beaver requires 0.5 to 2.5 kg of woody food per day. Adaptations for digesting cellulose and extracting protein from the woody diet include a prominent cardiac gland in the stomach that produces digestive enzymes and a large tri-lobed caecum in the large intestine containing microbes that ferment cellulose. The Beaver also re-ingests its feces to extract additional nutrients.

Mating is from January to March. Females produce three to five young after a gestation period of 105 to 107 days. Newborn kits are relatively well developed – their eyes are open, their teeth erupted, and they can swim or walk within minutes of birth. A kit weighs about 0.5 kg at birth, and it has light brown to nearly black fur. The nursing period lasts for about two months, but kits will eat leaves at 4 to 14 days and by one month they are weaned. The young kit first leaves the lodge on exploratory trips at about two weeks of age. Females produce only one litter a year. Males do not reach sexual maturity until their second year. Females are not capable of breeding until one year old, and most do not breed until they are in their second year, at 21 months of age.

Beavers can live up to 20 years, but few will reach 10 years. Mortality may be high among the young-of-the-year. The major predators of the Beaver in B.C. are the Grey Wolf and Cougar. Other sources of mortality include: starvation, accidents from falling trees, floods, injuries from fights, traps and tularemia (a bacterial infection that Beavers are especially susceptible to).

Range

The Beaver ranges across North America as far north as the treeline. In British Columbia it is found throughout the entire mainland and on at least 42 coastal islands including Vancouver Island. Populations only occur on islands larger than 15 km^2; 57% of the islands known to support this rodent are larger than 100 km^2. The most isolated islands with Beavers are the Dundas Group northwest of Prince Rupert, about 15 km from the mainland. This species is not native to the Queen Charlotte Islands. Beavers from the mainland were first transplanted there in 1936, and another introduction was made in 1949, using animals from Vancouver Island released at Gold Creek and Tlell River on Graham Island. The Beaver is now found on much of Graham and northern Moresby islands.

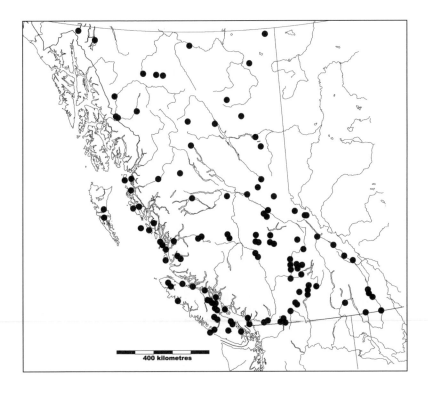

400 kilometres

Taxonomy

Twenty-four subspecies are listed for North America; four occur in British Columbia:

Castor canadensis belugae Taylor – from Cook Inlet Alaska along the coastal mainland and islands of B.C. to Dean Inlet.

Castor canadensis canadensis – across Canada from western Quebec to the Northwest Territories, Alberta and northeastern B.C.

Castor canadensis leucodontus Gray – Washington, Oregon, northern Idaho and B.C., where it occupies the southern interior, southwestern coastal mainland as far north as Rivers Inlet, Vancouver Island and larger associated islands.

Castor canadensis sagittatus Benson – the southeastern Yukon Territory and the northern and central interior of B.C. as far south as Kamloops and Glacier.

Conservation Status

This species is not of conservation concern in British Columbia. Managed as a fur-bearer, the trapping of Beavers is regulated under the provincial Wildlife Act. Over-trapping in the early 1900s lead to a provincial closure on trapping and hunting of this species and the development of some of the first regulations for managing wildlife in B.C. Thomas McCabe found that the Beaver had disappeared from a number of north coast islands in the 1930s and 1940s. He attributed their disappearance to over-trapping, but extinction from natural events followed by recolonization is probably a natural, ongoing process for Beavers living on smaller islands on B.C.'s coast.

Remarks

An important mammal in aboriginal culture, the Beaver was trapped for its fur and meat. Beaver pelts made excellent robes and other clothing. The large incisor teeth were used as carving tools, dice in gambling games, and decorations on leather shirts and on head-dresses worn by shamans. The Beaver appears in family crests used on totem poles, masks and ceremonial staffs.

The Beaver was one of the most important mammals in the historical fur trade. Fur traders prized Beaver pelts for their underfur, which was used to make felt hats. From 1919 to 1993, trappers declared about 1.14 million pelts in British Columbia, mostly from trap lines in the northern management regions.

Few fur-bearing mammals have been as well researched as the Beaver – the scientific literature on this species is enormous. Yet there has been remarkably little research done on this animal in British Columbia.

Selected References: Hatler 2002, Hill 1982, Jenkins and Busher 1979, McCabe 1948, Novak 1987, Slough and Sadleir 1977, Slough 1978.

Family Muridae:
Rats, Mice, Voles and Lemmings

Although their external appearance is variable, most species in this family tend to be mouse-like or rat-like: the skull has three upper and three lower pairs of cheek teeth and a V-shaped infraorbital opening. The 20 species found in British Columbia are classified into three subfamilies: Arvicolinae, Murinae and Sigmodontinae.

Subfamily Arvicolinae:
Voles, Lemmings and Muskrat

Members of this group are stocky rodents with a blunt muzzle, small eyes and ears, relatively short legs, and a tail that is usually shorter than the head and body length. The skull is angular in profile. The crowns of the molars have prisms arranged in alternating triangles (figure 59). The Muskrat (*Ondatra zibethicus*), red-backed voles (*Clethrionomys*) and Heather Vole (*Phenacomys intermedius*) have rooted molars; in other species the molars are ever-growing. Thirteen species occur in the province.

Southern Red-backed Vole
Clethrionomys gapperi

Other Common Names: Boreal Redback Vole, Gapper's Redback Vole.

Description

The Southern Red-backed Vole is a small vole with long, thick dorsal fur with a rufous or chestnut median stripe that extends from the head to the rump. The dorsal stripe is inconspicuous in some populations. The ventral pelage is white to grey; the hind feet are dark grey or black washed with silver. The tail, which is 22 to 32% of the total length, is slender and sparsely furred with grey or white hairs on the ventral surface. A dark colour phase found in some northern populations has a grey to black dorsal stripe, grey sides and dark tail. The eyes and ears are conspicuous. The rounded delicate skull has small rooted cheek teeth and a palate that terminates posteriorly in a simple transverse shelf in which the posterior edge is usually completely fused (figure 79).

Measurements:
total length: 133 (110-162) n=597
tail vertebrae: 36 (22-55) n=596
hind foot: 18 (14-22) n=597
ear: 14 (9-19) n=512
weight: 22.9 (15.0-45.0) n=597

Dental Formula:
incisors: 1/1
canines: 0/0
premolars: 0/0
molars: 3/3

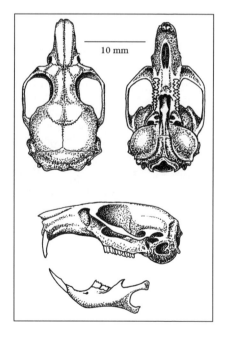

Identification:
Identification is problematic in parts of northern British Columbia where this species overlaps in its distribution with the Northern Red-

backed Vole (*Clethrionomys rutilus*). Adult Northern Red-backed Voles have a more densely furred and thicker tail that is yellow on the ventral surface. Its skull differs from that of the Southern Red-backed Vole in having a palate that is incompletely fused at its posterior edge (figure 80). Nevertheless, the fusion of the palate is related to growth and age. Some old Northern Red-backed Voles may have complete palates and the palate may not be completely fused in some immature Southern Red-backed Voles.

Natural History

In British Columbia, the Southern Red-backed Vole ranges from sea level on the coast to 2250 metres elevation in the southern Rocky Mountains and 2010 metres in the southern Coast Mountains. Primarily a forest species, it inhabits coniferous forests with abundant woody debris on the floor. In southern B.C., it lives in forests of Douglas-fir, Western Hemlock, Lodgepole Pine, Subalpine Fir, Engelmann Spruce and Western Larch. In northeastern parts of the province, it inhabits Trembling Aspen, Paper Birch, White Spruce and Black Spruce forests. Although several researchers have reported high populations in recent clear-cut habitats, the Southern Red-backed Vole generally is most abundant in mature forests with abundant shrub and ground cover. In spruce-fir forests of the Rocky Mountains in the western United States, this vole is considered to be an indicator of old-growth conditions. Nevertheless, in Western Hemlock forests of coastal Washington, Ernest Taylor found the Southern Red-backed Vole living in all successional stages, from old-growth forests to young-sapling and even grass-forb habitats. In spruce-fir forests of northern B.C., Thomas Sullivan and colleagues found this species rare in clear-cut forests and virtually absent from clear-cut sites that were burned.

Population densities can reach 48 animals per hectare in subalpine forests of the Rocky Mountains. In spruce-fir forests of northern B.C. and montane spruce forests in the south, average population densities are about 3 to 15 animals per hectare, although peak numbers can reach 84.5 per hectare. Population densities in coastal Western Hemlock forests of southwestern B.C. are lower, usually fewer than 2 animals per hectare. Populations peak in late summer or early autumn with recruitment of the young. They fluctuate from year-to-year, but there is no evidence that the Southern Red-backed Vole undergoes regular population cycles. There are no data on home-range size for B.C. In the Rocky Mountains of Colorado, home-range size was similar for males and females (0.01 to 0.06 ha). The Southern Red-backed Vole appears to have a solitary social structure (see page 15). Mature females maintain exclusive home ranges, though mature males have larger home ranges that overlap extensively with other mature males and females. The species is promiscuous, showing no strong bonds between males and females. A mother's association with her young ends when they are weaned at about 30 days.

The Southern Red-backed Vole is active day or night, but in summer it tends to be more active at night. It constructs nests about 75 to 100 mm in diameter from dried grass, leaves or moss, usually in underground tunnels, under stumps or logs. It maintains a home burrow as well as resting shelters that provide temporary protection from predators and extreme weather when feeding. Although it is not a strong climber, the Southern Red-backed Vole can climb and occasionally nests above ground. During a mammal survey in the Peace River area, Ian McTaggart Cowan found a Southern Red-backed Vole nest in an abandoned Crow's nest 3.4 metres high in a willow tree. The Southern Red-backed Vole is active throughout winter under the snow. Animals living at high elevations spend up to 7.5 months under the snow. In mid winter, it confines its activity to under the snow, but in the late-winter snowmelt, it may be active above ground. There is some anecdotal evidence that this species nests communally in small groups during winter.

The Southern Red-backed Vole is more omnivorous than most other voles in the province, eating green plant material, fungi, seeds, lichens and invertebrates. Lichens may dominate the spring diet and fungi the summer diet. In winter, this vole shifts to eating seeds. Captive animals store food throughout the year, but the extent that wild voles cache food is not known.

The breeding season is about seven months, from April to October. In harsh northern or subalpine environments, spring breeding often occurs under the snow. There is limited evidence that a few animals occasionally breed in mid winter under the snow. The gestation period is about 17 to 19 days. Embryo counts for 59 pregnant females from British Columbia averaged 4.9 (2-8). Females living in northern areas or at high elevations tend to produce larger litters. Older females surviving from the previous year can produce three litters in the breeding season; young-of-the-year females reach sexual maturity by early summer and produce one or two litters in their first summer. Newborns weigh about 1.3 grams; they are naked and blind. By 12 to 14 days, their eyes open and they begin to eat solid food. By 30 days the young are independent and ready to leave the nest.

A few Southern Red-backed Voles live up to 20 months in the wild, but most live only 10 to 12 months, rarely surviving two winters. This vole is a major prey species for many raptors and mammalian carnivores, including the Ermine and Marten.

Range

The Southern Red-backed Vole has an enormous range across the boreal regions of Canada and the United States. It is found across most of British Columbia, except in the extreme northwest where it is replaced by the Northern Red-backed Vole. The northern limits of its range are the lower Stikine River on the coast and the northern edge of the Interior Plateau (e.g., Bear Lake). In the northeast, it ranges as far west as the foothills and eastern slopes of the Rocky Mountains. The Southern Red-backed Vole occurs on only five islands in B.C.: Horsfall, Princess Royal and Athlone (Smythe) islands along the central coast; and Sonora and Hardwicke islands in Johnstone Strait.

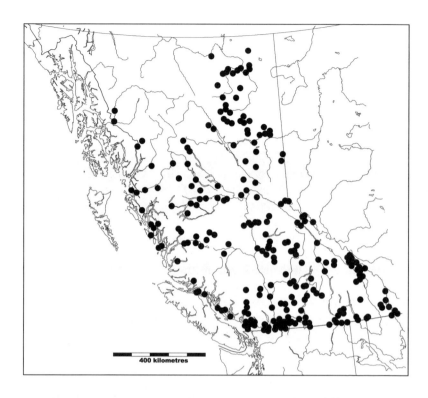

Taxonomy

In Alaska, the Yukon Territory, the Northwest Territories and northern British Columbia, *Clethrionomys gapperi* is replaced by *C. rutilus* (Northern Red-backed Vole); south of the Columbia River in the American Pacific Northwest, it is replaced by *C. californicus* (Western Red-backed Vole). Taxonomy of these three forms has been contentious but recent genetic studies suggest that they are distinct species. *C. gapperi* has 29 subspecies listed; 8 occur in B.C. The validity of many of these subspecies is questionable.

Clethrionomys gapperi athabascae (Preble) – the southern Northwest Territories, northeastern B.C., northern Alberta, Saskatchewan and Manitoba. Some authorities consider this race to be restricted to northeastern B.C., but in the original *Mammals of British Columbia*, Cowan and Guiguet showed the range extending down the Rocky Mountains as far south as Tornado Pass.

Clethrionomys gapperi cascadensis Booth – the Cascade and Coast mountains from Washington to Garibaldi Provincial Park in B.C.

Clethrionomys gapperi caurinus (Bailey) – a coastal race restricted to the province, where it ranges from Jervis Inlet to Lowe Inlet; it also inhabits several islands: Dufferin, Horsefall, Smythe, Princess Royal, Sonora and Hardwicke.

Clethrionomys gapperi galei (Merriam) – the Rocky Mountains from Colorado north to southern Alberta and extreme southeastern B.C. (Akamina-Kishinena area).

Clethrionomys gapperi occidentalis (Merriam) – the Olympic Peninsula and Puget Sound area of Washington; restricted to the lower Fraser River valley in B.C. Cowan and Guiguet (in the original *Mammals of British Columbia*), and Walter Sheppe treated *occidentalis* as a separate species, but it is now considered to be a subspecies of *C. gapperi*. The taxonomic status and geographic limits of this coastal form need to be assessed.

Clethrionomys gapperi phaeus (Swarth) – restricted to southeastern Alaska and western B.C. where it is confined to the vicinity of the Portland Canal and Portland Inlet.

Clethrionomys gapperi saturatus (Rhoads) – northern Washington, Idaho, Montana and B.C. where it occupies a broad region east of the coastal mountains and west of the Rocky Mountains.

Clethrionomys gapperi stikinensis Hall and Cockrum – Cleveland Peninsula of southeastern Alaska and the Stikine River area of B.C.

Conservation Status

Two subspecies are considered to be at risk provincially: *C. gapperi galei*, and *C. g. occidentalis*. *C. g. galei*, a widespread race found in the Rocky Mountains of the United States, barely enters the southeast corner of British Columbia. It appears on the provincial Blue List presumably because of few locality records, although the species seems abundant and widespread in the southern Canadian Rocky Mountains. Because it was described from only a few specimens on the basis of vague descriptive traits, the taxonomic validity of *C. g. galei* is questionable.

C. g. occidentalis is a dark coastal form that occupies the Puget Sound lowlands of Washington Sate and the extreme southwestern edge of the Fraser River valley. It appears on the provincial Red List. Its occurrence in the province was based on historical museum specimens from only three localities in the Vancouver region: Stanley Park, 1910; Vancouver, 1922; and Point Grey, 1946. It was thought to be extirpated in the lower Fraser River valley because of the loss of forested habitat, but in 1999, a small population was discovered in Burns Bog in Delta. The Burns Bog population, which is associated with Lodgepole Pine barrens, appears to be an isolated relict population surrounded by the urban areas of Delta and Surrey. The nearest known populations of Southern Red-backed Vole south of the Fraser River are near Chilliwack.

Remarks

The dark colour phase of this species, which is characterized by a black dorsal stripe and a grey dorsal pelage, is more prevalent in northern populations. Evidently, the dark pigmentation is most intense in young animals and with each successive moult the dark pigmentation becomes progressively lighter. The adaptive significance of this dark colour phase is unknown.

Selected References: Fuller 1969, Merritt 1981, Merritt and Merritt 1978, Sheppe 1960, Sullivan et al. 1999, Taylor 1999.

Northern Red-backed Vole

Northern Red-backed Vole
Clethrionomys rutilus

Other Common Names: Tundra Red-backed Vole.

Description

The Northern Red-backed Vole is a small vole with a rufous or chestnut median stripe on its back. The dorsal fur is long and thick; the undersides are buff to greyish. The dorsal surface of the hind feet are buff. The pelage colour is duller in young animals. A dark colour phase found in some northern populations has a grey or black dorsal stripe, grey sides and dark tail. The tail, which is 19 to 30% of the total length, is thick and densely furred, with yellowish fur on the ventral surface. The rounded delicate skull has small rooted cheek teeth; the palate terminates posteriorly in a simple transverse shelf with the posterior edge incompletely fused (figure 80).

Measurements:
total length: 129 (110-148) n=85
tail vertebrae: 31 (21-41) n=104
hind foot: 19 (15-21) n=103
ear: 14 (11-17) n=37
weight: 25.4 (18.0-31.8) n=25

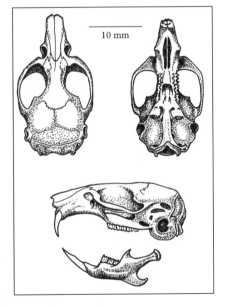

10 mm

Dental Formula:
incisors: 1/1
canines: 0/0
premolars: 0/0
molars: 3/3

Identification:
The Northern Red-backed Vole (*Clethrionomys rutilus*) can be distinguished from other voles – except the Southern Red-backed Vole (*C. gapperi*) – by its median dorsal stripe and a palate that terminates in a transverse shelf rather than a median spiny process. See the account for the Southern Red-backed Vole (page 240) for distinguishing characteristics.

Natural History

The Northern Red-backed Vole exploits a wide range of habitats, from tundra to forests. Population densities can reach 20 to 60 animals per hectare in late summer or early autumn. Although this vole is generally most abundant in mature forests, it will occupy habitats disturbed by fire or forest harvesting. Although it quickly invades an area following a forest fire, it delays breeding there until several years after the fire. In Alaska, Steve West and colleagues found population densities of Northern Red-backed Voles in clear-cut habitats were only about half the densities that occur in mature forests. In British Columbia, this vole's elevational range is from 150 to 1460 metres in the northern Rocky Mountains. Most researchers have concluded that this species does not undergo regular cycles in population. But Scott Gilbert and Charles Krebs used 17 years of population data to show that this vole has cyclic periods, though it has other periods with no cycles.

Little is known about this vole's social structure and there are no data on the size of its home range. Breeding females evidently have non-overlapping home ranges. This vole is active throughout winter under the snow. Although solitary most of the year, it may be social in winter, nesting communally in small groups. The Northern Red-backed Vole is dependent on blueberries, cranberries and Crowberries. The autumn berry crop persists under the snow and may be a major food source for this species throughout winter and early spring. Moss is an alternate food, eaten in summer when berries are unavailable. Lichens and invertebrates are minor food items.

The length of the breeding season varies from year to year, but breeding usually begins under the snow in late winter or early spring. In the taiga forests of the southern Northwest Territories, mature females that survive the winter produce their first litter from late May to early June and a second litter in late June to early July. A few females survive to produce a third or, rarely, a fourth litter in late summer. Females born in the first litters may produce one or two litters in their first summer. In the southern Yukon Territory, young-of-the-year first appear in early June. The breeding season can extend into September in some years. The gestation period is about 18 days. Embryo counts for 10 pregnant females from B.C. averaged 6.1 (range: 4 to 9). Newborns weigh about 1.7 grams; they are naked and blind. The Northern Red-backed Vole may survive up to 14 months in the wild. It is taken by various raptors and mammalian predators such as Ermine, Marten and Red Fox.

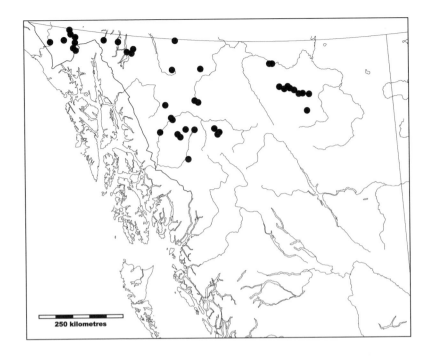

Range

The Northern Red-backed Vole ranges broadly across Eurasia and northwestern North America, including Alaska, the Yukon Territory, the Northwest Territories and northern British Columbia. It inhabits the extreme northwestern part of the province from the Haines Triangle east to Summit Lake Pass and Tuchodi Lakes in the northern Rocky Mountains, and as far south as the Spatzizi Plateau and Bob Quinn Lake on the Stewart Cassiar Highway.

Taxonomy

Because the Southern Red-backed Vole and Northern Red-backed Vole are similar in morphology some authorities have considered them to be conspecific. They differ in pelage colour, skull morphology, genetics (including DNA) and minor behavioural traits. Euan McPhee, who studied an area in the Northwest Territories where both voles live, found little evidence for hybridization and concluded that they are separate species. Eight subspecies of the Northern Red-backed Vole are recognized in North America; one inhabits British Columbia:

Clethrionomys rutilus dawsoni (Merriam) – broadly distributed across the Yukon Territory, the Northwest Territories and the northern part of the province.

Conservation Status
British Columbian populations are not of conservation concern.

Remarks
Ecologically similar, the Southern Red-backed Vole and Northern Red-backed Vole would be expected to compete in areas where they coexist. To what extent the two species overlap in their distributions in northern British Columbia is not clear. They have been taken at locations only 32 km apart in the Coast Mountains along the Stikine River. In northeastern B.C., the northern Rocky Mountains appear to delimit their ranges. Along the Alaska Highway, the Northern Red-backed Vole occurs as far east as Summit Lake in Summit Pass; specimens of the Southern Red-backed Vole have been taken only 24 km east at the Tetsa River. The two species should overlap in northern areas of the interior plateau, but because of spotty sampling in this region, the precise range limits of the two species are unknown.

Selected References: Fuller 1969; Gilbert and Krebs 1991; McPhee 1984; West 1977, 1982; West et al. 1980; Whitney 1976.

Brown Lemming *Lemmus trimucronatus*

Other Common Names: Siberian Lemming.

Description

The Brown Lemming is a stocky rodent with short limbs, a short tail (less than 20% of the total length) and small ears that are nearly hidden in the long, soft fur. The third, fourth and fifth digits on the front feet have short, curved claws; the tiny thumb has a distinctive large, flat nail. The dorsal pelage is greyish-brown in the head region and bright orange-brown in the rump area. The sides are bright orange-brown and the belly is light orange washed with grey. The dorsal surfaces of the hind feet are dark grey. The robust skull has broadly divergent zygomatic arches. The re-entrant angles on the upper molars are deeper on their outer side (figure 67), extending to the inner border of the tooth in the first and second upper molars.

Measurements:
total length: 142 (122-170) n=48
tail vertebrae: 17 (12-28) n=48
hind foot: 21 (19-24) n=47
ear: 11 (9-14) n=23
weight: 65.1 (42.5-98.4)
　　n=10

Dental Formula:
incisors: 1/1
canines: 0/0
premolars: 0/0
molars: 3/3

Identification:
No other British Columbian rodent is similar in external appearance. The skull of the Northern Bog Lemming (*Synaptomys borealis*) has similar enamel patterns on its upper molars, but is easily identified by its smaller size, shorter rostrum, grooved upper incisors and lower molars that lack closed triangles on their outer sides.

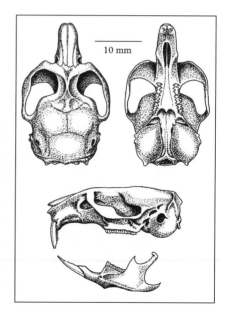

10 mm

The Brown Lemming's greenish fecal pellets, about 5 to 10 mm with rounded ends, also are quite distinctive. Most of our voles have smaller fecal pellets that are brown or black.

Natural History

Throughout arctic and subarctic regions, the Brown Lemming inhabits dense sedge-grass communities associated with open meadows, wetlands, and riparian habitats in the tundra or taiga forest. In British Columbia, it usually lives in subalpine or alpine areas. Although no habitat studies have been done on these high-elevation populations, anecdotal reports and information associated with museum specimens suggest that the Brown Lemming mostly inhabits open meadows, bogs or riparian habitats bordering alpine streams or lakes. Most records from the province range from 950 to 1830 metres elevation, but during population irruptions, the Brown Lemming will move into forested habitats at lower elevations.

Arctic populations undergo pronounced year-to-year cycles of abundance, reaching peaks every four to six years. As a result, this species has been used as a model by ecologists testing the various theories of population fluctuations in small rodents. At population peaks, Brown Lemmings can reach densities as high as 500 animals per hectare. No long-term monitoring has been done on alpine populations in B.C. to verify that they experience similar fluctuations. Nevertheless, anecdotal reports from early museum collectors, naturalists and wildlife biologists demonstrate that Brown Lemmings in B.C. undergo occasional irruptions.

A radio tracking study by Edwin Banks and colleagues revealed that male Brown Lemmings occupy larger home ranges than females (males 0.01 to 2.85 ha, females 0.02 to 1.68 ha). Males move around more frequently than females. Although the social structure of this species is not well known, males appear to be solitary and highly intolerant of other males in their home range. Brown Lemmings are extremely active in the summer, spending most of their time outside their burrows. Dispersing Brown Lemmings can travel long distances. A Brown Lemming was observed 16 km from shore on the sea ice in Alaska; a female released by James Bee and E. Raymond Hall travelled 2.4 km in about 90 minutes.

Brown Lemmings confine their movements to well-worn runways and trails that are conspicuous in dense grass or sedge habitats. Its distinctive fecal pellets are scattered in piles throughout its habitat. The burrow system consists of shallow tunnels that lead to an

enlarged nest chamber. Depth of the burrow system depends on the soil conditions. In substrates with permafrost, nest chambers are less than 10 cm below the surface. In soils that can be excavated, tunnels may be 5 to 100 cm under the surface and the nest chamber as deep as 60 cm. Temporary resting burrows have small, sparsely lined chambers, but the burrows used for rearing young have large nest chambers lined with dried grass or shredded leaves. In winter, Brown Lemmings construct elaborate nests under the snow on the surface of the ground. The winter nest has an inner nest made from dried grass and lemming fur surrounded by thick outer layers of dried grass. Well adapted to wet environments, the Brown Lemming has been seen swimming in pools of water created by the spring snowmelt.

Throughout the summer this rodent feeds mainly on the tender stems of grasses and sedges. It scatters grass and sedge cuttings along its runways, but unlike some species of *Microtus*, it does not cache large amounts of plant cuttings. Mosses are a minor food item in summer, but more important in autumn and winter as graminoid plants become less available. Although a number of diet studies in the arctic have shown that Brown Lemmings rarely eat woody plants, Thomas Sullivan and Wayne Martin attributed feeding damage on pine and spruce seedlings in central B.C. to this species.

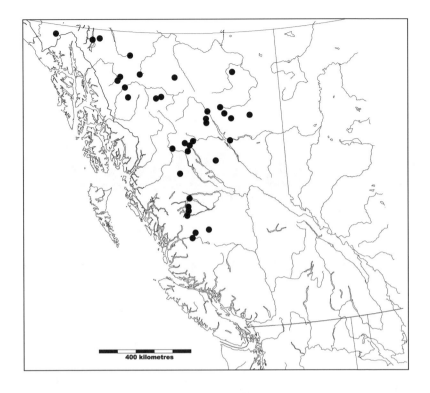

Brown Lemmings are capable of breeding under the winter snow in the arctic, but reproduction usually begins during the spring snowmelt. The initiation of the breeding season is closely synchronized with the first appearance of fresh grasses and sedges, the Brown Lemming's primary food. Evidently, chemicals present in these plants provide the essential cues to trigger reproduction. The gestation period is 21 to 23 days. The naked, newborn young weigh about 3.3 grams and are 33 mm in total length. Young lemmings grow rapidly, opening their eyes at 11 days, and leaving the nest by 16 to 17 days. Females can breed and produce their first litter when only three to four weeks old. Adult females produce as many as three litters in the breeding season; young-of-the-year produce one or two litters. Average embryo counts reported for arctic populations range from 4 to 12. The only breeding data for the B.C. population is derived information associated with museum specimens. Litter size, based on embryo counts from 19 pregnant animals collected from June 25 to September 2, averaged 6.0 (range: 3 to 10).

Potential predators in B.C. include the Coyote, Ermine, Least Weasel, Red Fox and various raptors. Four Brown Lemmings were found in the stomachs of Dolly Varden trout caught in the Driftwood River of central B.C.

Range

The Brown Lemming inhabits eastern Siberia and North America, where it ranges from Alaska, the Yukon Territory and the Northwest Territories east to Manitoba and south to British Columbia and Alberta. In B.C., it is sporadically distributed across northern and central areas of the province, reaching its southern limits at Mount Selwyn in the Rocky Mountains, the Ilgachuz Mountains in the Interior Plateau and the Rainbow Mountains on the eastern side of the Coast Mountains. In the Rocky Mountains, Mount Selwyn is the only known location for this species south of the Peace River, but the discovery of 21 Brown Lemming carcasses in a subalpine lake in the Willmore Wilderness Park of Alberta suggests that this species must range further south in the Rockies.

Taxonomy

Taxonomy and nomenclature of the genus *Lemmus* has been controversial, resulting in a number of species names in the taxonomic literature. Recent research, including a study of DNA, suggests that there are three species. North American and eastern Siberia populations are classified as *Lemmus trimucronatus*. In the original *Mammals of British Columbia*, Cowan and Guiguet classified the B.C. population as *Lemmus sibericus* (*Lemmus sibiricus*), but the scientific name *L. sibiricus* is now applied to a species found in western Siberia.

Five to nine subspecies of *L. trimucronatus* are recognized in North America, depending on the authority. But most of these races were defined on the basis of pelage colour, a trait that is highly variable in *L. trimucronatus*, even among individuals in the same population. James Bee and E. Raymond Hall, for example, described 11 distinct pelage types associated with age and seasonal variation in a single population from Barrow, Alaska. Genetic studies are required to resolve the subspecific taxonomy of this species. One of the North American subspecies occurs in British Columbia:

Lemmus trimucronatus helvolus (Richardson) – a brightly coloured form that inhabits the southern Yukon Territory, extreme southwestern Northwest Territories, Alberta and B.C.

Conservation Status
This species is not of conservation concern in British Columbia.

Remarks
One of the more fascinating aspects of this rodent's biology is its occasional population irruption followed by a complete crash in numbers. In the summer of 1987, Robert Cannings and Richard Hebda of the Royal BC Museum encountered large numbers of Brown Lemmings in the Ilgachuz Mountains at the southern limits of this species' range. They observed many lemmings moving through runways in forest and wetland areas well below the treeline. The following summer, I visited the same site and found none. The only evidence of Brown Lemmings from the past year were numerous old, worn runways and a few mummified carcasses. Thomas Sullivan and Wayne Martin's report of a population outbreak of Brown Lemmings during 1986-87 in the Morice region, southwest of Burns Lake, and several reports I received in 1987 of Brown Lemmings found drowned in northern lakes suggest that these occasional population irruptions may be synchronous across large areas of the province.

Selected References: Banks et al. 1975, Bee and Hall 1956, Negus and Berger 1998, Smith and Edmonds 1985, Sullivan and Martin 1991.

Long-tailed Vole   *Microtus longicaudus*

Other Common Names: None.

Description

The Long-tailed Vole is a medium-sized vole with a long tail that is 30 to 44% of its total length. The dorsal pelage ranges from greyish-brown to reddish-brown; the ventral pelage is grey. The dark, sparsely furred tail is slightly paler on its underside. The dorsal surface of the hind feet is grey or light brown. Populations from the interior tend to be paler with lighter bellies and more distinctly bicoloured tails than the coastal populations. The skull has a broad, incisive foramen with a sharply tapered anterior region. The incisors do not protrude beyond the nasal bones when the skull is viewed dorsally. The second upper molar lacks a posterior loop (figure 82), the third upper molar has four inner salient angles (figure 86), and the first lower molar has five or six closed triangles (figure 84).

Measurements:

total length: 182 (150-267) n=632
tail vertebrae: 66 (50-90) n=622
hind foot: 21 (16-27) n=627
ear: 13 (8-21) n=139
weight: 43.7 (20.0-85.0) n=146

Dental Formula:

incisors: 1/1
canines: 0/0
premolars: 0/0
molars: 3/3

Identification:

The Long-tailed Vole's range overlaps with six other *Microtus* species in British Columbia: the Creeping Vole (*M. oregoni*), Meadow Vole (*M. pennsylvanicus*), Montane Vole (*M. montanus*), Townsend's Vole (*M. townsendii*), Tundra Vole (*M. oeconomus*) and Water Vole (*M. richardsoni*). Externally, the Long-tailed Vole can easily be

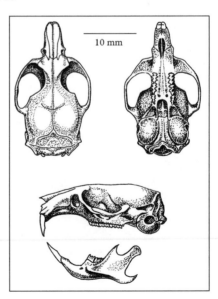

10 mm

discriminated from most of these species by its relatively longer tail (50 mm or more, and usually greater than 33% of the animal's total length). The Water Vole has a long tail, but it can be distinguished by its much larger size (weight > 90 g) and longer hind foot (> 28 mm).

The skulls of the Tundra, Creeping and Meadow voles can be discriminated from that of the Long-tailed Vole by their different molar enamel patterns (see those accounts on pages 268, 273 and 277). The Water Vole has the same molar enamel patterns, but its skull is much larger (maxillary toothrow length > 7 mm) and its upper incisors protrude more. The Montane Vole and Townsend's Vole also share the same molar enamel patterns and their skulls are similar in size. Various authorities have reported that these species differ from the Long-tailed Vole in having incisors that protrude beyond the nasal bones when their skulls are viewed dorsally and incisive foramen that are narrow and not tapered anteriorly. Nevertheless, these traits are variable and age-related; distinguishing the skulls of these species is difficult.

Natural History

With an elevational range from sea level to 2170 metres in British Columbia, the Long-tailed Vole lives in a wide range of habitats, from open meadows to shrubby riparian habitats and forest. But there are only a few detailed studies of its habitat requirements in B.C. In the Lower Mainland area near Vancouver, David Hawes found this species restricted to shrubby riparian habitats and early successional stages with dense shrubs. It is absent from grassy meadows and forests where it is replaced by species such as the Creeping Vole and Townsend's Vole. On the isolated Goose Islands, where it is the only vole species, Charles Guiguet found the Long-tailed Vole most abundant in grassy meadows, although it also inhabited bogs and forest. The most comprehensive habitat study is Beatrice van Horne's study on Prince of Wales Island, Alaska. She found that the optimum habitat for this vole was a clear-cut forest 7 to 20 years old with a mix of openings and shrub cover. Thomas Sullivan and colleagues reported similar results for spruce-fir forests in northern B.C., with this species most abundant in clear-cut sites. The Long-tailed Vole inhabits mature closed forest (70 years old) only if they have broken canopies creating small openings on the forest floor. In the dry grasslands, the Long-tailed Vole is commonly found in shrubby thickets of Common Snowberry. It also inhabits open grass-sedge meadows in subalpine and rocky alpine areas.

Long-tailed Voles rarely attain the high populations characteristic of many voles. Most population estimates for the Long-tailed Vole range from 1.1 to 12 animals per hectare, although one study reported densities as high as 120 animals per hectare. On Prince of Wales Island, Alaska, where the Long-tailed Vole is the only vole, density estimates ranged from 10 to 40 animals per hectare. In the spruce-fir forests of northern B.C., densities typically range from 1 to 20 animals per hectare but can reach 70 per hectare. Populations have a marked annual cycle that peaks in late August or September with the recruitment of the young. They also vary from year-to-year. A nine-year study done in Washington found evidence for a three-year population cycle, but no long term studies have been done in other parts of the range to establish if a similar pattern exists.

The social structure of the Long-tailed Vole is unknown. Estimates of average home-range size in summer range from 395 to 1615 m^2 for males and 353 to 2360 m^2 for females. Dispersing animals will move up to 1.1 km from their birthplace. Unlike most species of *Microtus*, the Long-tailed Vole does not form well-defined runways in its habitat. This species is mainly nocturnal with short bouts of activity in daytime. It is active throughout the winter constructing nests under the snow in close proximity to food sources. In summer the Long-tailed Vole eats forbs, graminoids, berries, truffles and arthropods. In winter, it feeds mostly on grasses or sedges buried under the snow. It selects the green parts of grass or sedge

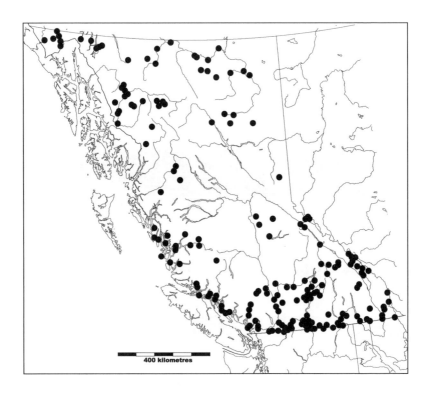

400 kilometres

tussocks for food, and may harvest the dried or dead parts of these plants for winter nesting material, but rarely eats them. When food is scarce in winter, the Long-tailed Vole eats the bark of trees and shrubs.

In B.C., the breeding season extends from late March to November. The average litter-size in the province is 5.0 (range: 3 to 9), based on embryo counts for 84 museum specimens collected between April 2 and November 11. The gestation period and development of the young have not been determined for this vole. Beatrice van Horne found that females produced no more than two litters in the breeding season. Most animals do not survive beyond their first breeding season. The oldest known individual in the wild was 13.5 months old. Potential predators include the Coyote, Ermine, Marten, Least Weasel, Long-tailed Weasel, Red Fox and various raptors.

Range

The Long-tailed Vole is associated with the western cordillera from Alaska, the Yukon Territory and the western Northwest Territories to the southwestern United States. In British Columbia it occupies the entire mainland. On the south coast, the Long-tailed Vole is absent from Vancouver Island and its associated smaller islands, which are occupied by Townsend's Vole. On the central and northern coast, where it is the only species of *Microtus*, the Long-tailed Vole has been recorded from only five coastal islands: Goose, King, Princess Royal, Stuart and Swindle. Except for the Goose Islands which lie about 38 km from the mainland, they are all near-shore islands.

Taxonomy

Fourteen subspecies are recognized with four occurring in British Columbia. But there has been no modern taxonomic study to verify the validity of these races – Christopher Conroy and Joseph Cook found little DNA genetic divergence among *Microtus longicaudus* populations from eastern Alaska, B.C. and Washington.

Microtus longicaudus littoralis Swarth – the mainland coast and Alexander Islands of southeastern Alaska and the coastal area of northwestern B.C.

Microtus longicaudus longicaudus (Merriam) – across western North America; in B.C., restricted to the southern dry interior as far north as Kamloops and Shuswap. In the original *Mammals of British Columbia*, Cowan and Guiguet classified these populations as *M. l. mordax*, but some authorities consider *M. l. mordax* and *M. l. longicaudus* indistinguishable.

Microtus longicaudus macrurus Merriam – coastal regions of western Washington and B.C., where it extends from Vancouver north to the Dean Channel. This subspecies includes the Goose Islands form, an isolated population that demonstrates gigantism, with some adults reaching 85 grams. This population also demonstrates some peculiar dental traits: of 22 skulls I examined from Goose Island, 10 (46%) had a distinct groove or shallow trough on the anterior face of one or both of their upper incisors; grooved upper incisors are rare or absent in other populations of the Long-tailed Vole.

Microtus longicaudus vellerosus Allen – widely distributed across eastern Alaska, the Yukon Territory, the western Northwest Territories, western Alberta, and northern, central and southeastern B.C., throughout the Selkirk, Purcell, Monashee and Rocky mountains.

Conservation Status
British Columbian populations are not of conservation concern.

Remarks
Although the Long-tailed Vole is one of the most widespread voles in the province, its biology has been little explored. In British Columbia, this species overlaps in its distribution with six species of *Microtus*. On the southwest coast it appears to be excluded from forest habitats by the Creeping Vole and from grassy meadows by Townsend's Vole; in the southern interior grasslands it is excluded from grassland habitats by the more aggressive Montane Vole. Its ecological interaction with other species of *Microtus* in the interior of B.C. is unknown.

Another subject requiring more study is the Long-tailed Vole's curious distribution on the islands along B.C.'s central coast. It has managed to reach the isolated Goose Islands, yet it is absent from most near-shore islands, while inhabiting the adjacent coastal mainland. The Long-tailed Vole is not known to occur on Vancouver Island in historical time, but there is evidence that it inhabited the island before the last glaciation: fossil teeth found in a sea cave near Port Eliza date from 16,000 to 18,000 years ago.

Selected References: Conroy and Cook 2000, Guiguet 1953, Hawes 1975, Smolen and Keller 1987, Sullivan et al. 1999, van Horne 1982.

Montane Vole   *Microtus montanus*

Other Common Names: Mountain Vole.

Description

The Montane Vole is a small vole with grizzled-grey ("salt-and-pepper") dorsal pelage and grey to white undersides. The fur on the dorsal surface of the feet is silver-grey. The tail, which is less than 30% of the animal's total length, is bicoloured with grey to greyish-brown above and whitish below. The skull has a short incisive foramen tapered in the posterior region, and the incisors protrude beyond the nasal bones when the skull is viewed dorsally. The second upper molar lacks a posterior loop (figure 82), and the first lower molar has five or six closed triangles (figure 84).

Measurements:
  total length: 142 (120-165) n=95
  tail vertebrae: 38 (29-51) n=95
  hind foot: 18 (16-22) n=92
  ear: 12 (9-14) n=11
  weight: 31.6 (22.0-45.0) n=14

Dental Formula:
  incisors: 1/1
  canines: 0/0
  premolars: 0/0
  molars: 3/3

Identification:

The range of the Montane Vole overlaps with the Meadow Vole (*Microtus pennsylvanicus*) and Long-tailed Vole (*M. longicaudus*) in British Columbia. It can also be confused with the Heather Vole (*P. intermedius*). See page 257 for the diagnostic traits of the Long-tailed Vole. The Meadow Vole does not have salt-and-pepper pelage, the dorsal surface of its hind feet are darker and its second upper molar has a distinct posterior

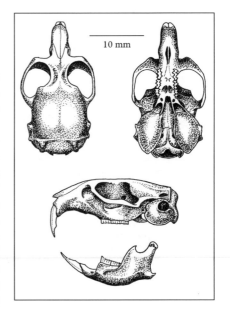

10 mm

loop (figure 81). The Heather Vole has softer, finer fur, more brownish dorsal pelage, a thinner more sparsely furred tail, and lower molars with the inner re-entrant angles deeper than the outer re-entrant angles (figure 75).

Natural History

In British Columbia, the Montane Vole lives in grassland habitats of the dry interior where it ranges from 300 to 1190 metres elevation. The few detailed habitat studies for the province show that it inhabits grassland-steppe communities with Bluebunch Wheatgrass, Antelope-Bush and Big Sagebrush; grasslands with Common Rabbit-brush, Prairie Sagewort or Common Snowberry thickets; and brushy habitats bordering lakeshores, grassy benches and Russian Thistle gulches. Montane Voles can be abundant in irrigated habitats such as orchards, hay fields and Alfalfa fields. This species co-occurs with the Meadow Vole and Long-tailed Vole throughout its range, though it generally avoids the wet meadows and riparian habitats occupied by the Meadow Vole and the shrubby thickets that seem to be the optimum habitat of the Long-tailed Vole.

Most Montane Vole populations undergo pronounced three-to-four-year cycles of abundance, reaching densities as high as 560 animals per hectare in some habitats. In the Okanagan Valley, Thomas Sullivan and colleagues estimated late summer population densities of up to 125 animals per hectare in old fields and 27 to 40 per hectare in apple orchards.

The Montane Vole has a complex social system, being monogamous when the population density is low but polygamous when it is high. Breeding males maintain home ranges that overlap minimally with those of other breeding males and significantly with those of several females. Males mark their territories with urine and feces, and defend them from other males by fighting and chasing intruders away. Territorial males may occupy the same general home range for up to nine months. Other males do not establish territories but reside in an area briefly and presumably play a minimal role in mating. Breeding females are also territorial, maintaining home ranges exclusive of other females. At low population densities, a female Montane Vole usually abandons her brood nest when the young are only 13 to 16 days old, moves to a new territory and builds a new nest. When the population is high and a breeding female is surrounded by other adult females, she will remain with her young and form an extended maternal family.

Although active in summer at any time of day, the Montane Vole shows distinct activity peaks in the morning and evening; it seems to be most active a few hours before sunset. Montane Voles construct conspicuous surface runways – well-worn trails three to six centimetres wide, radiating out from their burrows. They commonly scatter caches of plant cuttings along the runways and deposit feces in toilet areas at the intersections of several runways. Nests constructed from dried grass are located either in underground burrows or under the protective cover of woody debris or dense vegetation. Females usually place their brood nest underground in a burrow. The nest is about 13 cm in diameter with a hollow chamber 3.5 to 6.0 cm in diameter. Territorial males may build surface nests of grass within a few metres of a female's brood. Males also excavate burrows and will occupy the burrows of other mammals, such as the Northern Pocket Gopher.

Although grasses and sedges appear to be the main food of the Montane Vole, it has a flexible diet. A subalpine population in Colorado fed mostly on forbs such as vetch, agoseris, collomia and primrose in summer. A population living in a salt marsh in Utah relied almost entirely on Alkali Saltgrass throughout the entire year. In the Okanagan Valley, this vole eats the bark of fruit trees, particularly in winter, causing considerable damage.

Chemical compounds in the plants eaten by the Montane Vole provide the essential cues for breeding. Norman Negus and colleagues found that breeding in Wyoming and Utah was initiated in the spring by the appearance of green plants; timing of the first matings varied as much as a month from year-to-year. The gestation period is about 21 days. Litter size, which varies from year to year and with the mother's age, is from 1 to 10. Newborn Montane Voles are naked, blind and weigh about 2.2 grams. Their eyes open by 10 days, weaning begins by 12 days, and they finish nursing by 15 days. Growth rates and age at sexual maturity vary greatly. Animals born in early spring can breed at three to four weeks of age, but young born in later litters become sexually mature at seven to eight weeks. Young born in late summer may not breed until the following spring, although in some winters animals breed under the snow. A female can produce as many as five litters a year. Reproductive data for British Columbian populations is limited to data associated with museum specimens. Pregnant specimens have been taken from May 11 to September 15. Embryo counts for 18 museum specimens averaged 4.0 (range: 1 to 6).

Predators include the Burrowing Owl, Long-eared Owl, Western Screech-Owl, Northern Saw-whet Owl, Short-eared Owl, Badger, Ermine, Long-tailed Weasel, Striped Skunk and Western Rattlesnake.

Range

The Montane Vole occupies the cordillera and intermontane valleys of the western United States. In Canada, at the northern periphery of its range, it is restricted to the interior grasslands of southern British Columbia where it inhabits the Kettle, Nicola, Okanagan, Similkameen, Thompson and Fraser river valleys north to Westwick Lake and Riske Creek in the Chilcotin River valley.

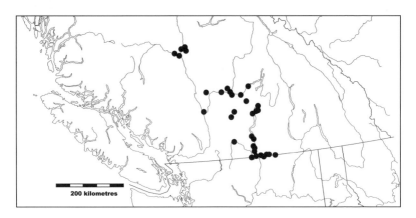

200 kilometres

Taxonomy

Fifteen subspecies are recognized; one is found in British Columbia:

Microtus montanus canescens Bailey – a small pale form found in eastern Washington and B.C.

Conservation Status

This species is not of conservation concern in British Columbia.

Remarks

Other than Thomas Sullivan's research on controlling populations in fruit orchards, the biology of this vole has received little attention in British Columbia. Given its association with native steppe-grassland habitats, it may a be a good indicator species for monitoring the impact of livestock grazing or alien plants such as Diffuse Knapweed and Spotted Knapweed on the province's grassland communities.

Selected References: Jannett 1980, Negus et al. 1977, Seidel and Booth 1960, Sullivan and Hogue 1987, Sullivan and Sullivan 1988.

Tundra Vole   *Microtus oeconomus*

Other Common Names: Root Vole.

Description

The Tundra Vole is a medium-sized vole with a short tail (less than 30% of the animal's total length). The dorsal pelage is grizzled brown with tinges of buff on the flanks and rump; the undersides are grey washed with buff. The tail is strongly bicoloured; the dorsal surface of the hind feet is pale brown. In the skull the second upper molar lacks a posterior loop (figure 82) and the first lower molar usually has only four closed triangles (figure 83), fewer than any other species of *Microtus* in British Columbia.

Measurements:

  total length: 156 (131-179) n=66
  tail vertebrae: 36 (26-47) n=81
  hind foot: 20 (16-23) n=81
  ear: 13 (10-16) n=56
  weight: 44.2 g (20.0-69.0) n=41

Dental Formula:

  incisors: 1/1
  canines: 0/0
  premolars: 0/0
  molars: 3/3

Identification:

The Tundra Vole co-occurs with the Meadow Vole (*Microtus pennsylvanicus*) and Long-tailed Vole (*M. longicaudus*) in northern B.C. See the Long-tailed Vole account (page 257) for diagnostic traits. The Meadow Vole can usually be identified by the lack of yellow on its flanks and rump, a grey ventral pelage and a tail that is not strongly bicoloured. Nevertheless, positive identification especially for immature animals may require a skull. The

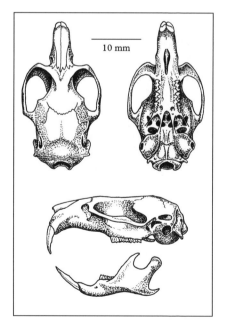

10 mm

lack of a posterior loop on the second upper molar and the presence of only four closed triangles on the first lower molar distinguish the Tundra Vole.

Natural History

In arctic regions, the Tundra Vole is associated with wet meadows and marshy areas bordering streams and lakes. During their five-year study in Kluane National Park, Yukon Territory, Charles Krebs and Irene Wingate captured this species in alpine tundra, subalpine shrub tundra and marshes. In 1983, I found Tundra Voles common in dense subalpine willow thickets, open subalpine and alpine habitats, sedge habitats bordering a stream and open grass-sedge meadows at Tats Lake in the Tatshenshini-Alsek Wilderness Park. It has also been captured by various museum collectors in willow thickets and subalpine meadows along the Haines Road. Charles Guiguet found this vole common in subalpine-alpine habitats at Mount McDame in the Cassiar Mountains. Its elevational range in British Columbia is from 820 to 1550 metres.

This vole undergoes pronounced three-to-four-year cycles in population density. During population highs, densities can reach 70 to 80 animals per hectare in summer. No data exist on its movements

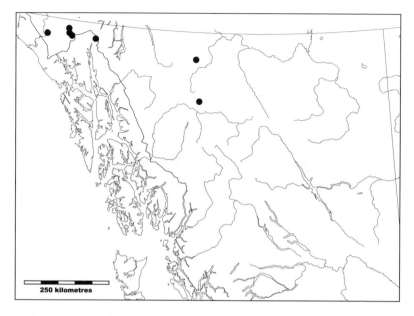

or home-range size and little is known about the social structure of this animal. In summer, Tundra Voles are solitary, but in winter they nest communally. In the northern Yukon Territory, I have captured Tundra Voles and Meadow Voles in the same runway. At Mount McDame, Charles Guiguet found both species sharing the same habitat. In the subalpine habitats at Tats Lake where the Meadow Vole is absent, the only vole found with the Tundra Vole was the Long-tailed Vole. Although they co-occurred in the same habitats, the Long-tailed Vole was most common in shrub thickets whereas the Tundra Vole mainly inhabited wet and dry meadows.

The Tundra Vole maintains a system of interconnecting runways, making piles of plant cuttings and mounds of fecal pellets throughout the system. It builds summer nests of grass and sedge on the surface or in excavated burrows and constructs winter nests on the ground under the snow. Tundra Voles feed mostly on graminoids, particularly sedges, which form much of the summer diet. Other foods include forbs, the shoots of woody shrubs and small amounts of seeds. In mid summer, locally abundant horsetails may be a major food for some populations. In autumn, this species eats roots and rhizomes. In Europe, the Tundra Vole relies mostly on the rhizomes of plants, particularly cottongrass, for winter food; but little is known about its winter diet in North America.

The breeding season begins in mid April to May, when this vole is still living under the snow, and usually ends in late September. Although winter breeding has been documented in northern Europe, it has not been reported for North American populations. The gestation period is 21 to 25 days. Estimates of litter size are based on embryo counts. Charles Krebs and Irene Wingate reported average counts of 4.9 to 5.4 for Kluane National Park in the Yukon Territory. In B.C. the average embryo count for 15 pregnant females was 5.3 (range: 3 to 7). Females may produce as many as four litters in a season.

Predators include various raptors, the Coyote, Ermine, Least Weasel and Red Fox.

Range

The Tundra Vole has a vast range across northern Europe, Asia and northwestern North America, where it is found in the Yukon Territory, the Northwest Territories and northwestern British Columbia. It has a small range in B.C., mostly in the Haines Triangle area; but it has also been found in White Pass on the Alaska border and in the Cassiar Mountains at Mount McDame and Gnat Lake (near Dease Lake).

Taxonomy

Ten subspecies are listed for North America with seven restricted to Alaskan islands. The validity of these subspecies is questionable. A genetic study by Ellen Lance and Joseph Cook revealed minor variation among Alaskan populations. British Columbian populations are situated in a transitional area between the mainland races *Microtus oeconomus macfarlani* and *M. o. yakutatensis*. In the original *Mammals of British Columbia*, Cowan and Guiguet assigned B.C. populations to *M. o. yakutatensis*, but genetic studies are required to resolve their taxonomic status.

Microtus oeconomus yakutatensis Merriam – a dark form associated with southeastern coastal Alaska and possibly extreme northwestern British Columbia.

Conservation Status

Several subspecies endemic to islands in Alaska are listed by the IUCN, but the Tundra Vole is not of conservation concern in British Columbia.

Remarks
Similar to the Collared Pika and Arctic Ground Squirrel, the Tundra Vole became isolated in Beringia during the last glaciation – this region in northern Alaska and the Yukon Territory was free of ice during the last glaciation. At the southern edge of the North American range, the British Columbian population is the result of a recent southward range expansion following the last ice age. More field work needs to be done to determine the precise distributional limits of the Tundra Vole in British Columbia. It appears to be absent from the northern Rocky Mountains, but it may be more widespread in the Cassiar Mountains than the few available records suggest.

Selected References: Batzli and Henttonen 1990, Krebs and Wingate 1985, Lance and Cook 1998.

Creeping Vole

Creeping Vole　　　*Microtus oregoni*

Other Common Names: Oregon Vole.

Description

Our smallest species of *Microtus*, the Creeping Vole weighs less than 30 grams. It has tiny eyes (< 4 mm in diameter) and a short tail that is less than 30% of the animal's total length. The fur is short, almost shrew-like. The dorsal pelage is a dark reddish-brown; the ventral pelage is dark grey washed with brown. The brownish tail is bicoloured. The dorsal surface of the hind feet is light brown. Although a number of authorities claim that this species has only five pads on the sole of its hind foot (most species of *Microtus* have six), this trait needs to be confirmed in a large series of live animals or fresh specimens. The skull is characterized by three inner salient angles on the third upper molar (figure 85), a pattern found in no other species of *Microtus* in British Columbia.

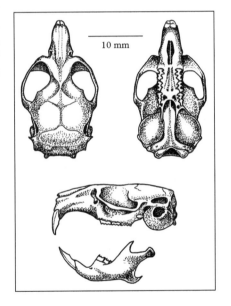

10 mm

Measurements:
total length: 135 (122-153) n=90
tail vertebrae: 34 (28-42) n=90
hind foot: 18 (16-20) n=89
ear: 10 (9-13) n=43
weight: 21.5 g (10.0-31.0) n=59

Dental Formula:
incisors: 1/1
canines: 0/0
premolars: 0/0
molars: 3/3

Identification:
In southwestern B.C., the Creeping Vole co-occurs with other voles: Long-tailed (*Microtus longicaudus*), Meadow (*M. pennsylvanicus*), Water (*M. richardsoni*) and Townsend's (*M. townsendii*). Diagnostic traits for identifying the Creeping Vole include its short, velvety fur, small eye and the distinctive pattern on its third upper molar.

Natural History

In the lower Fraser River valley, the Creeping Vole lives in coastal forests of Douglas-fir and Western Hemlock, shrubby habitats in recently logged forests, forest edges, riparian areas, grassy fields and abandoned agricultural land. Old fields and the early successional stages that follow forest harvesting seem to provide the ideal habitat. As shrubs and young trees replace grasses, sedges and herbaceous plants, these logged habitats become less suitable. The Creeping Vole also occupies dry subalpine forests and meadows up to 1980 metres in elevation on the eastern slopes of the Cascade Mountains in British Columbia.

For their study sites at Ladner and the University of British Columbia Research Forest, Thomas Sullivan and Charles Krebs found that population densities of Creeping Voles averaged about 32 animals per hectare at summer maximums and 7 per hectare during winter. They recorded the highest density, 72 per hectare, in abandoned farmland. Most researchers have found little evidence for year-to-year fluctuations in numbers. But Sullivan and Krebs concluded that three-to-four-year cycles of abundance occurred in some of their trapping grids set in old fields, shrub and forest habitats that lacked Townsend's Vole. Because it is a potential competitor, the presence of Townsend's Vole may prevent the Creeping Vole from reaching a sufficient population to undergo cycles. There are no estimates for this vole's home-range size in B.C.; in Oregon, they are from 0.05 to 0.38 ha for males and from 0.04 to 0.23 ha for females.

Little is known about the behaviour and basic natural history of this vole. In his study area in the Lower Mainland, David Hawes found that the presence of Townsend's Voles had an adverse effect on its populations, but that Creeping Voles avoided Townsend's Voles through their subterranean activity.

In summer, Creeping Voles presumably feed on green plant material; in coastal Oregon they also eat underground fungi. In winter, they eat the bark, twigs and buds of tree seedlings. Victor Scheffer described a stand of seedling Sitka Spruce on the Olympic Peninsula of Washington that was heavily damaged by Creeping Voles gnawing the main stems and clipping off branches. The Creeping Vole evidently makes surface runways in grassy meadows, but they tend to be less conspicuous than those made by the larger Townsend's Vole. A semi-fossorial rodent, the Creeping Vole constructs shallow tunnels in the soil. Several authorities have reported capturing Creeping Voles in mole tunnels.

Based on field studies and a captive laboratory colony, Ian McTaggart Cowan and Margaret Arsenault determined that the breeding season in B.C. generally begins in early March (although in mild winters it can begin as early as January) and usually ends about mid September (though it can extend into winter). The gestation period is about 23 days. The average litter size is 3.1 (range: 1 to 5), based on embryo counts for a sample of 26 females caught in B.C.; 28 litters born in captivity averaged 2.8 young (range: 1 to 5). Newborn Creeping Voles are naked and weigh about 1.7 grams. Their eyes open by 10 to 11.5 days, their molar teeth erupt by 11.5 days and weaning occurs at about 13 days. Female Creeping Voles can produce four or five litters a year. They are capable of coming into estrus as young as 22 days, but only the females born in the early litters will breed in the summer of their birth; females born after July delay breeding until the following spring.

Although a few Creeping Voles in the wild have survived 14 to 16 months, the average life span is much shorter – only a few animals survive beyond a year. Predators include the Barn Owl, Short-eared Owl, Northern Harrier, Rough-legged Hawk, Bobcat, Coyote, Ermine, Long-tailed Weasel and Western Spotted Skunk.

Range

A Pacific coast species, the Creeping Vole inhabits the coastal low-lands and coastal mountain ranges of North America from northern California to British Columbia, where it occupies the lower Fraser River valley and the Cascade Mountains. The eastern limit of its range in B.C. is on the eastern slopes of the Cascade Mountains in Manning Provincial Park. The northernmost record from the province is Fishblue Lake south of Lytton on the east side of the Fraser River.

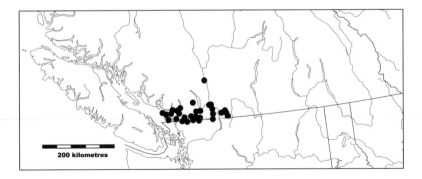

200 kilometres

Taxonomy
Four subspecies are recognized; one occurs in British Columbia:
Microtus oregoni serpens Merriam – B.C. and possibly north-western Washington.

Conservation Status
This species is not of conservation concern in British Columbia.

Remarks
Although the habitat requirements and population biology of the Creeping Vole have been well studied by Thomas Sullivan and David Hawes for several low-elevation coastal sites in the Fraser River valley, much remains to be learned about the basic biology of this inconspicuous mammal. The biology of Creeping Vole populations living in subalpine forests and meadows in the Cascade Mountains is essentially unknown.

Selected References: Carraway and Verts 1985, Cowan and Arsenault 1954, Hawes 1975, Scheffer 1995, Sheppe 1959, Sullivan and Krebs 1981.

Meadow Vole   *Microtus pennsylvanicus*

Other Common Names: None.

Description

The Meadow Vole is a medium-sized vole with variable dorsal pelage ranging from grey to rich brown; the undersides are silver-grey. The dorsal surface of the hind feet is grey to blackish. The tail, which is whitish-grey on the underside and brown on the dorsal surface, is usually less than 30 percent of the animal's total length (20 to 32%). The skull is characterized by a posterior loop on the second upper molar (figure 82), a trait found in no other British Columbian species of *Microtus*, and by five or six closed triangles on the first lower molar (figure 84).

Measurements:
total length: 147 (120-186) n=251
tail vertebrae: 38 (24-49) n=256
hind foot: 19 (16-260) n=259
ear: 12 (7-17) n=218
weight: 34.9 (20.0-63.5) n=262

Dental Formula:
incisors: 1/1
canines: 0/0
premolars: 0/0
molars: 3/3

Identification:

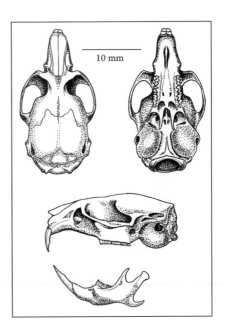

The Meadow Vole's distribution overlaps with five other *Microtus* species in British Columbia, but it can only be confused with the Montane (*M. montanus*), Tundra (*M. oeconomus*) and Townsend's (*M. townsendii*) voles – see those species accounts for distinguishing external traits. The posterior loop on the second upper molar identifies the Meadow Vole.

Natural History

The Meadow Vole is most abundant in open grass-sedge habitats associated with lakes, streams, dykes, sloughs and roadside ditches. It is also common in agricultural lands such as Alfalfa fields, hay meadows and orchards. Other habitats used by this vole include bogs, fens, lowland forests, recently burned or logged forests, sub-alpine forests, and subalpine or alpine meadows. Its elevational range in British Columbia is from about 300 to 2320 metres in the Rocky Mountains.

In the dry grasslands of the southern interior where this species competes with the Montane Vole, it usually confines itself to moist riparian habitats, whereas the Montane Vole occupies the more arid grasslands. Nevertheless, Thomas Sullivan and Eugene Hogue found both species coexisting in grassy habitats associated with apple orchards in the Okanagan Valley.

Meadow Vole populations vary widely throughout its range, reaching 600 animals per hectare in some habitats. Patterns of year-to-year variation in numbers also vary geographically: some populations display regular two-to-four-year cycles of abundance, but others exhibit only annual variations with no evidence of regular cycles. There have been no long-term population studies of this vole in B.C. Sullivan and Hogue estimated 22 to 60 animals per hectare in apple orchards in the Okanagan Valley. Populations in clear-cut habitats are much lower, usually less than 1 per hectare.

Males occupy larger home ranges than females, and both sexes have larger home ranges during the breeding season than in winter. Based on captures in live traps, Sullivan and Hogue estimated male home-range size at 405 to 3480 m^2 and female home-range size at 160 to 3115 m^2. But Dale Madison, who followed animals tagged with radio transmitters, found that Meadow Voles use much smaller areas during a 24 hour period – 192 m^2 for males and 69 m^2 for females. This vole is a strong swimmer – experiments have shown that it can swim up to 200 metres. I have captured Meadow Voles in marshes and beaver ponds on small, elevated tussocks completely surrounded by water.

The Meadow Vole's social behaviour varies seasonally. During the breeding season, sexually mature females are territorial, showing little overlap in their home ranges. In contrast, the home ranges of the breeding males overlap extensively with those of other males and females. Generally, breeding males live solitary lives, only coming into contact with other males when they compete for a breeding

female. They tend to be more aggressive in the breeding season and the presence of wounds on many males suggests that they compete for mates by fighting. During winter the Meadow Vole appears to be more social, nesting with others, presumably to conserve heat. As many as seven voles have been found in the same winter nest. Meadow Voles communicate with squeaks and various postural displays, including a threat display where an animal lunges at another squealing loudly with incisors bared. They appear to scent-mark their home ranges with fecal pellets and urine deposited in runways.

The Meadow Vole is active all day, but least active in the early morning. The grassy habitats it occupies are usually honeycombed with well-trampled runways. Nests are situated under woody debris or in dense grass. This vole eats many plants in summer, but mostly the stems, leaves and seed heads of grasses and sedges. It often makes conspicuous piles of stems and leaves in its runways. In agricultural areas, it readily exploits clover, Alfalfa, grains and garden vegetables, and it also feeds on invertebrates and truffles. For a winter food supply, the Meadow Vole makes caches of leaves, roots, rhizomes and corms on its runways under the snow. It also eats the inner bark of trees and shrubs during winter.

No information is available on the Meadow Vole's breeding season in British Columbia. In other parts of Canada, it begins in early spring, often under the snow, and ends in autumn. Winter breeding is common in the more temperate regions of North America, but is rare in the north. The gestation period is about 21 days. In B.C., the litter size averaged 5.8 (range: 2 to 10), based on embryo counts for 128 pregnant females. Newborn young are naked and weigh 1.6 to 3.0 grams. Their eyes open at 8 days and they eat solid food at 12 to 14 days. Females are capable of breeding at 25 days, males can breed at 35 to 45 days. Most Meadow Voles do not survive beyond a few months in the wild – only a few live to a second breeding season.

Range

The Meadow Vole has the most extensive distribution of any North American species of *Microtus*, ranging across Alaska, most of Canada and the northern United States. It is found throughout the entire mainland of British Columbia east of the coastal mountain ranges, although it penetrates into the Coast Mountains in some valleys. Westernmost records include: the Exchamsiks River in the Skeena River valley; Aiyansh in the Nass River valley; Dokdaon Creek on the Stikine River; Tunjony Lake, south of the Taku River; and Haines Road in extreme northwestern B.C.

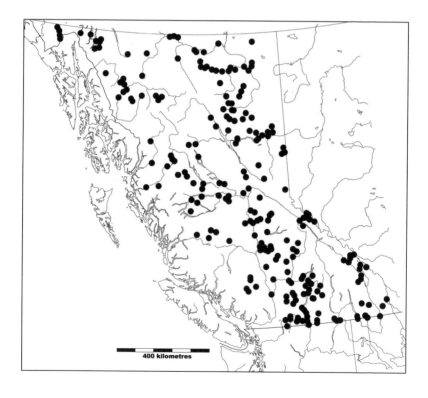

400 kilometres

Taxonomy

Although as many as 26 subspecies are recognized, they are based on pelage colour, size and skull features, highly variable traits in *Microtus pennsylvanicus*. Some authorities recognize four subspecies in British Columbia: *M. p. drummondii*, *M. p. microcephalus*, *M. p. funebris* and *M. p. rubridus*. But in the original *Mammals of British Columbia*, Cowan and Guiguet concluded that there are only two races in the province: *M. p. drummondii* and *M. p. modestus*. A modern genetic study is required to resolve the taxonomy of *Microtus pennsylvanicus*.

Microtus pennsylvanicus drummondii (Audubon and Bachman) – across Alaska, the Yukon Territory, the Northwest Territories, B.C., Alberta, Saskatchewan, Manitoba and Ontario. In B.C., it inhabits the entire mainland except for the extreme south.

Microtus pennsylvanicus modestus (Baird) – the Rocky Mountains of the United States from New Mexico to northern Montana, central Idaho and eastern Washington and the southern interior of B.C. from the extreme southern Okanagan to the Rocky Mountain Trench. According to Cowan and Guiguet, it is replaced by *M. p. drummondii* at higher elevations.

Conservation Status

This species is not of conservation concern in British Columbia.

Remarks

Throughout most of the British Columbia interior, this is the common vole in agricultural areas, where it can be found in hay fields and barn yards. Although it is regarded as a pest because it damages fruit trees and consumes agricultural crops (such as Alfalfa), the Meadow Vole is one of the dominant small mammals in many habitats, representing a major prey species for raptors and mammalian carnivores.

Selected References: Madison 1980, Reich 1981, Sullivan and Hogue 1987, Webster and Brooks 1981.

Water Vole   *Microtus richardsoni*

Other Common Names: None.

Description

The Water Vole is our largest *Microtus* species: its hind foot is longer than 29 mm, and most adults weigh more than 150 grams. Its dorsal pelage is greyish-brown, and the dense, woolly fur on the undersides is silver-grey. The long tail (more than 50 mm, 26 to 39% of the animal's total length) is thick, dark brown on the dorsal surface and whitish-grey on the underside. Some authorities claim that the sole of the hind foot has only five pads, and this trait has been used in a number of identification keys, but I have observed a small, vestigial sixth pad on some specimens. The skull is large (maxillary toothrow length > 7 mm) with the incisive foramina long and tapered posteriorly; the nasal bones protrude well beyond the incisors in dorsal view. The second upper molar lacks a posterior loop (figure 82) and the third upper molar has four inner salient angles (figure 86).

Measurements:
  total length: 239 (177-276) n=57
  tail vertebrae: 78 (54-98)
     n=56
  hind foot: 29 (24-34) n=58
  ear: 14 (12-20) n=58
  weight: 151.2 (124.0-
     177.3) n=12

Dental Formula:
  incisors: 1/1
  canines: 0/0
  premolars: 0/0
  molars: 3/3

Identification:

The Water Vole's distribution overlaps with those of the Creeping (*Microtus oregoni*), Meadow (*M. pennsylvanicus*) and Long-tailed (*M. longicaudus*) voles in British Columbia. See those accounts for diagnostic traits.

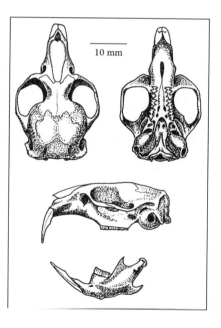

10 mm

Natural History

The Water Vole lives in subalpine-alpine meadows and riparian habitats that border cold mountain streams. Its elevational range in British Columbia is from about 1340 to 2320 metres. During population irruptions it may disperse to low elevations in the valley bottoms. For example, Ken Racey reported Water Voles at elevations as low as 183 metres in agricultural fields of the Pemberton Valley in the Coast Mountains. Specialized in its habitat requirements, the Water Vole occupies small patches of forb-grass meadow or alder-willow thickets adjacent to stream banks. In his study area in the Alberta Rocky Mountains, Daniel Ludwig captured Water Voles only within 5 to 10 metres of stream banks. Because this vole has specific habitat requirements, such as openings in stream banks with deep soil where it can construct burrows, it tends to be rather discontinuous in its distribution along a stream, with populations clumped in ideal sites and absent from unsuitable stretches. Ideal habitats may be occupied for many generations. The Water Vole also inhabits subalpine forests of Engelmann Spruce and Subalpine Fir, where it tends to be restricted to openings or clearings. In Oregon it has been found in recent clear-cut habitats.

Its sporadic distribution along streams prevents the Water Vole from reaching the high population densities reported for some other voles. Daniel Ludwig estimated densities of only 0.2 to 12.2 animals per hectare for riparian habitats in the Rocky Mountains of Alberta. With recruitment of the young, populations reach an annual peak in late summer. No long-term population studies have been done on this species to determine if it undergoes regular year-to-year cycles in abundance. Nevertheless, anecdotal observations suggest that the Water Vole undergoes occasional local irruptions. Ken Racey described three population outbreaks during 1927, 1949 and 1958 in the southern Coast Mountains. A population outbreak evidently occurred in Glacier National Park in the Columbia Mountains during 1982 when Water Voles invaded valley bottoms and even occupied the basement of the Glacier Park Lodge. In 1989, I observed a similar irruption at Blowdown Pass, west of Lillooet in the Coast Mountains. The subalpine meadows bordering a small stream were riddled with hundreds of runways and burrows, and much of the vegetation throughout the meadows showed the effect of heavy grazing. Water Voles also were common in the meadows and clearings of and Engelmann Spruce forest several hundred metres from the stream. When I revisited this area the following summer, I found no evidence of recent Water Vole activity, only the old runways and abandoned burrows from the previous year.

Males move greater distances and maintain much larger home ranges than females. Tracking animals tagged with radio transmitters over 72 hours, Daniel Ludwig estimated an average home-range size of 222 m² for females and 770 m² for males. From captures in live traps, he estimated that the maximum distances moved between captures over the breeding season was 94 metres for females and 332 metres for males. Adult females tend to remain in the same area throughout the breeding season, maintaining exclusive home ranges that overlap minimally with those of other adult females. Adult males travel more, but they remain close to the home ranges of one to four females, suggesting that Water Voles are polygamous.

Semi-aquatic, the Water Vole can swim on the surface, even against the current, and dive under water. Stream sides occupied by this vole are scarred with runways (5 to 7 cm wide) worn down to the surface soil. These paths run parallel to the stream and criss-cross its banks. Along the stream banks are small, trampled clearings unconnected to the runways – swimming voles use these clearings as temporary resting areas and haul-out sites.

Summer nests are in subterranean burrows, with entrances (about 8 cm in diameter) in stream banks or feeding areas close to streams. A burrow entrance leads to a branching network of tunnels one to three metres in length and 3 to 6 cm below the surface. Some tunnels are dead-ends used as temporary refuges from predators or as feeding areas. The others connect to a nest chamber (about 9.5 cm high, 15 cm long, 11 cm wide) that houses a large, domed nest made of short pieces of grass or sedge. Each nest is occupied by a single adult Water Vole.

Although active at any time of day, the Water Vole is most active after dark. During the population outbreak at Blowdown Pass, I observed Water Voles moving along runways in the daytime. This vole is territorial, but little is known about how it communicates. Odours from its feces, urine and glandular secretions deposited along runways presumably play some role. Large fecal droppings (3 x 6 mm) can be found scattered along runways, but this species does not deposit its droppings in large latrines in its runways as is typical of many voles. Adult males have prominent flank glands, 10 to 12 mm long. In a laboratory study, captive male Water Voles used these glands when confronted by a strange male; they rubbed their flank glands with the soles of their hind feet, then stomped the substrate of the nest box, presumably to leave a scent mark.

During the seven or eight months of winter in the high country, the Water Vole restricts its activity to snow tunnels. Daniel Ludwig found that activity above the snow surface ceased once the snow depth reached 6 cm. Temperatures in these snow tunnels are moderate, ranging from -1.0 to 1.0°C. Surface nests, often revealed in spring after the snow melts, may be used under the snow throughout winter or only during the spring snowmelt when subterranean tunnels and nests are flooded.

Water Voles feed on forbs and, to a lesser extent, grasses and sedges. They also eat seeds, invertebrates, and the buds and woody twigs of willows, as well as roots, rhizomes and corms, particularly in winter. In summer, they often deposit caches of fresh cuttings of forbs, graminoids and willow twigs in burrow entrances and in runways, presumably as stores for later consumption. Daniel Ludwig found no evidence that the Water Vole stores food for winter.

No data are available on the breeding season in British Columbia, but in the Rocky Mountains of Alberta, it begins in late May or early June and ends until early September; on exposed south-facing slopes, breeding begins as early as May 1. Males appear to become sexually active earlier than females. After a gestation period of 22 days, females give birth to about five young. The litter size in B.C., based on embryo counts for 16 museum specimens, averaged 5.2 (range: 3 to 8). Newborns are naked, blind and weigh about 5 grams. By 12 days their eyes open and by 21 days they eat solid food. They first appear out of the nest about mid July. Adult females produce only two litters in the breeding season. Although young-of-the-year are capable of breeding in the summer of their birth, most do not breed until the following year. A few individuals may survive two winters in the wild, but most Water Voles do not live beyond their second autumn or winter. Predators include various raptors, the Coyote, Ermine, Long-tailed Weasel and Marten.

Range

The Water Vole is associated with higher elevations in the cordillera of the western United States and Canada. In British Columbia, it inhabits the Cascade, Coast, Columbia and Rocky mountains, but not the lowlands of the Interior Plateau. Populations in the coastal mountain ranges are separated by a hiatus of some 200 km from the nearest populations in the Columbia Mountains. The northern limits of its range are Gold Bridge in the Coast Mountains and Trophy Mountain in the Columbia Mountains. Although a number of authorities, including Cowan and Guiguet, give Mount Robson as

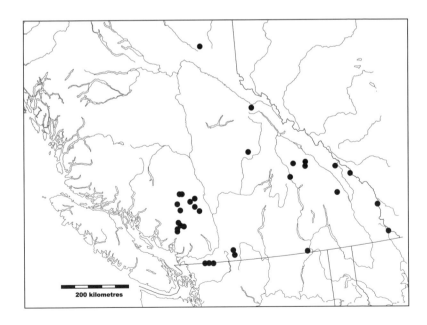

200 kilometres

the northern limit in the Rocky Mountains, there is a historical museum specimen from the head of the Parsnip River, nearly 150 km to the north.

Taxonomy

Some authorities recognize two geographically isolated races in British Columbia: *Microtus richardsoni arvicoloides* in the coastal mountain ranges and *M. r. richardsoni* in the Columbia and Rocky mountains. But Cowan and Guiguet in the original *Mammals of British Columbia*, considered these two forms indistinguishable, classifying them as *M. r. richardsoni*. Recent DNA studies by John Demboski and Jack Sullivan confirmed the genetic similarity of these two populations.

Microtus richardsoni richardsoni (DeKay) – a widespread race inhabiting the coastal mountain ranges north of the Columbia River and the Columbia and Rocky mountains of the western United States and Canada.

Conservation Status

This species is not of conservation concern in British Columbia.

Remarks

With adults weighing more than 100 grams and reaching nearly 28 cm in total length, this is an impressive vole. Of the British Columbian arvicoline rodents (voles and lemmings), only the Muskrat is larger. Living in subalpine-alpine meadows covered in snow up to eight months of the year, the Water Vole is able to cope with a short growing season and a limited breeding season. Water Vole fossils recovered from several caves in the foothills and Rocky Mountains of southwestern Alberta date from about 18,000 years ago, just before the last glaciation and from the early postglacial period. These fossils suggest that the Water Vole survived in the harsh environments that bordered the cordilleran ice-sheet during the height of the last glaciation.

Selected References: Anderson et al. 1976; Ludwig 1984a, 1984b, 1988; Racey 1960.

Townsend's Vole   *Microtus townsendii*

Other Common Names: None.

Description

Townsend's Vole is a large *Microtus* with a hind foot usually greater than 21 mm long, and a tail more than 50 mm long. The dorsal pelage is dark brown and the undersides are grey to white. The skull has a narrow incisive foramen that is not tapered; the incisors protrude slightly beyond the nasal bones when the skull is viewed dorsally. The second upper molar lacks a posterior loop (figure 82) and the third upper molar has four inner salient angles (figure 86).

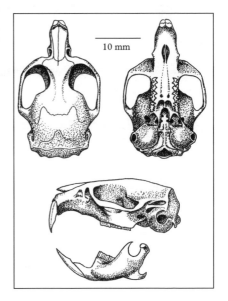

10 mm

Measurements:
  total length: 192 (155-235)
    n=186
  tail vertebrae: 58 (42-75)
    n=183
  hind foot: 24 (19-29)
    n=185
  ear: 14 (10-18) n=65
  weight: 59.6 g (40.0-103.0)
    n=89

Dental Formula:
  incisors: 1/1
  canines: 0/0
  premolars: 2/1
  molars: 3/3

Identification:
The distribution of Townsend's Vole overlaps that of the Long-tailed Vole (*Microtus longicaudus*) and Creeping Vole (*M. oregoni*) in the lower Fraser River valley. See those accounts for diagnostic traits.

Natural History

Townsend's Vole lives in non-forested habitats, such as marshes, wet meadows, salt marshes, hay fields, abandoned pastures, riparian areas bordering streams and lakes and alpine meadows. In the Fraser River valley of mainland British Columbia, it is restricted to elevations below 300 metres where it is most abundant in grassy

fields, including poorly drained fields partially flooded in winter. On the coastal islands where Townsend's Vole is the only vole species, it is occasionally found in forested habitats, especially recent clear-cuts in Douglas-fir forests. In Strathcona Provincial Park on Vancouver Island, I saw signs of Townsend's Vole (burrows, winter runways under the snow, fecal pellets) at elevations as high as 1740 metres in subalpine meadows dominated by heather plants and a sparse covering of sedges.

Townsend's Vole undergoes both annual fluctuations in population numbers and pronounced year-to-year cycles of abundance. At its peak, a population can reach densities of 525 to 880 animals per hectare, among the highest known for any North American species of *Microtus*. Average population densities during annual fluctuations range from 94 to 239 animals per hectare. Xavier Lambin and Charles Krebs used radio transmitters to track Townsend's Voles living in a grassy field near Ladner and found that males had larger home ranges than females. They estimated home-range sizes to be from 198 to 219 m² for breeding males and from 94 to 152 m² for breeding females. Males showed no seasonal differences in their home-range size, but females occupied smaller home ranges in summer. In spring, males and females maintain territories where they exclude other individuals of the same sex – many of the animals caught by Lambin and Krebs showed wounds, suggesting that these voles defend their territories aggressively. At this time of year, a male's territory overlaps with just one female, indicating that this species is monogamous in spring. In summer, as density increases because of the recruitment of young, territory sizes decrease and females overlap more in their home ranges. At this time, the mating system appears to be polygamous, because a male's territory now overlaps with those of females.

Townsend's Vole generally confines its movements to well-worn runways used over many generations. Piles of fecal droppings and fresh plant cuttings along runways are a sure sign of this vole. At high elevations on Vancouver Island, spring snowmelt often reveals runways and large latrines of fecal droppings deposited by Townsend's Voles active under the snow during the previous winter. In dry habitats, these voles maintain subterranean burrows, but in wet areas, such as flooded fields, they construct globular grass nests above ground. A strong swimmer, this rodent can occupy habitats completely inundated with water.

Several small rodents overlap extensively with Townsend's Vole in their distributions. The Deer Mouse co-occurs with Townsend's Vole on some of the coastal islands. In grassland habitats of the Fraser River valley, the Creeping Vole and Townsend's Vole often coexist. Removal experiments done by Charles Krebs and colleagues showed that Townsend's Vole excludes the Deer Mouse from grassland habitats through competition. Removing Townsend's Vole from a study area resulted in the Deer Mouse population increasing to high numbers. The Deer Mouse population remained high until Townsend's Voles re-established themselves.

Townsend's Voles feed on grasses, sedges and forbs, often storing cuttings of stems and leaves in caches scattered throughout their runway system. In winter, they live off bulbous roots stored under logs and grass piled in runways. Caches of mint bulbs found in Washington were as large as 13 litres. Lichens were found in the stomachs of a few individuals captured in a clear-cut. This vole also girdles the trunks of coniferous trees up to 19 cm in diameter.

The length of the breeding season varies with population abundance and winter climate. Usually the breeding season begins in January and continues until late autumn. In some years, however, there is winter breeding – nursing females can be found in all months. The gestation period is 21 to 24 days. Estimates for average litter size vary from 4.9 to 5.4, depending on the population cycle and the age of females. Newborns are naked and blind; they are weaned by 15 to 17 days. The average age of females at first conception is about 100 days, but some may breed as young as 20 days.

Because it reaches high population densities, Townsend's Vole is the dominant small mammal prey species in southwestern British Columbia. It is the principal prey of Barn Owls living in the lower Fraser River valley and Vancouver Island, accounting for 50 to 80% of prey items recovered from its regurgitated pellets. Other avian predators include the Great Horned Owl, Short-eared Owl, Northern Harrier, Red-tailed Hawk, Rough-legged Hawk, Snowy Owl and Great Blue Heron. Several mammalian carnivores prey on this vole, including as the Ermine, Long-tailed Weasel, Bobcat, Mink, Raccoon, Spotted Skink, Domestic Cat and Coyote. During haying season, this species is especially vulnerable, as avian predators concentrate on recently mowed hay fields.

Range

A Pacific coast species, Townsend's Vole ranges from northern California to southwestern British Columbia. In B.C.'s lower Fraser River valley, it ranges east to Chilliwack and north to Burrard Inlet. In the original *Mammals of British Columbia*, Cowan and Guiguet reported that this species did not occur north of Burrard Inlet, but there are a few recent records from the lower slopes of the Coast Mountains. Townsend's Vole occurs on 30 islands along B.C.'s south coast, including Vancouver Island, islands in the Strait of Georgia and Johnstone Strait and various islands adjacent to Vancouver Island. Breakwater Island, 20 ha in area, is the smallest of these. The most isolated population is on Triangle Island, the outermost of the Scott Islands, more than 40 km from Vancouver Island. In contrast to the ubiquitous Deer Mouse, Townsend's Vole has a spotty, insular distribution, presumably because many of B.C.'s coastal islands have insufficient open grassland habitat to support viable populations of this vole.

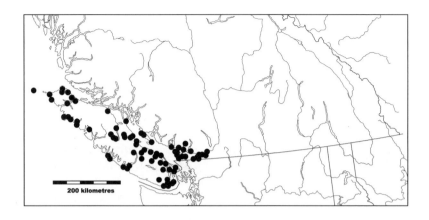

200 kilometres

Taxonomy

Six subspecies are recognized; four occur in British Columbia. Specimens from islands off the east and west coasts of Vancouver Island have never been formally described or assigned to a subspecies. Based on a study of DNA, William Thomas and Andrew Beckenbach demonstrated that Vancouver Island populations are strongly divergent from mainland populations. Moreover, they found evidence for two distinct lineages of *Microtus townsendii* on Vancouver Island.

Microtus townsendii cowani Guiguet – an insular race restricted to Triangle Island. Strongly differentiated from other populations, this race demonstrates gigantism, with some adults weighing more than 100 grams. This form has a peculiar coarse pelage with a high incidence of white markings on the head.

Microtus townsendii cummingi Hall – a weakly defined insular race restricted to Bowen and Texada islands. The Bowen Island population shows little genetic differentiation from populations in the Lower Mainland of B.C. and the Olympic Peninsula of Washington.

Microtus townsendii laingi Anderson and Rand – an insular race found on northern Vancouver Island and on Hope, Hurst and Nigei islands. Populations on Vargas, Moketas, Union and Little Bunsby islands off the northwest coast of Vancouver Island may be additional insular populations of *M. t. laingii*. Cowan and Guiguet classified the Little Bunsby and Vargas islands populations as *M. t. tetramerus*, although these islands are north of the range of this subspecies.

Microtus townsendii tetramerus (Rhoads) – an insular race found on southern Vancouver Island and the Gulf Islands. Populations on

Quadra, Denman and Hornby islands off the east coast of Vancouver Island, and East Turtle, Fleming, Tzartus, West Jarvis, Dodd and Sanford islands in Barkley Sound off the west coast of Vancouver Island have not been assigned to a subspecies, but they probably represent additional populations of this race.

Microtus townsendii townsendii (Bachman) – west of the Cascade Mountains in northern California, Oregon, Washington and southwestern B.C. where it ranges as far north as Burrard Inlet and east to Chilliwack.

Conservation Status

The Triangle Island subspecies (*M. t. cowani*) is on the province's Red List and was designated by the IUCN as Lower Risk (conservation dependent). This subspecies is endemic to Triangle Island, a small (1.07 km²) windswept and isolated island of the Scott Islands, 46 km from Cape Scott on the northern tip of Vancouver Island, where the nearest known populations of Townsend's Vole occur (figure 25). Although Triangle Island is a protected ecological reserve, the Townsend's Vole population there could be threatened by the introduction of rats or the arrival of a predator such as the Mink.

Remarks

The population biology and behaviour of Townsend's Vole populations living in the lower Fraser River valley have been studied for several decades by Charles Krebs and his students. Results of their research have provided considerable data for testing various theories of population fluctuations in small rodents. Townsend's Voles may damage reforested areas because they girdle the stems of tree seedlings. But this species is ecologically important, especially in the lower Fraser River valley where large populations of overwintering raptors rely on it as a food source.

Fossil teeth and bones of Townsend's Vole have been recovered from a cave at Port Eliza on Vancouver Island. They date from about 16,000 to 18,000 years ago, a period just before the ice of the last glaciation covered the south coast.

Selected References: Cornely and Verts 1988, Hawes 1975, Lambin and Krebs 1991, Taitt and Krebs 1985, Thomas and Beckenbach 1986.

Muskrat   *Ondatra zibethicus*

Other Common Names: None.

Description

Our largest vole, the Muskrat is a heavy, chunky rodent with small eyes and small ears. Its large hind feet are partially webbed and have stiff hairs on the edges, but the small front feet have no webbing. Its long tail, nearly the length of the body, is scaly and laterally flat with a keel. The Muskrat's rich pelage consists of long, shiny guard hairs and short, dense underfur; the dorsal pelage is dark brown; the undersides are paler. Its skull is massive and angular, with large incisor teeth and rooted cheek teeth.

Measurements:
  total length: 547 (448-685) n=140
  tail vertebrae: 242 (195-295) n=138
  hind foot: 76 (63-85) n=137
  ear: 23 (19-260 n=26
  weight: 1.02 kg (0.70-1.77) n=40

Dental Formula:
  incisors: 1/1
  canines: 0/0
  premolars: 0/0
  molars: 3/3

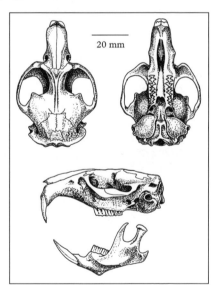

20 mm

Identification:
This species cannot be confused with any other rodent in British Columbia. Its flat tail, glossy fur and large size readily distinguish it from the Norway Rat (*Rattus norvegicus*) and Black Rat (*Rattus rattus*). The massive skull (length > 40 mm, maxillary toothrow length > 14 mm) is larger than any other vole skull.

Natural History

An aquatic rodent, the Muskrat inhabits small lakes, ponds, marshes, slow streams, sloughs, drainage ditches, dikes and brackish

estuaries. Its ideal habitat is a permanent wetland deep enough (0.5 to 3.0 metres) to not freeze solid in winter and with an abundance of cattails, bulrushes and sedges. In British Columbia, the Muskrat is generally confined to valleys and lowlands. High-elevation tarns and small lakes tend to freeze solid in winter and lack adequate vegetation. Productive areas for the Muskrat include: the Fraser River estuary, the Kootenay River floodplain in the Creston Valley, small lakes and sloughs of the Interior Plateau, and wetlands in the Peace River region.

There are no estimates of population densities for Muskrat populations in the province. In other regions of Canada, typical population densities range from about 7 to 60 animals per hectare. The Muskrat undergoes periodic fluctuations in numbers, although there has been some controversy about the regularity and length of these fluctuations. Recent research analysing fur returns to the Hudson Bay Company across Canada revealed that population peaks in the boreal forests and western cordillera regularly occur every eight to ten years. There appears to be a strong association with Mink populations, which show similar eight-to-ten-year cycles that lag two to three years behind the Muskrat trend. To what extent these cycles are the result of predator-prey interactions is unknown. Other explanations for these regular cycles of abundance include changes in food supply, or internal changes in reproduction and behaviour.

For most of the year, Muskrats are relatively sedentary. In summer they generally forage within 30 metres of their dwellings. There are no estimates of home-range size for B.C. habitats. Gilbert Proulx and Fred Gilbert estimated average home ranges of 302 to 1112 m^2 for Muskrats living in an Ontario marsh. During winter this animal usually feeds within 10 metres of its dwelling, rarely moving beyond 150 metres. In spring and autumn, when Muskrats disperse from their home area, they can move long distances.

A Muskrat moves slowly and awkwardly on land, but swims very well, up to five kilometres per hour on the water's surface. When swimming, it holds the front feet under the chin and propels itself by paddling with the hind feet and moving the tail from side to side; the tail also acts as a stabilizer. An excellent diver, the Muskrat can remain underwater up to 20 minutes. Adaptations for diving include the ability to lower its heart rate during a dive and a high tolerance for carbon dioxide. This species is so dependent on swimming that, during periods of drought or low water levels, it will excavate canals up to 30 cm deep connecting its lodge to open water.

Although biologists have occasionally observed more than one breeding male or female occupying the same lodge or burrow, they consider the Muskrat monogamous, with the basic social unit consisting of a breeding pair and their associated young. During the breeding season adult pairs are aggressively territorial. Young-of-the-year remain within the vicinity of their parents until they disperse in autumn or the following spring. During winter, as many as six Muskrats (presumably a family group) will share the winter lodge.

Muskrats construct two basic types of shelters. They excavate burrows in the banks of sloughs, ditches, dikes, streams or lakes. A burrow can be 8 to 150 cm deep in the soil with its entrance below the water's surface and leading to a labyrinth of interconnecting chambers and tunnels. In wetlands with abundant plant material and deep water, Muskrats construct lodges from roots and stems. The cone-shaped lodge may be several metres in diameter and up to a metre in height. Its floor is well above the water level and the Muskrat enters through one of several underwater tunnels. It has one or two nest chambers used for rearing the young. Before winter, the Muskrat adds more mud and vegetation to the lodge to increase its insulation; the winter dwelling lodge usually has larger chambers, presumably to accommodate more animals. Muskrats also build smaller lodges as shelters for feeding in the winter. Another winter structure made by the Muskrat is the "push-up", a dome-like mound of frozen vegetation created by pushing plant material through a hole in the ice.

The Muskrat is active throughout the winter. Its thick pelage effectively insulates the Muskrat while swimming in freezing water – its underfur keeps dry even when the animal is submerged. With its thick walls and an insulating blanket of snow, the winter dwelling lodge protects this animal from extreme winter temperatures while resting or feeding out of water. Muskrats may huddle together in the nest chamber to conserve heat. They further reduce their winter energy demands by foraging and feeding close to the lodge.

Muskrats feed on the stems, leaves, shoots, roots and tubers of aquatic plants such as cattails, bulrushes, pondweeds, arrowheads, water lilies and horsetails; in agricultural areas they eat cultivated crops. The diet varies greatly from place to place, reflecting differences in local plant abundance and individual preferences. Although generally herbivorous, the Muskrat will occasionally eat snails, mussels, crustaceans and young birds. In marshes, it usually constructs elevated feeding platforms or rafts from vegetation, enabling it to feed out of the water.

The Muskrat is a prolific breeder. The breeding season in British Columbia has not been studied, but mating probably occurs in April. The gestation period is 28 to 30 days. The average litter size ranges from three to nine, with northern populations having larger litters. Although the Muskrat may breed throughout the year in parts of the United States, in most regions of Canada it produces two litters. Newborns are blind, nearly hairless and weigh about 21 grams. By one week they have a covering of coarse hair; by two to three weeks they begin to swim. They are usually weaned at about four weeks. Young-of-the-year do not breed until the following year.

Muskrats can live up to three or four years in the wild, but the mortality rate, especially among young animals, is high during the summer and over winter. Estimates for overwinter survival for all ages range from about 20 to 68%. Mammalian predators include the Bobcat, Coyote, Marten, Mink, Racoon and Red Fox. Avian predators include the Bald Eagle, Great Horned Owl and Northern Harrier.

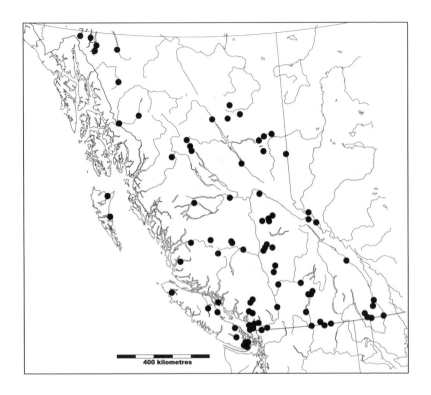

Range

The Muskrat is distributed across most of North America. It occupies the entire mainland of British Columbia, as well as Vancouver Island and Graham Island, where it was introduced. Although Cowan and Guiguet, in the original *Mammals of British Columbia*, show this rodent as absent from much of the southern coastal mainland, it appears to be spottily distributed throughout that region. The Muskrat is known to occur on 12 islands between the mainland and Vancouver Island.

Taxonomy

Sixteen subspecies are recognized across North America; two are found in British Columbia:

Ondatra zibethicus osoyoosensis (Lord) – the western United States, southwestern Alberta and southern B.C. Populations of this race were introduced to Vancouver and Graham islands.

Ondatra zibethicus spatula (Osgood) – Alaska, the Yukon Territory, the Northwest Territories and north-central B.C.

Conservation Status
This species is not of conservation concern in British Columbia. Managed as a fur-bearer, the trapping of Muskrats is regulated under the provincial Wildlife Act. In southern B.C., some habitat loss has resulted from the flooding of valleys and wetlands by hydroelectric dams and the drainage of wetlands for farmland or urban development.

Remarks
The Muskrat was trapped by several aboriginal groups for its fur and meat. Historically, it has been one of the most important fur-bearing mammals in the province. Although the value of its pelt is generally low, the rodent is abundant and easy to trap. From 1919 to 1993, nearly 3.5 million Muskrat pelts were declared from trap-lines in British Columbia, most from the Cariboo, Lower Mainland and Omineca-Peace management regions.

The Vancouver Island population originated from animals transplanted from the Lower Mainland in the 1920s. The first were released at Cowichan Lake in 1922, and then from 1924 to 1925, more animals were released at Comox, Merville (Hopkins Lake), Jordan River, Shaw Creek, Port Alice and Ucluelet. The Graham Island population arose from about 15 animals transplanted from the Fraser River delta to the Masset area in 1924 or 1925. From these introductions Muskrats have spread rapidly across both islands.

Selected References: Erb et al. 2000, MacArthur and Aleksiuk 1979, Perry 1982, Proulx and Gilbert 1983, Willner et al. 1980.

Heather Vole   *Phenacomys intermedius*

Other Common Names: None.

Description

The Heather Vole is a small vole with soft, fine fur. Its dorsal pelage ranges from brown to greyish-brown; the undersides are grey or whitish. Populations in northern British Columbian have yellow on the nose, but others do not. The dorsal surface of the hind feet is whitish. The tail is distinctly bicoloured, short (less than 30% of the animal's total length), thin and sparsely furred. The mature skull is angular with prominent ridges. The cheek teeth are rooted; on the lower molars the inner re-entrant angles are deeper than the outer re-entrant angles (figure 75).

Measurements:
   total length: 137 (120-164) n=82
   tail vertebrae: 33 (21-46) n=84
   hind foot: 18 (15-22) n=83
   ear: 14 (10-16) n=24
   weight: 32.1 g (24.0-50.0) n=21

Dental Formula:
   incisors: 1/1
   canines: 0/0
   premolars: 0/0
   molars: 3/3

Identification:

In external appearance, this vole is difficult to distinguish from some *Microtus* species, especially immature animals. The Heather Vole has softer, finer fur, the dorsal surface of its hind feet are white or silvery rather than black, and the tail is thin with a sparse covering of hairs. The enamel patterns on the lower molars are diagnostic – no *Microtus* species has lower molars with deeper inner re-entrant angles.

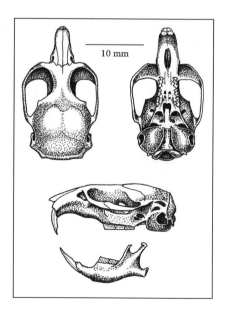

10 mm

Natural History

In western North America, the Heather Vole lives in open forest, stunted treeline forests and alder-willow thickets. In the Rocky and Columbia mountains, it is associated with Lodgepole Pine, Subalpine Fir, Western Larch, Subalpine Larch, Engelmann Spruce and Whitebark Pine communities. In the coastal mountain ranges, it inhabits forests with Subalpine Fir, Western Hemlock, Amabilis Fir and Yellow-cedar. Throughout the Interior Plateau of central British Columbia, this species is most common in dry Lodgepole Pine forests. At the northwestern limits of its range in the Yukon Territory, it is associated with willow thickets, closed and open spruce forests and Trembling Aspen stands. The Heather Vole also inhabits subalpine tree islands and alpine meadows. The characteristics most often mentioned in habitat descriptions are abundant woody debris on the forest floor and a dense shrub layer. In alpine areas, its occurrence is correlated with the presence of rocks or vegetative cover and nearby water. The Heather Vole colonizes habitats disturbed by fire or forest harvesting. For example, a small mammal study done in Glacier and Revelstoke national parks in the 1980s found that the highest captures of Heather Voles were in recent subalpine burns.

Estimates of population density range from 0.5 to 4.3 animals per hectare for subalpine communities in Alberta's Kananaskis Valley and from to 0.2 to 10 per hectare in alpine communities of Colorado. Thomas Sullivan and colleagues recorded average densities of less than one animal per hectare for montane spruce forests in B.C.'s southern interior, low compared to those recorded for some species of *Microtus*. Research on long-term population fluctuations is lacking, but anecdotal observations suggest that local populations of Heather Voles may experience periodic irruptions.

Data on movements are scanty, presumably because individuals are rarely recaptured in traps. Bristol Foster recaptured a marked adult male about 600 metres from its original capture location. No estimates exist for home-range size. Heather Voles are active mostly at twilight or night but they may feed during the day. Several researchers describe this vole as docile and easy to handle when captured. Its social behaviour is essentially unknown. Because adults are aggressive during the breeding season and only one adult Heather Vole occupies the summer nest, it is probably solitary in summer.

The Heather Vole constructs distinct summer and winter nests. Winter activity is largely confined to runways under the snow. Winter nests, located on the surface of the ground under snow, are thick-walled and about 15 cm in diameter; they are made of twigs, leaves, grass, and lichen, and usually lined with shredded grass. The Heather Vole deposits its feces outside the nest where it accumulates in huge piles over the winter – these large latrines are revealed with the spring snowmelt. Summer nests are in shallow underground burrows (not more than 20 cm deep) with a nest chamber, a separate toilet chamber, several short tunnels (less than a metre long) and one or more burrow openings. Each burrow has only one nest, about 10 cm in diameter, made of dead grass or moss. Burrow openings are located at the base of trees, shrubs or rocks. Although the Heather Vole makes runways near its burrow, it will use the runways of other voles, such as the Northern Bog Lemming and *Microtus* species.

Information on the diet of the Heather Vole is derived mostly from plant cuttings found piled near its burrow. Made at night, these summer food piles presumably allow this vole to feed during the day near the protection of its burrow. Throughout its range, the Heather Vole's dominant summer foods are the leaves, stems and berries of Soapberry and various species of heather (Ericaceae), including Sheep Laurel. In B.C., its food piles contain Kinnikinnick, cinquefoil, Black Huckleberry, Soapberry and white heather species. Curiously, the Heather Vole does not eat Pink Mountain-heather, a common plant in some alpine areas. Other plants found in summer middens include willow species, Scrub Birch, Trembling Aspen, Cloudberry, Sweet Gale, wild rose, grasses, sedges, and forbs such as lousewort, vetch, strawberry and hawkweed. Fungi have been identified in stomach contents and feces. In winter, the Heather Vole feeds predominately on the bark of trees and shrubs such as Scrub Birch, willow, alder and white heather, depending on their availability; evidently, it does not eat the bark of coniferous plants. This vole gnaws bark on the stems and branches of shrubs and fallen trees; it also cuts small branches and twigs (1 to 6 mm in diameter) into short lengths (20 to 80 mm), and stores them in caches. These winter caches may contain more than a thousand twigs.

The breeding season extends from May to September. The gestation period is 19 to 24 days. Females produce as many as three litters in the breeding season, bearing two to seven young in each. Adult females produce larger litters than young-of-the-year females. For a population in the Alberta Rocky Mountains, Duncan Innes and John Millar estimated average litter sizes of 4.4 for adults and 3.6 for young-of-the-year. Both sexes can breed in their first summer, but young born late in the season may not breed until the following year.

The Heather Vole appears to have a slower developmental rate than other voles. Newborns are naked, blind and weigh about 1.9 grams. By 13 to 15 days their eyes open, and by 14 to 18 days they begin to eat solid food. Females are capable of breeding at four to six weeks. The molar teeth become fully erupted at 8 to 10 months and the skull reaches its full size and development at about one year. This vole may live up to three years in the wild, but little is known about its survival rates. Duncan Innes and John Millar recaptured only two (6.3%) of the Heather Voles that they had marked and released the previous year in their study area.

Known avian predators include the Short-eared Owl, Great Horned Owl, Northern Hawk Owl, Northern Saw-whet Owl and Rough-legged Hawk. The Heather Vole is rarely a major prey for raptors, but in a study of two Northern Saw-whet Owl nests at about 1,525 metres elevation in the Okanagan Valley, Richard Cannings, found that Heather Voles represented about 15% of the prey. A major mammalian predator appears to be the Marten, a small carnivore that feeds extensively on voles. Other possible mammalian predators include the Ermine and Long-tailed Weasel.

Range

A boreal species, the Heather Vole is widely distributed across Canada from Labrador to British Columbia and the Yukon Territory; in the United States it is restricted to the western cordillera and

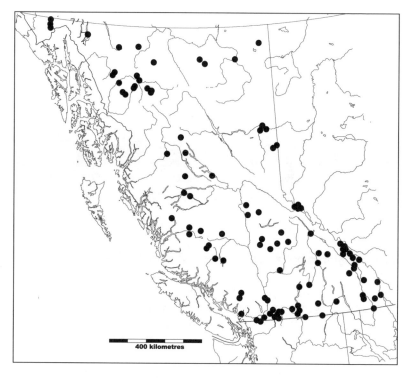

400 kilometres

Pacific coast region. Although Cowan and Guiguet, in the original *Mammals of British Columbia*, showed the range as spotty with large gaps separating most of the subspecies in the province, this was probably from inadequate sampling and few museum specimens. The available data suggest that the Heather Vole is distributed continuously throughout most of B.C.'s mainland, though it is largely absent from lowland coastal regions. The only coastal lowland record is from Cornice Creek near the mouth of the Kimsquit River. The Heather Vole has not been found on any B.C. island.

Taxonomy

Taxonomy of the Heather Vole is contentious, particularly the relationship between the eastern and western populations. Some taxonomists, including Cowan and Guiguet, have treated these two populations as distinct species: the Western Heather Vole (*Phenacomys intermedius*) and the Eastern Heather Vole (*P. ungava*). Other taxonomists consider the two forms to intergrade in areas where they are in contact and treat them as a single species, *P. intermedius*.

Small sample sizes and large geographic distances separating museum specimens has prohibited a rigorous study of geographic variation across the range. A modern genetic study is required to resolve the taxonomy.

Although the northern populations in British Columbia (i.e., the subspecies *P. i. mackenzii*) often have a yellow nose and are associated with the eastern *ungava* group, I chose to treat all the Heather Voles in B.C. as a single species. Other than the presence of a yellowish nose, a trait that appears to be somewhat variable, I can find no clear morphological differences for discriminating the two forms.

Nine subspecies are recognized in North America; five inhabit the province:

Phenacomys intermedius intermedius Merriam – cordillera of the western United States and Canada; in B.C. it ranges throughout the central interior east to the Rocky Mountain Trench and west to the Coast Mountains.

Phenacomys intermedius laingi Anderson – restricted to the province, where it occupies the central Coast Mountains and east to the Cariboo Mountains.

Phenacomys intermedius levis Howell – the Rocky Mountains of Alberta, northern Montana and B.C., where it ranges as far north as the Peace River. In the original *Mammals of British Columbia*, Cowan and Guiguet considered this race to be restricted to the extreme southern Rocky Mountains (Tornado Pass).

Phenacomys intermedius mackenzii Preble – broadly distributed across the Yukon Territory, the Northwest Territories, Alberta, Saskatchewan, Manitoba and northern parts of B.C.

Phenacomys intermedius oramontis Rhoads – the coastal mountain ranges from northern Oregon and western Washington to B.C. where it extends as far north as Garibaldi Provincial Park.

Conservation Status
This species is not of conservation concern in British Columbia.

Remarks
One of the least known rodents in the province, the Heather Vole is a fascinating small mammal that offers many research opportunities. A genetic study applying modern DNA techniques is essential to determine if there is more than one species in Canada. This vole's habit of cutting and storing twigs in winter caches raises a number of questions about its social behaviour. If these twig caches represent

food hoarding for winter, then the Heather Vole could be social in winter, living in communal nests. Its diet suggests several areas for study. For example: why does this vole not eat the needles or bark of coniferous plants but consumes Sheep Laurel, a plant known to be toxic to many vertebrates?

There are no historical records of the Heather Vole from any British Columbian island. But a tooth recently found in a sea cave near Port Eliza is evidence that this vole inhabited Vancouver Island in the late Pleistocene, just before ice of the last glaciation covered the island.

Selected References: Edwards 1955, Foster 1961, Innes and Millar 1982, McAllister and Hoffmann 1988, Nagorsen 1987, Racey 1928, Sullivan et al. 2000.

Northern Bog Lemming

Northern Bog Lemming
Synaptomys borealis

Other Common Names: Lemming Mouse.

Description

The Northern Bog Lemming is a small vole with a short tail (less than 20% of the animal's total length) that is equal to or slightly longer than its hind foot. The dorsal pelage is grizzled brown or greyish-brown; the undersides are grey. The underside of the tail is sparsely covered with whitish or greyish hairs. The angular skull has a short rostrum with faintly grooved upper incisors. The re-entrant angles on the upper molars are deeper on the outer side and in the first and second molars extend to the inner border (figure 67). The lower molars lack closed triangles on the outer sides (figure 73).

Measurements:
total length: 123 (102-142) n=91
tail vertebrae: 22 (17-30) n=90
hind foot: 19 (14-22) n=89
ear: 12 (6-15) n=20
weight: 26.4 g (21.7-48.0) n= 13

Dental Formula:
incisors: 1/1
canines: 0/0
premolars: 0/0
molars: 3/3

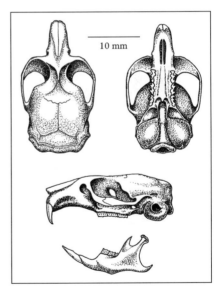

Identification:
The short tail and distinctive enamel patterns on its upper and lower molars distinguish this species from any other British Columbian vole.

Natural History

In British Columbia, the Northern Bog Lemming ranges from sea level on the central and north coast to 2230 metres elevation in the Rocky Mountains. It usually occurs in wetlands, such as meadows, bogs and fens, although it can be found in other habitats. Habitat descriptions from museum specimens and small-mammal surveys in B.C. include: patches of sedge bordering small mountain streams, alpine meadows, dry alpine slopes covered with heather and sedge among boulders, runways in blueberries, a sphagnum-willow swamp bordering an alpine meadow, grassy margins of lakes and beaver ponds, bog-birch or willow wetlands, and hummocks in moss fens. In southern B.C., this species is usually restricted to high elevations in small patches of wetland surrounded by forests of Subalpine Fir and Engelmann Spruce. The Northern Bog Lemming and Meadow Vole overlap in their habitats – a number of biologists have reported that the Northern Bog Lemming uses Meadow Vole runways.

There are no population estimates for this species, but most biologists agree that the Northern Bog Lemming is rare. It appears to have distinct summer and winter shelters. In winter, it occupies nests under the snow made from fine grass and sedges. In Mount Revelstoke National Park, Ian McTaggart Cowan and James Munro examined 15 old winter nests left by Northern Bog Lemmings, all adjacent to meadows in dry sites covered with heather: 14 consisted

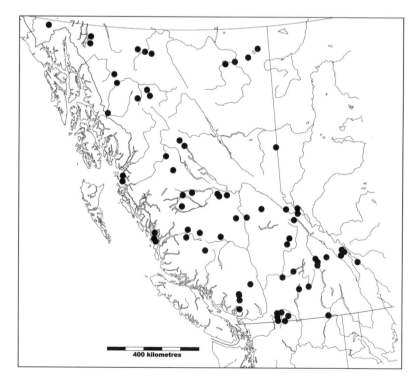

entirely of bound sedges lined with finely cut sedge material; the other nest was made from White Mountain-heather, sedge and moss. In summer, the Northern Bog Lemming occupies underground burrows. Areas adjacent to its burrows are usually honeycombed with small surface runways that contain piles of plant cuttings and scattered fecal pellets. Plants found in food piles include sedges, grasses, hawkweed, Western Bog-laurel (a plant toxic to many vertebrates), Fan-leaved Cinquefoil and saxifrage.

The scanty reproductive data for this species are limited to information from a few museum specimens. Fourteen pregnant females taken in B.C. contained an average of 5.2 embryos (range: 3 to 9). These pregnant females were captured from May 15 to September 23, suggesting that females have several litters in a year. Little is known about the Northern Bog Lemming's predators, but owls and various small mammalian carnivores are likely candidates. Remains of this vole have been found in Marten feces.

Range

This is a boreal rodent, distributed from Alaska and Canada, to extreme northern Minnesota, Idaho, Montana and Washington. Although specimen records from British Columbia are few and widely scattered, the Northern Bog Lemming appears to be found throughout most of the mainland. It also inhabits Campbell and Yeo islands, off B.C.'s central coast.

Taxonomy

Nine poorly differentiated subspecies are recognized; five occur in British Columbia:

Synaptomys borealis artemisiae Anderson – restricted to the eastern Cascade Mountains and possibly the Okanagan Valley in B.C., and extreme northern Washington.

Synaptomys borealis borealis (Richardson) – the Northwest Territories, northern Alberta and Saskatchewan and northeastern B.C.

Synaptomys borealis chapmani Allen – Rocky Mountains of Alberta, northwestern Montana, northeastern Washington, and the Rocky, Purcell and Selkirk mountains in B.C.

Synaptomys borealis dalli Merriam – Alaska, the Yukon Territory, and the northern interior of B.C. east of the Coast Mountains as far south as Quesnel.

Synaptomys borealis truei Merriam – southeastern Alaska including the Alexander Archipelago, the entire coastal mainland of B.C. and extreme northwestern Washington. There are also records in the province for Campbell and Yeo islands. In the original *Mammals of British Columbia*, Cowan and Guiguet assigned this coastal race to *S. b. wrangeli*.

Conservation Status

The subspecies *S. b. artemisiae* is on the provincial Red List, and was designated as Lower Risk (least concern) by the IUCN. It is listed provincially because it appears to be rare (known from only five localities in British Columbia and three from northern Washington) and is found in atypical habitat. The type locality (Stevenson Creek, near Princeton) is in low-elevation (730 metres) arid grasslands and sage, unusual habitat for the Northern Bog Lemming. This led Rudolph Anderson to apply the common name Sagebrush Northern Bog Lemming and the scientific name *S. b. artemisiae* for the subspecies when he first described it in 1933. The taxonomic validity of this race, however, is far from clear. Moreover, four of the five sites where it has been found in B.C. are in high-elevation wet meadows or willow-birch thickets, typical habitat for the species. The unusual habitat of the type locality may be the result of Northern Bog Lemmings dispersing into atypical habitats during a population irruption.

Remarks

The Northern Bog Lemming is the least known rodent in British Columbia. Every aspect of its biology, including distribution, taxonomy and ecology, needs more research.

Selected References: Anderson 1933, Cowan and Munro 1946, Racey and Cowan 1935.

Subfamily Murinae: Old World Rats and Mice

All three species of Old World rats and mice in British Columbia have been introduced from Europe or Asia. These are commensal species that live in association with humans, although feral populations exist in B.C. They have naked tails with visible annulations; their skulls have rooted molars, and the cusps of the first two upper molars are arranged in three longitudinal rows (figure 93).

House Mouse   *Mus musculus*

Other Common Names: None.

Description

The House Mouse is a small rodent with a long, naked, scaly tail, relatively large ears and coarse fur. Pelage colour is highly variable but most wild animals have a dull grey-brown dorsal pelage and grey ventral fur. Animals with a blond pelage or white markings have been found in British Columbia. A number of colour morphs, including melanic and albino forms, exist in House Mice used as laboratory animals or sold as pets. The small and delicate skull has upper incisors that are notched in side view (figure 95), and a tiny third molar.

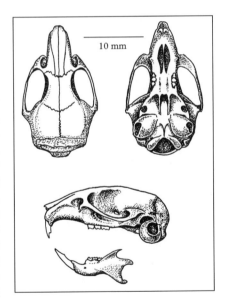

10 mm

Measurements:
  total length: 163 (142-197) n=41
  tail vertebrae: 79 (64-90) n=48
  hind foot: 18 (15-23) n=45
  ear: 13 (11-18) n=15
  weight: 21.4 (17.0-26.0) n=19
Dental Formula:
  incisors: 1/1
  canines: 0/0
  premolars: 0/0
  molars: 3/3
Identification:
This species could be confused with the Deer Mouse (*Peromyscus maniculatus*) and Western Harvest Mouse (*Reithrodontomys megalotis*). The Deer Mouse has brown dorsal pelage and white ventral fur, the tail is haired, and the cusps of the first two upper molars are arranged in two longitudinal rows, not three. The Western Harvest Mouse is smaller (adults weigh < 15 g) than the House Mouse, the tail is furred, the upper incisors are distinctly grooved and the cusps of the first two upper molars are arranged in two longitudinal rows. The House Mouse emits a strong odour so distinct that one can easily identify this rodent in a live-trap, even before opening it.

Natural History
Most occurrences of the House Mouse in British Columbia are from urban or agricultural areas where it is associated with human dwellings, barns and buildings with animal feed. It has also been found in isolated settlements such as Telegraph Creek in northern B.C., where it was presumably transported with food or livestock feed. In the Okanagan and Similkameen valleys and on Pender Island, feral populations live in agricultural fields. Charles Guiguet captured a number of House Mice in uncultivated orchard lands near Osoyoos in the early 1950s. During a survey of Western Harvest Mice in 1990 and 1991, I captured large numbers of House Mice in the grassy borders of hay fields near Keremeos in the southern Similkameen River valley and on the O' Keefe Ranch near Vernon. It coexisted with several small rodents, including the Deer Mouse, Meadow Vole and Long-tailed Vole; in many sites, the House Mouse was the dominant small mammal species. On Pender Island, off the southeast coast of Vancouver Island, I captured large numbers of House Mice in hay fields, where it coexisted with Townsend's

Vole. It is difficult to determine if these House Mice are year-round residents in agricultural fields, because all of these sites were within a few hundred metres of barns, which the House Mouse could occupy during winter.

No population estimates exist for British Columbia. In some parts of North America, this species has reached "plague" numbers. For example, E. Raymond Hall reported an outbreak in the San Joaquin Valley, California, where the House Mouse population reached more than 200,000 per hectare. In fields away from buildings, population densities typically range from 50 to 100 animals per hectare. Its social structure varies, but the House Mouse usually lives in small family groups that consist of a dominant male, several females and subordinate males. The dominant males are highly territorial, although females also will exclude females from other groups.

Long-distance movements of one kilometre have been recorded, but this is a relatively sedentary mouse, especially in buildings. John Reimer and Michael Petras found that the distance between captures in a barn averaged only 6 metres, whereas the average distance between captures for populations living in fields was about 30 metres. House Mice make nests in buildings, structures or underground burrows. Their burrows are complex, with a number of entrance holes and tunnels, usually 3 to 5 cm below the surface. The diverse diet includes plant and animal material. Unlike the Black

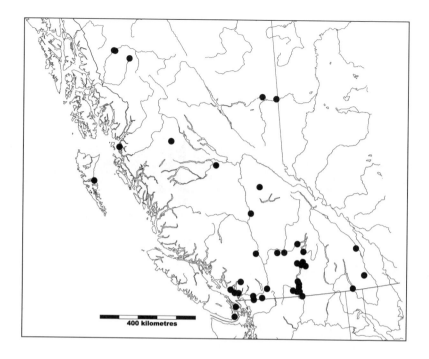

400 kilometres

Rat and Norway Rat (the other Old World rodents), the House Mouse does not prey on birds or bird eggs.

No information is available on the length or timing of the breeding season in British Columbia. The gestation period is 18 to 21 days. At birth, the young are naked and blind. Their eyes open at 2 to 3 days; they are weaned by 14 days and sexually mature by five to six weeks. Litter size usually ranges from four to eight, although in buildings House Mice may produce larger litters. (The only reproductive data for B.C. are from three specimens that had four or five embryos each.) Although this rodent can live several years, it rarely survives beyond 18 months in the wild. In urban and agricultural areas, the Domestic Cat is probably the major mammalian predator. In a study of Barn Owl predation in southwestern B.C., Wayne Campbell and colleagues found House Mouse remains in about 43% of the owl pellets collected from the city of Vancouver, but it was not present in pellets collected from other parts of the Fraser River valley and Vancouver Island.

Range
The House Mouse was originally restricted to Eurasia, but various commensal forms associated with humans have been transported to all the continents and numerous islands throughout the world. In British Columbia, there are sporadic records across the mainland as far north as the Peace River region and Telegraph Creek on the Stikine River. The House Mouse has been found on only three islands: Vancouver, Pender (in the Gulf Islands) and Moresby.

Taxonomy
Populations of the House Mouse in Canada presumably derive from the commensal race *Mus musculus domesticus*. Some authorities consider this subspecies a distinct species *Mus domesticus*.

Mus musculus domesticus Rutty – Europe and introduced to North America.

Conservation Status
This species is not of conservation concern.

Remarks
The history of the House Mouse in British Columbia is poorly documented. It probably first arrived on ships of the early explorers and fur traders, and spread to various regions through human transport. The earliest known records are museum specimens collected from Kamloops in 1889.

Selected References: Berry 1981, Campbell et al. 1987, Guiguet 1952, Hall 1927, Nagorsen 1995, Reimer and Petras 1968.

Norway Rat   *Rattus norvegicus*

Other Common Names: Brown Rat, Common Rat, Sewer Rat.

Description

The Norway Rat is stout with a thick, naked tail that is shorter than the head and body length (70 to 90%). It has small, thick ears that do not cover the eye when pushed forward, a blunt snout, and a thick, heavy rump. The dorsal pelage is brown and the undersides pale grey. The skull has a rectangular braincase with straight temporal ridges (figure 96); the upper incisors are not notched in side view.

Measurements:
  total length: 385 (334-436) n=14
  tail vertebrae: 175 (143-202) n=14
  hind foot: 42 (34-44) n=14
  ear: 20 (18-23) n=5
  weight: 354.4 (256.5-434.5) n=5

Dental Formula:
  incisors: 1/1
  canines: 0/0
  premolars: 0/0
  molars: 3/3

Identification:
Norway Rats can be confused with Black Rats (*Rattus rattus*), especially if immature. But the Black Rat's tail is longer than its head and body, and it has a slender body and rump; it also has large, thin, sparsely haired ears that cover the eyes when pushed forward, a pointed snout, and a skull with a rounded braincase and curved temporal ridges.

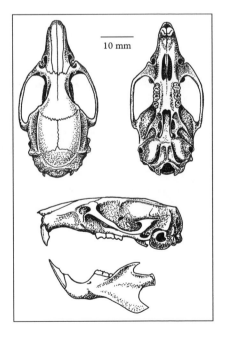

10 mm

Natural History

The Norway Rat inhabits urban and agricultural areas in human dwellings, garbage dumps, landfill sites, food-storage facilities, warehouses, barns and buildings with livestock feed. Feral populations occur on parts of Vancouver Island, some nearby smaller islands and the Queen Charlotte Islands. In productive environments with abundant food, Norway Rats can reach high population densities. For example, as many as 10 burrows per square metre were found in a landfill site in Richmond. The only information on wild Norway Rats in British Columbia is from research by Mark Drever and colleagues on a population on Langara Island in the Queen Charlotte Islands. Before the population was eradicated, this island of about 3200 ha supported about a thousand Norway Rats with an overall density of about one rat per hectare. They were most abundant in coastal areas with plentiful marine resources, particularly burrowing seabirds. In the interior of the island, Norway Rats lived in shrub thickets of dense Salal and huckleberry.

Norway Rats have a complex social structure. They live in colonies consisting of smaller social units comprised of a mated pair, or a male with a harem of several breeding females. Members of a social unit are highly territorial, defending their burrow from neighbouring rats. When populations are high, a dominance hierarchy is established among neighbouring social units, with dominant males and their females taking the most productive habitats. Daily movements depend on the proximity of food sources. If abundant food is nearby, these rats limit their movements to a few hundred metres. But if food is scarce they may forage as far as 3.3 km. On Langara Island, three Norway Rats captured inland had marine amphipods in their stomachs, indicating that they had foraged in coastal areas some 500 metres from their burrows. Although they climb in buildings, Norway Rats rarely climb trees. They are capable of swimming up to 300 metres across narrow ocean straits.

Norway Rats eat a great variety of plant and animal material. In urban areas, they feed on waste foods, often in garbage dumps; in agricultural areas they eat cereal and roots crops, animal feed and fruits. The stomachs of wild Norway Rats trapped on Langara Island contained 34 types of food, including fruits, seeds, plant shoots, fungi, terrestrial and marine invertebrates, fish, and birds. Most of the birds eaten were Ancient Murrelets, a seabird that nests in burrows on the island. Drever and colleagues estimated that Norway Rats living near the seabird colony ate one Ancient Murrelet every 2.4 days. Norway Rats also eat the eggs of seabirds.

In the San Francisco area of California, Tracey Storer and David Davis found that Norway Rats living in buildings bred throughout the year. Nothing is known about the breeding season in British Columbia, but it is possible that here also they breed all year. The gestation period is 21 to 24 days. The average litter size in the San Francisco population, based on embryo counts, was 9 (range: 8.3 to 10.3); Storer and Davis estimated that an adult female could produce as many as 42 young in a year. Young females produce smaller litters. The newborn are naked and blind; by six days their eyes open. By 11 weeks, females are sexually mature. Few Norway Rats survive beyond one year. In urban and agricultural areas, Domestic Cats are the major mammalian predator of young rats, but they rarely kill the larger adults (weighing over 200 g). From their study of Barn Owl predation in southwestern B.C., Wayne Campbell and colleagues found Norway Rat remains in 26.5% of the owl pellets collected from the city of Vancouver, but only in trace amounts in owl pellets collected from other parts of the Fraser River valley, and none in pellets collected from Vancouver Island.

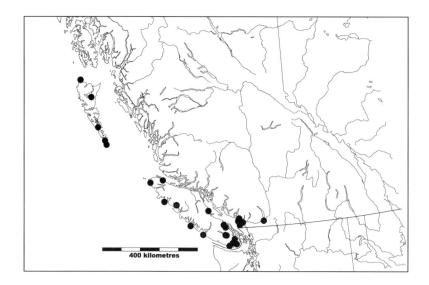

Range

Originally an inhabitant of Asia, this Old World rat has been introduced widely throughout the world. In British Columbia, it inhabits greater Vancouver, the lower Fraser River valley, the Queen Charlotte Islands, Vancouver Island and some small islands along the south coast. In the Queen Charlottes, the Norway Rat is found on Graham, Kunghit and Moresby islands; in the past decade it has been eradicated from Bischof, Cox, Langara, Lucy and St James islands.

Taxonomy

Several races have been described in the Old World, but the taxonomy of North American populations is obscured by multiple introductions.

Conservation Status

An alien species, the British Columbian populations are not of conservation concern, except for their potential impact on native species, particularly nesting seabirds on the Queen Charlotte Islands.

Remarks

The Norway Rat probably came from Europe with explorers and fur traders, but its history in British Columbia is obscure. Allan Brooks noted that, by the late 1880s, it was common in Vancouver, New Westminster and Victoria, but had yet to reach Chilliwack. The Norway Rat only arrived in the Queen Charlotte Islands in the early 1980s. Before 1980, the only rat on this archipelago was the Black Rat. The source of Norway Rats on the Queen Charlotte Islands is unknown. Although there are no established populations in B.C.'s interior, Norway Rats have turned up from time to time in towns such as Field, Kamloops and Vernon. These animals likely arrived on railway cars.

Introduced Norway Rats displaced the Black Rat on some of the Queen Charlotte Islands and caused serious population declines in ground-nesting seabirds such as the Ancient Murrelet. On Langara Island, Ancient Murrelet remains were found in the stomachs of 53% of the Norway Rats trapped near the colony. In 1995, the Canadian Wildlife Service used poisoned baits to eradicate Norway Rats from Langara Island and nearby Lucy and Cox islands. Parks Canada biologists eliminated this species from St James Island (1999) and Bischoff Island (2003) in Gwaii Haanas National Park Reserve. The Norway Rat is also suspected as the cause of the disappearance of Keen's Mouse from Bischof and Langara islands.

Selected References: Bertram 1995, Bertram and Nagorsen 1995, Brooks 1947, Drever 1997, Drever and Harestad 1998, Storer and Davis 1953.

Black Rat   *Rattus rattus*

Other Common Names: Alexandrine Rat, Roof Rat, Ship Rat.

Description

The Black Rat is a slender rat with a thin, naked tail that is longer than the head and body (100 to 160%). It has large ears that cover the eye when pushed forward, a pointed snout, and a slender body and rump. Three distinct colour morphs can be found in British Columbia, sometimes all at the same place: (1) *alexandrinus* morph – grey-brown dorsal pelage with slate-grey belly; (2) *rattus* morph – black dorsal pelage with a slate-grey belly; and (3) *frugivorous* morph – grey-brown dorsal pelage with a white to yellow belly. The skull has a rounded braincase with curved temporal ridges (figure 97); the upper incisors are not notched in side view.

Measurements:
  total length: 393 (330-455) n=60
  tail vertebrae: 216 (172-255) n=59
  hind foot: 37 (32-42) n=60
  ear: 23 (21-25) n=14
  weight: 185.6 (119.0-267.8) n=23

Dental Formula:
  incisors: 1/1
  canines: 0/0
  premolars: 0/0
  molars: 3/3

Identification:
Black Rats can be confused with Norway Rats (*Rattus norvegicus*), especially if immature; see the Norway Rat account for distinguishing traits (page 316).

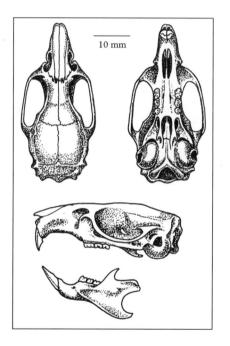

10 mm

Natural History

The Black Rat lives in urban and agricultural areas – many of the early records from the Vancouver region come from dumps and waterfront areas. But on southern Vancouver Island and in North

Vancouver it is also found in forested habitats well away from human dwellings. On the Queen Charlotte Islands, it inhabits isolated islands with seabird colonies.

This species appears to be less common than the Norway Rat, but there are no estimates of population densities for any British Columbian population. The social structure of the Black Rat consists of a small group dominated by an adult male who fought his way to the top – the members of a group apparently establish a hierarchy by fighting. The social group is highly territorial, excluding all other rats. There are no data on movements or home-range size for the B.C. populations. In California, the average home-range size for a population in a riparian habitat was 0.20 ha for males and 0.16 ha for females. In buildings, the Black Rat occupies attics, wall cavities or false ceilings. In the wild, it excavates burrows and constructs nests above ground in trees. Black Rats are superb climbers. I have observed them climbing trees to reach bird feeders and running along narrow power lines. This omnivorous rat eats a variety of plant and animal material. In urban areas, compost heaps are an attractive food source. On the Queen Charlotte Islands, Black Rats feed on the eggs and chicks of ground-nesting seabirds such as the Ancient Murrelet.

In many temperate regions, Black Rat populations declined after the arrival of the larger, more aggressive Norway Rat. On the Queen Charlotte Islands, the Norway Rat, which only appeared in the 1980s, has displaced the Black Rat from Langara Island and, possibly, Kunghit Island. In the Lower Mainland, the Norway Rat is the more common species and the Black Rat keeps to waterfront areas and forested habitats in North Vancouver. On Vancouver Island, the Norway Rat is the common rat in cities such as Victoria, but both species coexist in rural and agricultural areas.

In the San Francisco area of California, Tracey Storer and David Davis found that Black Rats living in buildings bred throughout the year. Nothing is known about the breeding season in British Columbia. Museum specimens noted as carrying embryos were collected between May and August. The gestation period is 21 to 29 days. The average litter size for the San Francisco population, based on embryo counts, was 7.3 (range: 5.4 to 9.0). Storer and Davis estimated that an adult female could produce as many as 40 young in a year. The young are weaned at 17 to 23 days.

In urban and agricultural areas, Domestic Cats are the major mammalian predator of Black Rats. Martens also prey on them on

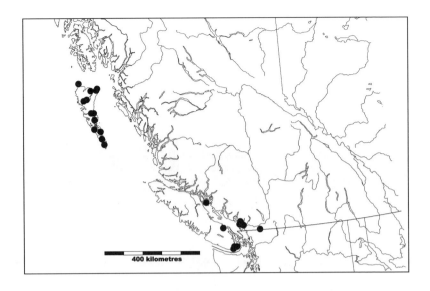

the Queen Charlotte Islands. In a study of Barn Owl predation in southwestern B.C., Wayne Campbell and colleagues found Black Rat remains in 6.5% of the owl pellets collected from Vancouver Island, but only trace amounts in pellets collected from the city of Vancouver and other parts of the Fraser River valley.

Range
Originally a species of India and southern Asia, the Black Rat has been introduced throughout the world, including many isolated islands. There are historical records from southern Ontario, but the only established populations of this rodent in Canada are in British Columbia where it inhabits Vancouver, the lower Fraser River valley, the Queen Charlotte Islands and Vancouver Island. Black Rats are absent from islands adjacent to Vancouver Island, except for Cortes Island. On the Queen Charlotte Islands it has been found on Burnaby, Graham, Huxley, Kingfisher (Swan), Kunghit, Kwaikans, Langara, Lyell, Moresby and Murchison islands.

Taxonomy
The Black Rat demonstrates an enormous range of colour, morphometric and genetic variation with distinct wild and commensal forms. In the original *Mammals of British Columbia,* Cowan and

Guiguet listed two subspecies in the province: *Rattus rattus alexandrinus* and *R. r. rattus*. A third colour form, with a white or yellow belly, also occurs in the province – it has been treated as distinct subspecies *R. r. frugivorous* in the older literature. Most authorities now treat all three as colour phases in a polymorphic species that do not warrant recognition as subspecies.

Conservation Status
As an alien species in British Columbia, the Black Rat is not of conservation concern, except for its potential impact on native species, particularly nesting seabirds on the Queen Charlotte Islands.

Remarks
The history of the Black Rat in British Columbia is not well documented. As with the House Mouse, it probably first arrived on ships of the early explorers and fur traders. The earliest record for the province comprises specimens collected on Vancouver Island in 1858 by the British naturalist John Keast Lord and housed at the Museum of Natural History in London, England. The earliest record from the Queen Charlotte Islands is a museum specimen taken on Graham Island in 1919. The date of this rat's arrival on the Lower Mainland is unknown, but there are museum specimens dating as early as 1921, and by the 1940s it was well established in the Vancouver area.

Every November, when the winter rains begin, young-of-the-year Black Rats colonize the attic of my house in Victoria. Presumably, they climb up the drain pipes and enter the attic through small openings near the chimney. Over several years, I have removed all three colour morphs from my attic.

Selected References: Bertram 1995, Bertram and Nagorsen 1995, Brooks 1947, Campbell 1968, Campbell et al. 1987, Storer and Davis 1953.

Subfamily Sigmondontinae: New World Rats and Mice

New World rats and mice demonstrate a diversity of body forms, but all four British Columbian species have a long, furred tail, slender body (figure 43), and conspicuous ears and eyes. Their molars are rooted and, except for those of the Bushy-tailed Woodrat, the molar crowns have two rows of cusps (figure 94).

Bushy-tailed Woodrat

Bushy-tailed Woodrat   *Neotoma cinerea*

Other Common Names: Packrat.

Description

Superficially, the Bushy-tailed Woodrat resembles a squirrel with a long, bushy tail and long vibrissae (extending to the shoulders). The dorsal pelage is grey to brown; the ventral pelage and dorsal surface of the feet are white. The soles of the feet are furred. The skull is large (skull length > 40 mm); the crowns of molars resemble those of a vole, with triangles or prisms of enamel (figure 68). Males are conspicuously larger than females, and they have a ventral gland that deposits a brown stain on their chest and belly fur.

Measurements:
  total length: 383 (318-468) n=223
  tail vertebrae: 166 (128-205) n=217
  hind foot: 45 (37-57) n=222
  ear: 30 (23-40) n=43
  weight: males 374.7 (315.0-425.0) n=7
        females 326.9 (292.0-394.6) n=13

Dental Formula:
  incisors: 1/1
  canines: 0/0
  premolars: 0/0
  molars: 3/3

Identification:
At a glance this species could be confused with a squirrel, but its pelage colour, long vibrissae and five digits on the front feet distinguish it from any of our squirrel (Sciuridae) species. The V-shaped infraorbital opening and prismatic molars readily distinguish the skull from any squirrel skull.

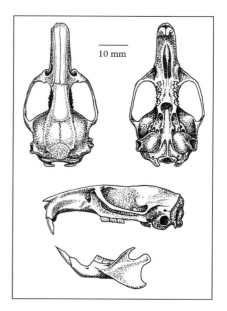

10 mm

Natural History

Ranging in British Columbia from sea level to 2380 metres elevation, the Bushy-tailed Woodrat lives in cliffs, rocky outcrops, rock crevices, caves, abandoned mine adits and taluses. It also occupies old buildings, such as abandoned homesteads, cabins and barns. It is associated with all of the province's ecosystems, including arid grassland, forest and open alpine. The woodrat's most critical habitat requirement is a sheltered den site to protect it from predators and the extreme temperatures of summer and winter. In coastal Oregon, woodrat dens have been found high up in Douglas-fir and Sitka Spruce trees. Although there are a number of historic museum specimens of this rodent from the coastal inlets of B.C., all were taken in rock slides or the abandoned buildings of logging camps.

There are few estimates of population density. In coastal areas and the Cascade Mountains of Washington, they range from 1.5 to 2 animals per hectare. But density may not be meaningful for this rodent because of its irregular and patchy distribution. In parts of the western United States, outbreaks of bubonic plague cause occasional dramatic population declines in local populations. In the late 1980s, I saw evidence of a population crash in Bushy-tailed Woodrat colonies in the Hat Creek valley of B.C. caused by plague.

Studies in the United States suggest that this species has a relatively small home range; but a radio-tracking study in the Rocky Mountains of Alberta revealed average home-range sizes of 6.12 ha (range: 1.6 to 11.2 ha) for males and 3.56 ha (range: 0.13 to 10.44 ha) for females. Females usually forage within a few hundred metres of their dens, but occasionally they venture as far as 470 metres. Young animals dispersing from their birth site move long distances. Peter Esherich reported that a female in California travelled 3.2 km in 28 days and that a young male moved 2.2 km in 10 days.

The social structure of the Bushy-tailed Woodrat is variable. Males are territorial and highly aggressive to other adult males during the breeding season. Escherich's study in California concluded that this rodent was polygamous: males maintained harems of one to three adult females, and excluded all other adult males. But woodrats in the Rocky Mountains of Alberta are less social and the home ranges of males and females overlap. A typical group there comprises an adult male with several adult females and their offspring. A female may mate with different males during the breeding season, but she has a strong connection with her offspring.

A male Bushy-tailed Woodrat has a skin gland on his chest that emits a strong, musky odour (unmistakable, if you have ever encountered Bushy-tailed Woodrats living in a building). He marks his territory by rubbing his chest on objects, and also by urinating on rocks. These urine spots are used for many generations. Fresh urine stains are yellow to dark brown, but after rain washes away the organic material, there remains a distinctive white, chalky material that resembles bird guano. In a protected site, such as a cave in an arid climate, many generations of woodrat urine forms a thick dark-brown or blackish veneer known as *amberat*. Woodrats display other behaviours presumably associated with social interactions, such as thumping with the hind feet and chattering with the teeth.

The Bushy-tailed Woodrat is a superb climber. I have watched individuals released from a live trap scamper up the face of a near vertical cliff. It can climb up to 15 metres in trees. Strictly noctur-nal, this rodent emerges from its den well after dark and spends sev-eral hours foraging. It is most active on nights with intermediate moonlight and less active on dark or bright nights, presumably to avoid predators. Woodrats memorize escape routes in their home range enabling them to find their den at night or in the dim light of a cave. Their long facial vibrissae help them navigate in the dark.

A Bushy-tailed Woodrat's den, constructed from branches and other vegetation, is usually wedged into a rock crevice or cliff over-hang (figure 10). In the Hat Creek valley, Douglas-fir, Common Juniper and Big Sage were the most common plants used for dens. A woodrat can carry sticks up to a metre long and three centimetres thick. It carries material in its mouth and may use the front feet to help put each part of the den in place. This animal is notorious for packing its den and food caches with items such as mammal bones and skulls, owl pellets, tin can lids and dried animal droppings.

A typical den houses one or more cup-shaped nests made from fine material such as shredded bark, grass, moss or dried sagebrush leaves. The den protects the Bushy-tailed Woodrat from predators and the extreme temperatures of winter and summer. Although it inhabits low-elevation valleys in the dry interior, this rodent has no special adaptations for coping with heat other than a well-sheltered dwelling. Because it is active throughout winter, the woodrat needs a den with thick nest insulation. In sheltered buildings or caves where this rodent occupies the "twilight zone" inside the entrance, the nest often has little or no stick cover.

The Bushy-tailed Woodrat has a broad diet. Robert Finely identi-fied some 80 plants eaten by woodrats in Colorado. It mostly eats the foliage of plants, but also conifer needles, bark, seeds and fungi. It stores plant material at or near the den site, apparently drying some before storage. The woodrat makes small caches in spring and summer, enabling it to remain close to its den on bright or dark nights when it is too risky to forage. In late summer or early autumn, it makes large caches, some weighing up to 50 kg, which are vital for its winter survival. Because the plants that the Bushy-tailed Woodrat eats are low in nutrition, it must consume a lot of them, and it re-ingests its fecal pellets to extract additional nutrients. The pellets are large and conspicuous, usually deposited near the den in distinct latrine areas built up, over time, with deep layers.

The breeding season begins in spring, with yearlings breeding later than adults. Year-to-year variation in spring climate and the first appearance of green vegetation causes variation in the timing of breeding in a local area. The gestation period is about 30 days. No breeding data are available for B.C. populations. In Alberta's Rocky Mountains, females usually produce three or four young. Females can breed again just 24 to 48 hours after giving birth to their first litter; but in the Alberta population only 30 to 62% of the females

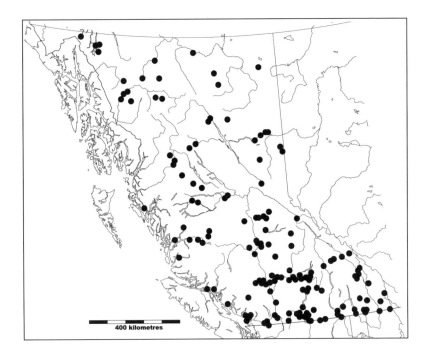

400 kilometres

produced a second litter. Newborn Bushy-tailed Woodrats are nearly naked and weigh about 14 to 15 grams, with males distinctly heavier than females. By 15 days their eyes open and they begin to eat solid food. The nursing period lasts for about a month. Young-of-the-year do not breed until the following spring. Shortly after they finish nursing, the young-of-the-year begin to disperse from their birth site. In Alberta, Richard Moses and John Millar found that most young females remained near their birth site, but most young males dispersed.

The Bushy-tailed Woodrat may live as long as three years, but only 31 to 34% of animals in the wild survive more than a year. Avian predators include large owls, such as the Barred Owl, Great Horned Owl and Great Gray Owl; in coastal Oregon and Washington, the Bushy-tailed Woodrat is a major prey species of the endangered Spotted Owl. Mammalian predators include the Coyote, Ermine, Long-tailed Weasel and Marten.

Range

The Bushy-tailed Woodrat inhabits most of western North America where it ranges from the southwestern United States to the southern Yukon Territory and the Northwest Territories. It is found throughout the entire mainland of British Columbia. Although it occurs at sea level on the coastal mainland, the Bushy-tailed Woodrat is absent from the coastal islands.

Taxonomy

Thirteen subspecies are recognized; three are found in British Columbia:

Neotoma cinerea cinerea Ord – Montana, Idaho, Wyoming, Alberta and B.C., where it is found in the extreme southeast (Newgate-Akamina Pass).

Neotoma cinerea drummondii (Richardson) – Alberta, the Yukon Territory, the Northwest Territories and B.C., where it ranges across the east from the Kootenays to the Yukon border.

Neotoma cinerea occidentalis Baird – Oregon, Washington, B.C., the Yukon Territory and the Northwest Territories; throughout B.C.'s entire central and western mainland.

Conservation Status

British Columbian populations are not of conservation concern.

Remarks

Because of its strong odour and tendency to inhabit buildings, this rodent is generally viewed as a nuisance species by ranchers and farmers. Yet it is a fascinating mammal demonstrating intriguing behaviours, such as a complex social structure and food hoarding tendencies. In dry environments of the southwestern United States, ancient plant remains and pollen preserved in Bushy-tailed Woodrat plant middens and amberat have provided insights into past climates and vegetation. Richard Hebda and colleagues found that amberat recovered from woodrat middens in B.C.'s Hat Creek valley and Bull Canyon yielded pollen, plant fragments and insect remains dating from 700 to 1100 years ago.

Selected References: Escherich 1981, Finley 1958, Hebda et al. 1990, Moses and Millar 1992, Smith 1997, Topping and Millar 1966.

Keen's Mouse

Keen's Mouse   *Peromyscus keeni*

Other Common Names: Cascade Deer Mouse, Sitka Mouse.

Description

Keen's Mouse has a long, furred tail, conspicuous eyes and long vibrissae. Its body size and fur colour vary considerably, particularly among island populations. The typical dorsal pelage is brown to dark grey, but some island populations are very dark with a nearly black dorsal band. The undersides are generally white to grey, but some island populations have reddish-brown markings on the chest and belly. The dorsal surface of the hind feet varies from white or silver to dark brown. Isolated island populations often demonstrate gigantism. The largest Keen's Mice in British Columbia live on Triangle Island, where adults average 44 grams and a few weigh more than 50 grams. The skull is characterized by molars with cusps and incisors that lack grooves.

Measurements:
  total length: 198 (140-263) n=773
  tail vertebrae: 102 (70-126) n=772
  hind foot: 24 (19-32) n=773
  ear: 18 (12-26) n=327
  weight: 26.0 (15.0-52.4)
    n=774

Dental Formula:
  incisors: 1/1
  canines: 0/0
  premolars: 0/0
  molars: 3/3

Identification:
The only similar species is the Deer Mouse (*Peromyscus maniculatus*), whose distribution overlaps with Keen's Mouse on the southern coastal mainland, the eastern slopes of the Cascade and Coast mountains, Vancouver Island and two associated islands (Balaklava and Malcom). The two species differ genetically,

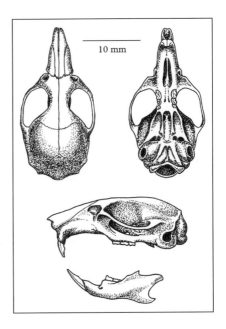

10 mm

but they are difficult to distinguish by external morphology in these regions. Adult Deer Mice usually have a shorter tail (< 98 mm), but this measurement overlaps to some extent among some populations of both species. Although Keen's Mouse averages larger cranial measurements than the Deer Mouse, the two species cannot be discriminated by any single skull measurement. Using specimens from western Washington, and islands and coastal areas of southwestern British Columbia that were identified from genetic traits, Marc Allard and Ira Greenbaum developed a formula based on tail length and the length of the mandibular diastema (the gap between the incisors and cheek teeth) that reliably discriminates the two species. Unfortunately, their formula cannot be used on live animals and its reliability needs to be assessed in other parts of the range where the two species co-occur (e.g., the eastern slopes of the coastal mountain ranges). Immature *Peromyscus* cannot be identified to species from morphology – they can only be identified from genetic traits.

Natural History

Ranging from sea level to 2010 metres elevation in the Coast and Cascade mountains, Keen's Mouse occupies many habitats. In the coastal lowlands it lives in forests of Western Hemlock, Western Redcedar and Sitka Spruce. At higher elevations it inhabits Mountain Hemlock, Engelmann Spruce, Lodgepole Pine and Amabilis Fir forests. It also occurs in subalpine and alpine habitats. On some of the smaller coastal islands, Keen's Mouse occupies shrub thickets and piles of driftwood adjacent to beach intertidal areas. On the dry eastern slopes of the Cascade Mountains at the eastern limits of its range, it is restricted to small pockets of moist Engelmann Spruce and Douglas-fir forests.

Keen's Mouse is abundant enough to be the dominant small rodent in a community, but there are no estimates of population density for British Columbia. In floodplain habitats of southeastern Alaska, Thomas Hanley and Jeffrey Barnard estimated densities as high as 96 animals per hectare, but these densities varied strikingly from year to year among different habitats and even within habitat types. The greatest variations occurred from year to year and they were consistent among different habitats. These fluctuations may be linked to variations in food supply and over winter survival. Because it is so adaptable, it is difficult to define specific habitat requirements. On some of the small islands off the north coast of Vancouver

Island, Barry Thomas found that Keen's Mouse was most abundant on beaches or forest edges, and rare or absent from inland forests. Its dependence on old-growth forests is not clear. On Prince of Wales Island in southeastern Alaska, Keen's Mouse was most abundant in young forests (23 years old); but several studies done in forests on the Olympic Peninsula of Washington found population densities nearly twice as high in continuous old-growth stands as in clear-cut habitats.

Few data are available on the movements of this species. The average home-range size of Keen's Mouse in southeastern Alaska is about 0.25 ha; dispersal distances range from 248 to 381 metres. With a long tail to help it balance, Keen's Mouse can climb up to 15 metres in trees. It probably makes dens in tree cavities as well as in logs and under woody debris. No one has described this rodent's social system; in other species of *Peromyscus*, males and females overlap home ranges but defend them from others of the same sex.

The distributions of Keen's Mouse and the similar Deer Mouse overlap extensively in southwestern B.C. Walter Sheppe demonstrated that, in isolation, the two species occupy similar habitats in the Coast and Cascade mountains, but when they co-occur they have more restricted habitats. Subtle differences in microhabitat may permit these similar species to coexist; for example, Keen's Mouse is more arboreal and seems to have a stronger association with mature forests. Nevertheless, the ecological distribution of these rodents is complex. On Mount Seymour near Vancouver, the two species appear to be separated by elevation: the Deer Mouse lives below 884 metres and Keen's Mouse above 762 metres; in the narrow overlap zone (from about 762 to 884 metres), they share the same habitat. Except for Balaklava, Malcom and Vancouver islands, the two species do not coexist on the islands of the south coast. On Vancouver Island, Keen's Mouse is the only species found in some subalpine-alpine habitats, but it co-occurs with the Deer Mouse at low elevations on some parts of the island. How Keen's Mouse and the Deer Mouse coexist on Balaklava Island, which is only six square kilometres in area, is unknown.

The diverse diet of Keen's Mouse includes high-energy seeds, fruits and arthropods. Captive mice have eaten mostly the seeds and fruits of Salmonberry, wild currants, Sitka Spruce and Western Hemlock. On small islands, this mouse forages mainly on beaches and in the intertidal zone, feeding on marine invertebrates such as amphipods. On Triangle Island, which supports more than a million

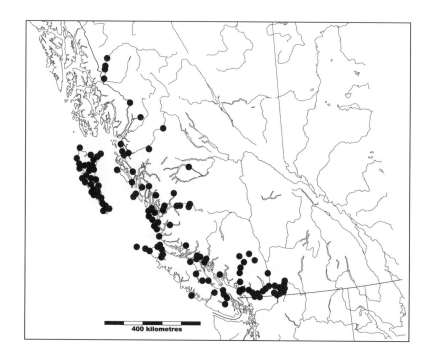

400 kilometres

seabirds, Keen's Mouse eats the eggs of burrowing seabirds such as Cassin's Auklet and the Rhinoceros Auklet. Mark Drever and colleagues attributed this island's large mouse population to its huge seabird population. Although coastal populations of Keen's Mouse have access to salmon carcasses during the annual salmon runs, captive animals showed no interest in salmon meat.

In the Cascade Mountains, a few females breed in late March, but most breed from April through June, and breeding ceases by the end of July. In coastal regions, the breeding season may be longer. Nursing females have been captured in mid April and pregnant females as late as mid August. The gestation period is 23 to 25 days. Walter Sheppe calculated an average litter size of 6.1 (range: 5 to 9) for the Cascade Mountains subspecies (*P. keeni oreas*). The litter size averaged 4.8 (range: 3 to 8), based on embryo counts for 81 museum specimens from across British Columbia. Females produce several litters per year. Newborns are pink and naked with closed eyes. The birth weight varies with subspecies; for the various B.C. races, it ranges from 1.6 to 3.4 grams. By 13 to 21 days the newborn's eyes open and it begins to eat solid food. After 30 days of age, it is capable of breeding, but young born in later litters may delay breeding until the following year.

Predators include various raptors and mammals, such as the Bobcat, Coyote, Ermine, Long-tailed Weasel and Marten. On many of the coastal islands where Keen's Mouse is the dominant mouse or vole species, it is the major small mammal prey for small carnivores such as the Marten. Black Rats and Norway Rats introduced to smaller islands in the Queen Charlotte Islands archipelago appear to have reduced Keen's Mouse populations either through competition for food or direct predation.

Range

This species ranges from coastal Washington to southeast Alaska. In British Columbia it inhabits the entire coastal mainland, as well as Vancouver Island and the islands off its north and west coast, and many central and north coast islands including the Queen Charlotte Islands. The eastern limits of its range are on the eastern slopes of the coastal mountain ranges. In the Cascade Mountains, the easternmost occurrence is at Bromley Creek near Princeton. Eastern records in the Coast Mountains include Blowdown Pass west of Lillooet, Stuie in the Bella Coola valley, Roche Deboule Mountain in the Skeena River valley and Dokdaon Creek on the Stikine River.

Taxonomy

Researchers have long recognized the presence of a long-tailed form of *Peromyscus* on the northwest coast of North America. Its taxonomic status and relationship with the Deer Mouse (*P. maniculatus*) was contentious until the genetic and morphological research by Ira Greenbaum and colleagues clearly demonstrated that this form represents a distinct species, Keen's Mouse (*P. keeni*). *P. keeni* includes populations previously classified as *P. oreas* and *P. sitkensis* and a number of coastal subspecies originally assigned to the species *P. maniculatus*. Thirteen subspecies are recognized, with eleven occurring in British Columbia. The subspecies *algidus, cancrivorous, maritimus, pluvialis* and *rubriventer* were assigned to the species *P. maniculatus* in the original *Mammals of British Columbia*. Although the genetics of these subspecies were not analysed in the studies done by Ira Greenbaum, their long tails and proximity to *P. keeni* on the adjacent mainland, suggest that these subspecies are members of the species *P. keeni*.

Peromyscus keeni algidus Osgood – southwestern Yukon Territory, the Haines Triangle region of B.C. and a small area of the Alaska panhandle. Based on a DNA analysis of samples from Skagway, Alaska, Kelly Hogan and colleagues proposed that this subspecies be classified as *P. keeni*. A genetic analysis is needed to determine the precise range of this subspecies. Specimens from further east in the Haines Triangle have short tails suggesting they may be *P. maniculatus*.

Peromyscus keeni cancrivorous McCabe and Cowan – an insular race restricted to Table Island.

Peromyscus keeni interdictus Anderson – an insular race inhabiting western and northern Vancouver Island, and possibly some islands off the west coast of Vancouver Island. In the original *Mammals of British Columbia*, Cowan and Guiguet listed this taxon as *P. maniculatus interdictus*.

Peromyscus keeni isolatus Cowan – an insular subspecies restricted to islands in Queen Charlotte Strait (Balaklava, Bell, Doyle, Duncan, Heard, Hurst, Malcom, Nigei and Pine) and the Scott Islands (Beresford, Cox, Lanz, Sartine and Triangle). *P. k. isolatus* coexists with *P. maniculatus austerus* on Balaklava and Malcom islands. This race has had a complicated taxonomic history. Originally described as a race of *P. sitkensis* (*P. sitkensis isolatus*) restricted to Pine Island, Cowan and Guiguet reclassified it as a subspecies of *P. maniculatus* (*P. maniculatus isolatus*) that occurs on Pine and Nigei islands. Research by Ira Greenbaum has demonstrated that this population as well as populations from the Scott Islands and various insular populations in Queen Charlotte Strait are a race of *P. keeni*.

Peromyscus keeni keenii (Rhoads) – an insular race restricted to the larger Queen Charlotte Islands.

Peromyscus keeni macrorhinus (Rhoads) – mainland coast and adjacent islands of southern Alaska and northwestern B.C. from Rivers Inlet to Wrangell Island.

Peromyscus keeni maritimus McCabe and Cowan – an insular race restricted to the Moore Islands.

Peromyscus keeni oreas Bangs – coastal lowlands, Cascade Mountains and Coast Mountains from Washington to Rivers Inlet, B.C.

Peromyscus keeni pluvialis McCabe and Cowan – an insular race restricted to the Goose Islands.

Peromyscus keeni prevostensis Osgood – an insular race restricted to the small, isolated outer islands of the Queen Charlotte Islands. In the original *Mammals of British Columbia*, Cowan and Guiguet classified this taxon as *P. maniculatus sitkensis*.

Peromyscus keeni rubriventer McCabe and Cowan – an insular race restricted to Reginald, Smythe, Townsend, Horsfall, Dufferin, Campbell, Hecate, Chatfield and Hunter islands.

Conservation Status
British Columbian populations are not of conservation concern.

Remarks
Except for the Deer Mouse, Keen's Mouse is the best represented rodent on British Columbia's islands occupying about 80 of them. The smallest islands known to support populations of this species are about 20 ha in area. The most isolated islands with Keen's Mice are the Queen Charlotte Islands that lie about 70 km from the B.C. mainland, and the Scott Islands (Cox, Beresford, Lanz, Sartine and Triangle), 10 to 46 km off the northern tip of Vancouver Island. Other isolated islands with Keen's Mouse populations are the Goose Islands and Moore Islands, 10 to 12 km from the nearest island with another population of Keen's Mouse and more than 30 km from the mainland coast. In contrast to the Deer Mouse, there has been little study of the biology of these island populations of Keen's Mouse.

More field research is needed to determine how this species co-exists with the similar Deer Mouse in southwestern B.C. Such studies will require the development of reliable field identification traits.

Selected References: Allard and Greenbaum 1988; Drever et al. 2000; Hanley and Barnard 1999; Hogan et al. 1993; McCabe and Cowan 1945; Sheppe 1961, 1963; Thomas 1971.

Deer Mouse   *Peromyscus maniculatus*

Other Common Names: White-footed Mouse.

Description

The Deer Mouse has a long, furred tail, conspicuous eyes and long vibrissae. The dorsal fur is brown and the belly white; the dorsal surface of the hind feet is also white. Some island populations of the coastal subspecies *Peromyscus maniculatus austerus* tend to be very dark, whereas populations of the *P. m. artemisiae* in the dry interior are pale. But this species has less colour variation than Keen's Mouse (*P. keeni*). The delicate skull has molars with cusps and the incisors lack grooves.

Measurements:
  total length: 170 (125-209) n=1519
  tail vertebrae: 80 (50-1040 n=1521
  hind foot: 21 (15-28) n=1521
  ear: 17 (10-22) n=662
  weight: 21.7 (15.0-41.8) n=1521

Dental Formula:
  incisors: 1/1
  canines: 0/0
  premolars: 0/0
  molars: 3/3

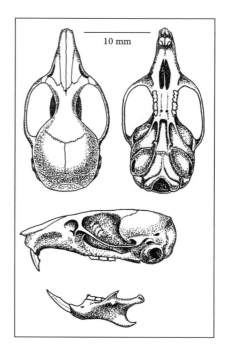

10 mm

Identification:
On the southern coast, Vancouver Island and the east slopes of the coastal mountain ranges, distinguishing this species from Keen's Mouse (*P. keeni*) is problematic (see its account, page 333). In the Okanagan, the Deer Mouse (*P. maniculatus*) may be confused with the Western Harvest Mouse (*Reithrodontomys megalotis*). *R. megalotis* is smaller – adults are less than 150 mm long and weigh less than 15 grams. Any Deer Mouse similar in size to a

Western Harvest Mouse is obviously a juvenile, discriminated by its dull grey pelage. The skull of a Western Harvest Mouse is smaller (skull length < 23 mm, maxillary toothrow length < 3.8 mm) and the upper incisors have distinct grooves on their anterior face.

Natural History

One of the most widespread and abundant rodents in the province, the Deer Mouse inhabits all of the province's biogeoclimatic zones, where it lives in a broad range of forested, shrub-steppe, arid-grassland and rocky alpine habitats. On smaller coastal islands that lack trees, the Deer Mouse inhabits shrub thickets of Common Snowberry and wild rose, and even grassy areas. This species is a common inhabitant of human dwellings in rural areas of British Columbia's interior, living in old cabins and homesteads in association with the Bushy-tailed Woodrat. It also occupies caves – its droppings and skeletal remains have been found in a number of caves on Vancouver Island. The Deer Mouse's elevational range in B.C. is from sea level to 2200 metres in the Columbia Mountains.

Detailed population studies on the Deer Mouse in B.C. are limited to coastal regions. In mature and logged Western Hemlock, Western Redcedar and Douglas-fir habitats in the Research Forest of the University of British Columbia in the Lower Mainland, average population densities range from 15 to 30 Deer Mice per hectare

Densities in Douglas-fir forests on some coastal islands are considerably higher; for example, Thomas Sullivan reported average densities of 43.5 animals per hectare on Saturna Island. In the spruce-fir forests of northern B.C., Sullivan and colleagues reported populations of 6.7 to 34.9 animals per hectare. In the boreal forest of the southern Yukon Territory, the northern limits of the Deer Mouse's range, summer densities are from 6 to 18 per hectare. There are no estimates for population density in montane forests or the shrub-steppe habitats of B.C.'s interior grasslands, but the Deer Mouse is abundant in these habitats, and it is often the dominant small rodent.

In contrast to Keen's Mouse, which is usually more abundant in old-growth forest, the Deer Mouse tends to have higher populations in logged habitats. Populations in forests and clear-cuts show a similar annual pattern related to recruitment of the young: they are lowest in the spring and, with breeding, increase throughout summer, reaching a peak in autumn; over winter, they decline to their spring low. Late-summer populations are higher in clear-cuts than in older forests because of an influx of juveniles, but by spring the densities are the same in both habitats. Although Deer Mouse populations fluctuate from year to year, there is no evidence for regular population cycles.

Estimates of home-range size from various habitats across North America range from 120 to 3000 m^2. Home-range size in B.C. has only been studied in coastal habitats. In Douglas-fir forests in the University of British Columbia Endowment Lands, Mary Taitt calculated home ranges of about 500 to 4000 m^2 for males and 500 to 2500 m^2 for females. Home-range size is linked to food resources – providing supplemental food will produce significantly smaller home ranges. In the Kananaskis Valley of the Alberta Rocky Mountains, homing experiments revealed that about 51% of adult Deer Mice were able to return to their home area after being moved more than 1500 metres. One animal returned a distance of 1980 metres.

The social system of the Deer Mouse is flexible. Some populations appear to be polygamous, living in small groups consisting of a mature male, a few mature females and their young. The adult males defend their territories from other mature males, but overlap with the home ranges of adult females. Other populations may be more promiscuous, because the home ranges of both species overlap. The Deer Mouse is an excellent climber. They make dens in tree cavities 3 to 10 metres above ground, and they can be attracted to artificial nest boxes. Deer Mice also build dens under logs, stumps, woody debris or at the base of trees. In shrub-steppe habitats where there are no trees, they burrow under woody shrubs such as Big Sage. Some Deer Mice maintain several active dens, switching between them over a period of a few weeks.

The Deer Mouse can enter brief periods of daily torpor when food is scarce, and some northern populations appear to enter torpor in response to shortening daylight. But this rodent does not hibernate – its periods of torpor usually last just four to nine hours. The Deer Mouse enters torpor near dawn, then arouses by nightfall, before its normal activity period. While in torpor, the Deer Mouse conserves energy by lowering its body temperature, but rarely below 13°C. In winter, this species relies on burrows insulated under deep snow or dens in tree cavities to escape extreme temperatures; it also nests communally and uses huddling to conserve heat.

The Deer Mouse is omnivorous. In late summer and autumn, it eats mostly the seeds of trees (coniferous and deciduous), shrubs and grasses, and it may cache some seeds; this mouse's seed harvesting can have a negative impact on reforestation and the seeding of range land. The Deer Mouse also eats invertebrates. On some of the smaller coastal islands, it forages in the intertidal areas and beaches feeding on marine invertebrates such as amphipods. In coastal forests, the Deer Mouse also consumes fungi, including truffles.

The timing of the breeding season varies greatly in response to climate, length of daylight and the availability of food resources. Timing and duration of the breeding season even varies from year-to-year in the same population. On the south coast of B.C., the length of the breeding season is 12 to 34 weeks; some animals begin breeding in late winter, but the main breeding season extends from May to late autumn. On islands in the Strait of Georgia and Barkley Sound the breeding season lasts from mid April to December or January. Mainland populations living in northern regions or at high elevations have a shorter breeding season of only 8 to 10 weeks. The gestation period is 22 to 23 days. Size of the young at birth varies with subspecies. Newborns of the subspecies *Peromyscus maniculatus artemisiae* from B.C.'s interior weigh about 1.3 to 2.2 grams. Newborns are pink, naked and blind. Hair begins to appear by the second day, their eyes open by 15 days and they are weaned by 25 days. Litter size for this species varies. The average litter size for populations in the University of British Columbia Endowment Lands, ranged from 3.0 to 5.3. Thomas Herman found that Deer Mice living on small islands in Barkley Sound (Helby, Diana and Fleming) had smaller average litter sizes than Deer Mice on nearby Vancouver Island. Females may breed as early as 30 days of age; in northern populations young-of-the-year females usually delay breeding until the following year.

Few Deer Mice survive beyond their second summer. This species is eaten by many raptors and mammalian predators. In the Okanagan Valley, the Deer Mouse is the dominant small-mammal prey of the Western Rattlesnake.

Range

The Deer Mouse ranges from Mexico, through much of the United States, to Canada as far north as the treeline. It inhabits the entire mainland of British Columbia, except for the central and northern coastal mainland and associated islands, where it is replaced by Keen's Mouse. It occurs on 97 islands on the southern coast, including Texada Island, Vancouver Island, many in the Strait of

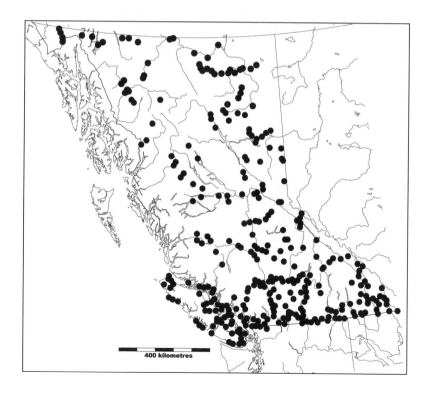

400 kilometres

Georgia and Johnstone Strait, and those off the west coast of Vancouver Island, including islands of the Broken Group and Deer Group in Barkley Sound.

Taxonomy

The taxonomy of the genus *Peromyscus* has been extensively revised since the original *Mammals of British Columbia*. Many coastal forms Cowan and Guiguet treated as subspecies of the Deer Mouse (*Peromyscus maniculatus*) are now classified as Keen's Mouse (*P. keeni*) – see the previous account (page 337). Moreover, some of the insular races of *P. maniculatus* from the southwest coast have been synonymized. Four subspecies are now recognized in British Columbia:

Peromyscus maniculatus alpinus Cowan – a weakly defined race restricted to high elevations in B.C.'s Selkirk Mountains.

Peromyscus maniculatus artemisiae (Rhoads) – Oregon, Washington, Idaho, Montana, Wyoming and the south-central part of B.C. as far north as Prince George.

Peromyscus maniculatus austerus (Baird) – Washington and the southwestern coast of B.C., where it inhabits the mainland as far north as Rivers Inlet, as well as many islands: Vancouver Island; Malcom, Balaklava, Hope and Vansittart islands in Queen Charlotte Strait; and islands in Johnstone Strait and the Strait of Georgia. Marc Allard and Ira Greenbaum assigned the four coastal subspecies listed in the original *Mammals of British Columbia* – *P. m. angustus*, *P. m. balaclavae*, *P. m. georgiensis* and *P. m. saturatus* – to the race *P. m. austerus*. Populations from various islands off the west coast of Vancouver Island have never been formally assigned to a subspecies and they may represent additional populations of *P. m. austerus*.

Peromyscus maniculatus borealis Mearns – the Yukon Territory, the Northwest Territories, Alberta and the central and northern part of B.C. east of the Coast Mountains.

Conservation Status

British Columbian populations are not of conservation concern.

Remarks

The Deer Mouse occurs on more islands than any other small mammal in B.C. From his study of islands in the Strait of Georgia, J. Redfield concluded that this species did not inhabit islands smaller than 25 ha in area regardless of their isolation. But on the Broken Islands in Barkley Sound, the Deer Mouse occurs on islands as small as 4 ha. In contrast to Keen's Mouse, which has managed to reach some of the most isolated islands on the coast, the Deer Mouse occupies islands that are relatively close to a source population. Of the 97 islands with Deer Mice, 76% are less than two kilometres from the source population on the mainland or nearby islands.

Some of the early published ecological literature on the Deer Mouse from southwestern B.C., including the Lower Mainland and Vancouver Island, should be regarded with caution. Because researchers were unaware of the presence of two species of *Peromyscus* this is region, some of their study animals could have been Keen's Mice.

Selected References: Herman 1979; King 1968; Kirkland and Layne 1989; Redfield 1976; Sullivan 1977, 1979; Sullivan et al. 2000; Taitt 1981.

Western Harvest Mouse

Western Harvest Mouse
Reithrodontomys megalotis
Other Common Names: None.

Description
The smallest rodent in British Columbia, the Western Harvest Mouse has prominent ears and a thin, sparsely-haired bicoloured tail. The dorsal pelage is a dull reddish-brown with long, scattered greyish to blackish guard hairs and shorter hairs that tend to be bicoloured with a buff base and a dark tip. The underparts are whitish; the dorsal surface of the hind feet are white. The small delicate skull has molars with two rows of cusps, and upper incisors with a prominent groove on their anterior face (figure 49).
Measurements:
  total length: 136 (116-151) n=42
  tail vertebrae: 67 (54-86)
    n=42
  hind foot: 16 (12-19) n=42
  ear: 14 (12-17) n=14
  weight: 11.0 g (8.0-15.0)
    n=38
Dental Formula:
  incisors: 1/1
  canines: 0/0
  premolars: 0/0
  molars: 3/3
Identification:
This species could be confused with the House Mouse (*Mus musculus*) and Deer Mouse (*Peromyscus maniculatus*). See those accounts for diagnostic traits.

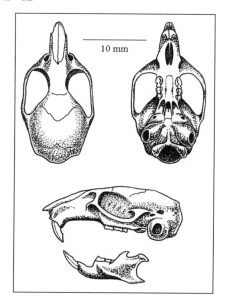

Natural History
In British Columbia, the Western Harvest Mouse is associated with the intermontane grasslands where it inhabits shrub-steppe rangeland, old fields and grassy areas bordering cultivated fields and roadsides. The elevational range extends from 300 to 780 metres

with most occurrences below 600 metres. In shrub-steppe grass-lands it lives in Antelope-Bush, Big Sage, Common Rabbit-brush, and grasses such as Bluebunch Wheatgrass and needle grass.

Although it occurs in shrub-steppe rangeland, abandoned fields, pastures, organic orchards and grassy roadsides, this species tends to be most common in edge habitats or dry gullies with an abundant shrub cover of wild rose, Mock-orange, Saskatoon, Choke Cherry, Squaw Currant, Common Snowberry and Black Hawthorn. It also exploits undisturbed edge habitats with high grass and alien weeds. In the Similkameen River valley, I found the Western Harvest Mouse in undisturbed edge habitats of brome grass, thistles and goldenrod that bordered hay and Alfalfa fields. In the north Okanagan, James Munro captured this rodent in weedy areas bordering roadsides. The most important habitat features are high grasses and shrubs.

In the United States, the Western Harvest Mouse may be a domi-nant small mammal in grassland communities, with densities as high as 60 animals per hectare, but in B.C., it is rarely a dominant species. During spring and summer of 1994 and 1995, Walt Klenner captured this species in a number of his study grids in the south Okanagan. In 1994, population densities in some grids reached 40 animals per hectare. In most of his grids, Western Harvest Mice showed striking seasonal variation in density, peaking in July. In some, it was absent in spring and early summer, but then suddenly appeared in large numbers in mid summer. Population densities were much lower in 1995, just 10 to 14 animals per hectare, although other rodents, such as the Great Basin Pocket Mouse, showed no similar decline. The cause of these year-to-year fluctua-tions in Western Harvest Mouse numbers is not known.

There are no estimates of Western Harvest Mouse home-range size in B.C. In the southwestern United States, estimates range from 0.95 to 1.12 ha. Maximum distances between capture sites for indi-viduals are usually less than 300 metres, but when population densi-ty is high a few individuals will move long distances (375 to 3200 metres). A series of experiments testing homing ability demonstrat-ed that Western Harvest Mice displaced as far as 300 metres were able to return to their home areas. The Western Harvest Mouse can climb as high as one metre in shrubs and nests have been found one metre above the ground in shrubs.

Steve Thompson induced captive Western Harvest Mice starved for 12 hours to enter shallow torpor when they were exposed to temperatures of 10 to 30°C. The bouts of torpor rarely lasted more than four hours. Below 10°C, Western Harvest Mice could not be aroused completely from torpor unless they were given food. It appears that the use of torpor varies among subspecies with some rarely entering torpor. Torpor would be expected to be critical for the survival of Canadian populations, because they are at the extreme northern limits of the range where they may be exposed to cool winter temperatures and periodic food shortages. Although a study in Nevada concluded that this rodent hibernated because it was not captured in winter months, the Western Harvest Mouse's ability to hibernate is unknown. Thomas Sullivan captured this species throughout the year on his trapping grids at Summerland in the Okanagan Valley.

There have been no comprehensive behavioural studies on wild Western Harvest Mice – their social structure is unknown. Experiments on captive animals showed a strong dominance hierarchy among males. It appears that a breeding male may nest with several females. Wild and captive animals are strictly nocturnal with activity greatest on moonless or rainy nights. Several researchers have noted this species' tendency to use vole runways. The Western Harvest Mouse constructs spherical or cup-shaped nests on the ground or in the branches of shrubs above ground. Evidently, it does not construct burrows, but will occupy the burrows made by other small mammals.

The Western Harvest Mouse eats seeds and invertebrates (especially Lepidoptera larvae), as well as flowers, herbaceous material and fungi. The Western Harvest Mouse's arboreal activity in shrubs is probably related to foraging for seeds, flowers and invertebrates. There is no evidence that the Western Harvest Mouse caches food in or near its nest or burrow.

Females breed throughout the year in southern parts of the range; in northern regions, the breeding season extends from April to October. Pregnant or lactating females have been captured between June and September in B.C. Although females may produce as many as 14 litters per year in captivity, wild females in Canada probably produce two or three litters. The gestation period is 23 to 24 days. The litter size averaged 2.6 (range: 1 to 7). Newborns weigh 1.0 to 1.2 grams, and are blind and naked. By 11 days their eyes open; they are weaned at about three weeks. Females are capable of breeding at about four months age; reproductive senility begins at about 45 weeks of age.

Western Harvest Mice can live up to 18 months in the wild, but

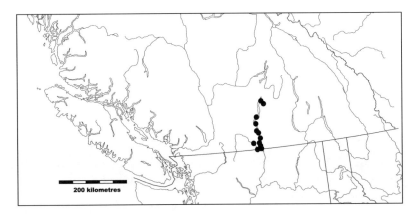

200 kilometres

only a few survive beyond 6 months. In the Okanagan Valley, remains of the Western Harvest Mouse have been recovered in the pellets of the Northern Saw-whet Owl; other owl species presumably also prey on this rodent. Potential mammalian predators include the Ermine, Long-tailed Weasel and Striped Skunk.

Range
Western Harvest Mice have an extensive range throughout western and central North America and Mexico. In Canada, where it is at the northern periphery of its range, it is restricted to southern British Columbia and southeastern Alberta. In B.C. it is found throughout most of the Okanagan Valley as far north as Vernon and in the Similkameen River valley north to Keremeos. The Western Harvest Mouse is absent from the central Okanagan Valley in the Kelowna area, and from the Thompson River and Kettle River valleys.

Taxonomy
Thirteen subspecies are recognized in North America; one is found in British Columbia:

Reithrodontomys megalotis megalotis (Baird) – the western United States, Mexico, and B.C.

Conservation Status
The Western Harvest Mouse appears on the provincial Blue List and the British Columbian population is listed nationally by COSEWIC as Special Concern. The growth of urban centres such as Vernon, Kelowna and Penticton has resulted in significant habitat loss and

contributed to the fragmentation of this rodent's range. The conversion of steppe-grassland to vineyards has also eliminated habitat. The Western Harvest Mouse population in the north Okanagan at Vernon is separated from the nearest known population in the south Okanagan by nearly 50 km, making it isolated and vulnerable to habitat loss.

Remarks

Although naturalist-collectors such as Hamilton Mack Laing collected small mammals in the south Okanagan and Similkameen River valleys during the 1920s and 1930s, they took no Western Harvest Mice, and this species is not mentioned in their field notes. The first documented records for British Columbia were animals captured by George Holland at the north end of Osoyoos Lake and at Skaha Lake near Penticton in 1942. James Munro, who trapped six at Okanagan Landing near Vernon in 1956, speculated that the Western Harvest Mouse was a recent invader to the province. If this species expanded its range into B.C. from Washington in the 1940s, then it spread rapidly throughout the Okanagan and Similkameen valleys. Curiously, Cowan and Guiguet, in the original *Mammals of British Columbia*, were not aware of this species' presence in the north Okanagan.

Selected References: Holland 1942, Munro 1958, Nagorsen 1995, Thompson 1985, Webster and Jones 1982.

Family Dipodidae:
Jumping Mice and Jerboas

This family includes the jerboas and jumping mice. Some authorities treat the jumping mice as a distinct family: Zapodidae. The North American jumping mice are small mouse-like rodents adapted for jumping with elongated hind limbs and hind feet, and extremely long, thin, nearly naked tails. Their skull has a large oval infraorbital opening, upper incisors with a deep groove on their anterior face and rooted cheek teeth that are high crowned with transverse ridges. Three species are found in British Columbia.

Meadow Jumping Mouse

Meadow Jumping Mouse   *Zapus hudsonius*

Other Common Names: None.

Description

The smallest of the three jumping mice in British Columbia, the Meadow Jumping Mouse has bright yellow to buff sides, a grey dorsal band and whitish undersides typically with buff patches on the belly, chest and throat. The skull is small (maxillary toothrow length < 3.7 mm; incisive foramen length < 4.2 mm) and the upper premolar is simple, lacking a crescent fold at the base of the main cusp (figure 65).

Measurements:
  total length: 212 (185-237) n=113
  tail vertebrae: 130 (112-149) n=111
  hind foot: 30 (27-34) n=110
  ear: 13 (10-16) n=32
  weight: 16.7 (10.0-25.0) n=34

Dental Formula:
  incisors: 1/1
  canines: 0/0
  premolars: 1/0
  molars: 3/3

Identification:

Throughout most of its range in British Columbia, this species co-occurs with the Western Jumping Mouse (*Zapus princeps*) and can even occupy the same habitat. In most areas where they co-occur, the Western Jumping Mouse is slightly larger and has duller sides that range from yellow to grey, and white undersides that lack a buff wash or buff patches. Nevertheless, because they overlap in size and show considerable variation in colour, distinguishing these species

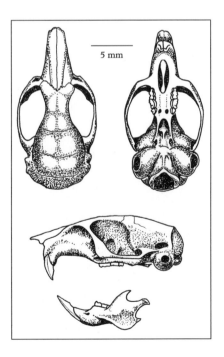

5 mm

from external traits is exceedingly difficult in some regions. Positive identification can only be made from the skull – the skull of the Western Jumping Mouse is larger (incisive foramen length > 4.2 mm, maxillary toothrow length > 3.7 mm).

Natural History

The Meadow Jumping Mouse lives in grassy fields and dense riparian thickets bordering streams or ponds. Meadows dominated by grasses appear to be the most productive habitats. Habitat descriptions recorded on museum specimens from British Columbia include: willow or alder thickets; grassy meadows and thickets bordering lakes, beaver ponds or streams; and open clearings in Trembling Aspen forest. The elevational range in B.C. is from 300 to 1200 metres. Some researchers have concluded that the Meadow Jumping Mouse prefers habitats that are moist or near water bodies, but John Whitaker found that soil moisture and distance to water had little influence on the presence or abundance of this species in his study area. Population estimates from various habitats in eastern North America are typically 2 to 4 animals per hectare, although densities can reach 37 per hectare. Competition with the Meadow Vole appears to influence the Meadow Jumping Mouse – its populations are higher when this vole is absent.

The Meadow Jumping Mouse is primarily nocturnal, though sometimes active during the day. Its social structure is essentially unknown, except that the home ranges of males and females overlap extensively. Home-range sizes (0.08 to 1.10 ha) are similar for males and females. Several researchers suspected that this species is transient, because tagged individuals suddenly disappeared from their study areas. But Rudy Boonstra and James Hoyle concluded that the Meadow Jumping Mouse is simply trap shy – once captured in a live-trap it rarely enters one again.

Meadow Jumping Mice build summer nests from grass, moss or dried leaves in underground burrows or on the ground under logs, woody debris, or clumps of grass or sedge. They either dig their own shallow burrows or occupy burrows made by other small mammals. The Meadow Jumping Mouse typically hibernates in underground burrows. The dark band on its back probably provides camouflage against a background of dense vegetation. It does not construct runways in the grass, but moves about randomly.

Normally, this species moves by short hops that rarely exceed 15 cm, or by slowly crawling on all four limbs. When alarmed, it takes several long hops to escape and then freezes. Some biologists have said that this mouse can jump three metres, but the evidence shows that these claims are exaggerated. Don Quimby and John Whitaker separately carried out extensive field studies on this mouse and found that its maximum jumping distance was no more than a metre. It readily climbs grass stems to cut seed heads, and it is a strong swimmer that can remain in water up to five minutes.

The Meadow Jumping Mouse's diet includes invertebrates, seeds and underground fungi. In spring, its main foods are caterpillars and beetle larvae. In summer and fall, it eats the seeds of various grasses, touch-me-not, oxalis and cinquefoil, the fruits of various shrubs, and underground fungi or truffles. It does not store food in its burrow for winter.

The Meadow Jumping Mouse hibernates for six to eight months. Little else is known about its hibernation in British Columbia, but in eastern North America this mouse typically enters hibernation between mid September and late October, although it has been seen active as late as mid November. Evidently, this species rapidly gains weight during the two weeks just before it enters hibernation. Emergence dates range from late April to early May, with males usually appearing before females.

Breeding information is also lacking for western populations. In eastern North America, the breeding season begins as soon as the Meadow Jumping Mouse emerges from hibernation. Young are born between mid May and early September, most in June. The gestation period is 17 to 21 days. Litter sizes, estimated from embryo counts, average 5.5 (range: 2 to 9). Many females produce two litters per year and some have three litters. The scanty information for B.C. is limited to reproductive data recorded on museum specimens, but it is consistent with eastern populations: seven pregnant females collected between June 15 and July 24 had 4 to 7 embryos. Newborns are naked, blind and weigh about 0.7 to 0.9 grams. By 17 days they have all their hair. Their eyes open between 22 and 25 days. The upper incisor teeth erupt at 11 days and all teeth erupt by 33 days. The young reach adult weight by about 60 days. Some females born in the early litters may breed in their first summer. Mortality is high in this species, but some can survive two years in the wild. Known predators include raptors such as the Great Horned Owl, Long-eared owl, Western Screech-Owl and Northern Harrier, and mammalian carnivores such as the Red Fox and weasels.

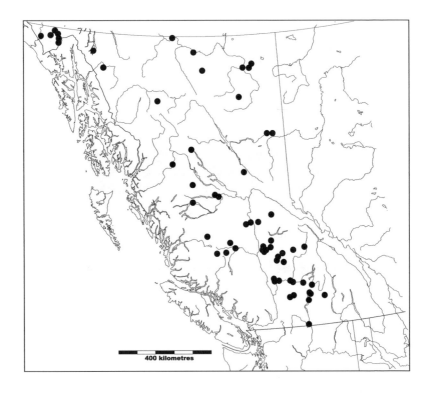

400 kilometres

Range
A boreal species, the Meadow Jumping Mouse ranges across Alaska, Canada and the central and eastern United States. It is found across most of northern and southern British Columbia as far south as Osoyoos Lake and east to the Monashee Mountains. It appears to be absent from coastal regions and the southern Rocky Mountains.

Taxonomy
Most authorities recognize 11 subspecies with 3 occurring in British Columbia. In a study based largely on dental morphology, Gwilyn Jones concluded that no subspecies should be recognized for *Zapus hudsonius*, but his taxonomy was never published and it has not been adopted by mammalogists.

Zapus hudsonius alascensis Merriam – Alaskan Peninsula, the coastal mainland of southern Alaska and the Haines Triangle area of extreme northwestern B.C.

Zapus hudsonius hudsonius (Zimmerman) – central Alaska and northern Canada from the Yukon Territory to Ontario; the northern half of B.C., except for the Haines Triangle.

Zapus hudsonius tenellus Merriam – restricted to central and south central B.C. from Burns Lake and Prince George south to the Okanagan Valley.

Conservation Status

The subspecies *Zapus hudsonius alascensis* is on the provincial Blue List, probably because of its limited range and few occurrences. But it is known from 11 localities in the Haines Triangle. No data are available on the size of the population, but its habitat is not at risk and most occurrences are within the Tashenshini-Alsek Wilderness Provincial Park, a large protected area.

Remarks

Almost all of the research on the Meadow Jumping Mouse has been done in eastern North America. Much remains to be learned about the biology of this species in British Columbia.

Selected References: Boonstra and Hoyle 1986; Jones 1981; Quimby 1951; Whitaker 1963, 1972.

Western Jumping Mouse *Zapus princeps*

Other Common Names: None.

Description

The Western Jumping Mouse has a grey band on its back, and its sides are dull yellow to grey. The undersides are white, except for some populations of the subspecies *Zapus princeps kootenayensis*, which have a buff wash and buff patches on the throat, chest or abdomen. The skull is large (incisive foramen length > 4.2 mm, maxillary toothrow length > 3.7 mm) and the upper premolar is simple, lacking a crescent fold at the base of the main cusp (figure 65).

Measurements:
   total length: 235 (209-265)
      n=218
   tail vertebrae: 142 (122-
      162) n=218
   hind foot: 32 (27-34)
      n=220
   ear: 12 (10-16) n=81
   weight: 24.6 g (17.7-37.5)
      n=68

Dental Formula:
   incisors: 1/1
   canines: 0/0
   premolars: 1/0
   molars: 3/3

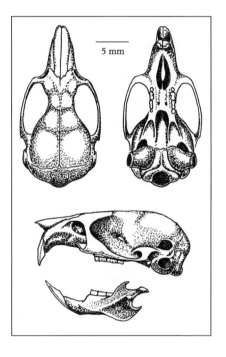

5 mm

Identification:
To distinguish this species from the Meadow Jumping Mouse (*Zapus hudsonius*) see that species account (page 353). Identification is problematic in southwestern British Columbia, where this species coexists with the Pacific Jumping Mouse (*Z. trinotatus*) in a narrow zone along the crest of the Cascade Range and the Coast Mountains. The Pacific Jumping Mouse tends to be more brightly coloured, with a buff wash and buff patches on the ventral pelage, and its upper premolar has a crescent fold at the base of the main cusp, but these traits tend to be variable.

Natural History

The Western Jumping Mouse lives in willow or birch thickets, shrubby or grassy environments bordering streams or lakes, subalpine or alpine meadows, mature forests and recent burns. Although it is often found in sites dominated by grasses or sedges, these habitats do not support high populations. The most productive habitats for this species are rich meadows with abundant forbs. The elevational range in British Columbia is from sea level to 2100 metres in the Coast Mountains and 2300 metres in the Rocky Mountains.

Estimates of population densities range from 1.9 to 35.0 per hectare, with males maintaining larger home ranges than females. Abundance increases markedly in summer when juveniles enter the population. Densities also vary from year to year and among different habitats. Estimates of average home-range size are 0.17 ha (range: 0.03 to 0.33 ha) for males and 0.10 ha (range: 0.07 to 0.13 ha) for females. The social structure of this species is unknown, but females appear to be territorial, showing little overlap in their home ranges, while the home ranges of males overlap extensively.

The Western Jumping Mouse hibernates for 7 to 10 months of year. It accumulates fat for about a month before entering hibernation. In the Rocky Mountains of Utah, Jack Cranford found that this species entered hibernation in September or October, and emerged from May 1 to July 22, the males about a week before the females. Hibernation has not been studied in B.C., but collecting dates for museum specimens range from May 15 (at Newgate) to October 4 (at Kelley Lake). Although some researchers have reported that elevation influences the length of hibernation, the evidence shows that year-to-year differences in spring climate have more effect on hibernation and the timing of spring emergence. Because the cue for arousal occurs when the soil temperature reaches 8.5 to 9.0°C, there can be considerable year-to-year variation depending on spring climate. During his three year study, Jack Cranford found that spring arousal dates varied by as much as 30 days at one of his study sites.

Hibernation nests are usually in chambers about 60 cm below ground. A single entrance tunnel leads from the surface to the nest. In autumn, the entrance is usually plugged with soil. The nest chamber is about 16 cm in diameter; it houses a round nest constructed from plant material such as forb stems, bark, spruce needles, dried grass and leaves. No food is stored in the hibernacula. After spring snowmelt, summer nests have been found under logs and stumps.

The Western Jumping Mouse eats seeds, fruits, fungi, green plant material and invertebrates. Gwilyn Jones and colleagues dissected the stomachs of 26 specimens collected in B.C. and found that they contained mostly Salmonberry (28%), grass seed (23%), underground fungi (13%) and blueberries (10%), along with small amounts of unidentified seeds and fruit, and invertebrate fragments. The Western Jumping Mouse evidently shifts its diet to high-energy

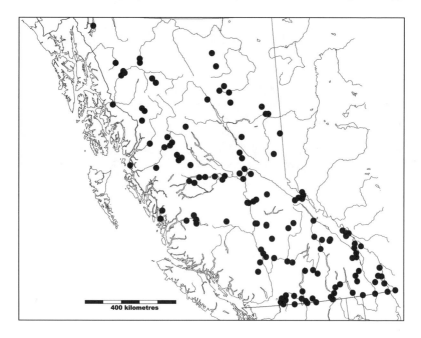

400 kilometres

seeds later in the season as it builds up fat reserves for hibernation.

Mating occurs immediately after animals emerge from hibernation in spring. Reproductive data for B.C. are limited to information recorded for museum specimens. Embryo counts for 21 pregnant females collected from June 14 to July 19 averaged 4.9 (range: 3 to 6); 6 nursing females were taken between July 11 and August 9. Females produce one litter per year. The gestation period is 18 days; the nursing period is about 33 days. The young begin to grow fur by 9 days and open their eyes by 25 days. They remain in the nest until about 30 days; after 35 days they are independent. Young-of-the-year are smaller and lighter than adults when they enter hibernation in autumn – they reach full adult size by the end of their second summer. The Western Jumping Mouse can live more than three years in the wild, with a few individuals surviving up to five years; the average life span is 16.5 months. Predators include owls and the Coyote, Marten, Badger, Ermine, Long-tailed Weasel and Least Weasel.

Range
The Western Jumping Mouse inhabits the Great Plains and mountains of western Canada and the United States. It is found throughout the entire mainland of British Columbia except for the extreme southwest where it is replaced by the Pacific Jumping Mouse. The Western Jumping Mouse is not known to occur on any of the province's coastal islands.

Taxonomy
Eleven subspecies are generally recognized with three occurring in British Columbia. Gwilyn Jones proposed radical revisions to the taxonomy of *Zapus princeps*, treating all B.C. populations of *Z. princeps* and *Z. trinotatus* as a single subspecies, *Z. princeps princeps*, but his taxonomy has not been adopted by most authorities.

Zapus princeps idahoensis Davis – northern Idaho, Montana, Wyoming, southwestern Alberta and extreme southeastern B.C., where it occupies the Rocky Mountains and Rocky Mountain trench as far north as Mount Assiniboine.

Zapus princeps kootenayensis Anderson – extreme northern Idaho, northeastern Washington and southern B.C. as far north as Barkerville and west to Allison Pass in Manning Provincial Park.

Zapus princeps saltator Allen – southeastern Alaska, the south-central Yukon Territory, western Alberta and the entire northern and central regions of B.C.

Conservation Status
British Columbian populations are not of conservation concern.

Remarks
Our knowledge of the biology of the Western Jumping Mouse is based almost entirely on studies done in the Rocky Mountains, an area with long winters, unpredictable spring snowmelt and a short growing season. This species' adaptations to such a harsh, unpredictable environment include: a short active period above ground, a lengthy hibernation period, the production of only one litter per year and a long life span. Comparative life history studies of Western Jumping Mice living in more moderate or stable environments at low elevations would be interesting.

Selected References: Cranford 1978, 1983; Jones 1981; Jones et al. 1978.

Pacific Jumping Mouse

Pacific Jumping Mouse   *Zapus trinotatus*

Other Common Names: None.

Description

The Pacific Jumping Mouse is our largest and most brightly coloured jumping mouse. Its dorsal pelage has bright yellow to light orange sides that contrast sharply with the dark dorsal band. The undersides are usually white, washed with buff; buff patches often occur on the throat, chest and abdomen. The skull is large (incisive foramen length > 4.2 mm, maxillary toothrow length > 3.7 mm) and the upper premolar is complex, with a crescent fold at the base of the main cusp (figure 64).

Measurements:

total length: 238 (215-258) n=87

tail vertebrae: 145 (120-162) n=86

hind foot: 32 (29-35) n=86

ear: 14 (12-16) n=10

weight: 24.2 (17.0-36.0) n=20

Dental Formula:

incisors: 1/1

canines: 0/0

premolars: 1/0

molars: 3/3

Identification:

See the previous account for distinguishing this species from the Western Jumping Mouse (*Zapus princeps*).

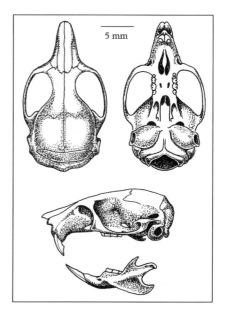

5 mm

Natural History

The Pacific Jumping Mouse inhabits forests, wet meadows, forest clearings, marshy areas, riparian areas along streams and lakes, sub-alpine willow-alder thickets and alpine meadows. The elevational range in British Columbia is from sea level to 1580 metres in the southern Coast Mountains and 1500 metres in the Cascade Mountains. Researchers working in Oregon captured this species more frequently in riparian stream-side habitats than in upland areas. The Pacific Jumping Mouse has been found in clear-cut forests, but the impact of forestry practices on this species is not clear. This mouse appears to more abundant in shrubby communities than forests.

There are no data on this animal's population densities, movements or social structure. Although it does not make runways in vegetation, it occasionally uses the runways of other mammals, such as the Mountain Beaver and Shrew-mole. The Pacific Jumping Mouse is a docile animal, but when alarmed, it can jump 0.9 to 1.8 metres. In summer it constructs spherical nests from grasses, sedges or mosses, usually on the ground. A nest that contained small young was located in the crook of a tree 1.4 metres above ground. The Pacific Jumping Mouse hibernates in underground burrows that have a nest chamber lined with dried grass. It accumulates fat in September and October for hibernation, but there is little specific information on the length of the hibernation period or the dates it enters and emerges from hibernation. Collecting dates for museum specimens taken in B.C. range from March 8 (at Vancouver) to November 20 (at Alta Lake).

The Pacific Jumping Mouse eats seeds, fruits, fungi and animal matter. Gwilyn Jones and colleagues dissected the stomachs of 27 specimens collected in B.C. and found that they contained mostly grass seeds (47%) and Thimbleberry (31%), along with small amounts of underground fungi, plant material and invertebrate remains. Lichens and conifer seeds have also been found in stomach remains of Pacific Jumping Mice. This mouse usually eats the fruits that fall to the ground, but it will climb into the branches of low shrubs, such as Salal, to collect fruits.

The breeding season extends from May or June through September. In the western United States, females have four to eight young with most births in July or August. The gestation period is 18 to 23 days. Females evidently produce only one litter in a year. Breeding data for B.C. are limited to two pregnant females collected on July 8 and August 9 and a nursing female taken on August 8. Newborns are blind, naked and weigh only 0.7 to 0.9 grams. They reach maturity just before they enter hibernation, but do not breed until the following spring. Owls are the major predators – the Barn Owl, Great Horned Owl, Long-eared Owl and Spotted Owl. Some mammalian carnivores also occasionally prey on the Pacific Jumping Mouse, such as the Long-tailed Weasel, Ermine, Western Spotted Skunk and Bobcat.

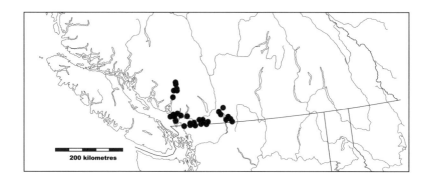

Range

A coastal species, the Pacific Jumping Mouse inhabits California, Oregon, Washington, and extreme southwestern British Columbia as far north as Garibaldi Provincial Park in the southern Coast Mountains and east to Manning Provincial Park and Treasure Mountain on the crest of the Cascade Mountains.

Taxonomy

Gwilyn Jones reported intergradation in dental morphology among the Western Jumping Mouse (*Zapus princeps*) and Pacific Jumping Mouse (*Z. trinotatus*) – he interpreted this as evidence that the two forms interbreed. Most authorities, however, treat the Pacific Jumping Mouse as a separate species. Four subspecies are recognized; one occurs in British Columbia:

Zapus trinotatus trinotatus Rhoads – a coastal race associated with northern California and Oregon, Washington, and extreme southwestern B.C.

Conservation Status

British Columbian populations are not of conservation concern.

Remarks

Surprisingly little is known about this intriguing rodent and more research is needed on its distribution, population dynamics, taxonomy and basic natural history in British Columbia.

Selected References: Gannon 1988, Jones 1981, Jones et al. 1978.

Family Erethizontidae:
New World Porcupines

There are 10 species of New World porcupines; one occurs in North America.

Porcupine

Porcupine   *Erethizon dorsatum*

Other Common Names: None.

Description
A large rodent, the Porcupine has a heavy body with a small head, short limbs and a short thick tail. Its pelage, unique among British Columbian mammals, consists of long, thick barbed quills, stiff guard hairs and a woolly underfur. The longest hairs are the stiff guard hairs that cover the dorsal parts of the body. Distributed throughout the dorsum, including the tail, are the shorter protective quills; the longest quills are on the rump. No quills occur on the undersides. The overall pelage colour ranges from dark brown to blackish. Individual quills may be completely white or bicoloured with a whitish or yellow base and black tip. The front feet have four toes and the hind feet have five. Adapted for an arboreal lifestyle, the feet have large curved claws and their naked soles have coarse pads. The skull is large and robust, with a round infraorbital opening that is larger than the foramen magnum. The incisors are deep yellow or orange.

Measurements:
total length: 795 (657-932) n=34
tail vertebrae: 231 (170-330) n=86
hind foot: 107 (84-124) n=86
ear: 24 (15-34) n=6
weight: 9.1 kg (4.3-12.3) n=18 males; 5.6 (3.3-8.4) n=9 females

Dental Formula:
incisors: 1/1
canines: 0/0
premolars: 1/1
molars: 3/3

Identification:
This species cannot be confused with any other rodent.

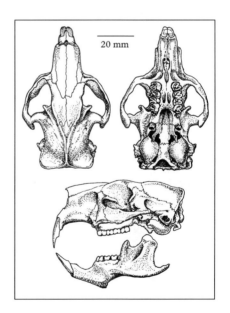

20 mm

Natural History

Associated with forests, the Porcupine ranges from sea level to sub-alpine areas. The most essential habitat requirements are trees for food and suitable den sites. Because it mostly exploits pole-sized trees for food, the Porcupine reaches its highest populations in early successional forests that develop after forest harvesting or fire. The only population density data for British Columbia are John Krebs' estimates of 4.3 to 14.3 animals per hectare for second-growth forest in the Kalum Valley near Terrace in the Coast Mountains. Porcupines undergo short-term fluctuations in numbers in response to food supply, and long-term fluctuations spanning many decades in response to forest succession or vegetational changes resulting from shifts in climate.

Porcupines build dens under fallen trees, brush piles or stumps, in rock crevices, small caves or talus, and in the burrows of other animals. Todd Zimmerling, who studied 46 winter dens in north-western B.C. located by radio-tracking, found that a Porcupine used 2 or 3 dens throughout a winter. Although the animals in his study rested occasionally in the branches of trees, none of them spent more than a day away from their den site. They seemed to prefer the dens located in rock crevices and under stumps, possibly because these sites provided the best insulation from extreme temperatures.

The Porcupine has a solitary social structure, although males may live in loose associations of related males. A male's home range usually overlaps with those of several females and other males. Females appear to be highly territorial, their home ranges not overlapping with other females. The extent that Porcupines remain solitary in winter depends on the availability of suitable den sites. In northwestern B.C., where winter dens are abundant, Porcupines keep to themselves; Todd Zimmerling found only one instance of two occupying the same den in that region. But in areas where winter dens are scarce, several will share a den. Porcupines feed close to their dens in winter. In northwestern B.C., John Krebs estimated average winter home-range sizes of 11.2 ha for males and 5.7 ha for females, although daily movements were only 63 to 295 metres. The summer home range is larger, from 2.4 to 83.5 ha. In late summer or autumn, young-of-the-year females disperse from their mother's territory, but young males tend to remain at their birth site. Adult Porcupines occasionally travel 8 to 10 km in search of a new territory.

An adult Porcupine is armed with as many as 30,000 quills. These modified guard hairs are thicker than normal hairs, and have a spongy filling and a narrow barbed tip. Porcupine quills are one of nature's smart weapons, designed to intimidate or injure a threatening predator. Normally the quills lie flat against the body, but when the Porcupine is aroused to defend itself, they stand up, pointing in all directions. While erect, the quills separate easily from their roots, ensuring that any contact by a predator will release them. Once a quill has entered the skin of an unfortunate victim, the barbed tip quickly pulls it deep into the flesh – a quill seven millimetres long was observed to move at the rate of one millimetre an hour through the muscle of a human arm. The quills on a Porcupine's back and tail are covered with a lubricating fat that ensures deep penetration.

Quills are used as a last resort. As with most animals armed with defensive weapons, the Porcupine has several early warning signals for predators. When threatened, it makes clacking sounds with its teeth and emits a strong odour from glands near the base of its tail. The conspicuous white markings created by the contrast of whitish quills against the animal's black back signal danger to a predator, in the same way as a skunk's white stripes. A disadvantage of quills as a defence is that a Porcupine falling out of a tree risks impaling itself with its own quills.

The Porcupine is clumsy and slow, but well equipped for climbing, with long, curved, sharp claws, gripping foot pads and powerful thigh muscles. When climbing up or down a tree, the Porcupine spreads itself flat against the trunk, with its head facing upwards, and drags its belly over the bark. Bristles on the underside of the tail prevent the animal from slipping. While feeding, the Porcupine can support itself with its powerful hind feet, leaving the front feet free to handle food. This animal has been seen swimming to lily pads, plants rich in salt.

Strictly herbivorous, the Porcupine eats grasses, forbs, leaves and bark. In summer, it feeds on a wide range of succulent ground plants, and climbs Trembling Aspens and birches to eat the buds, catkins and leaves; in agricultural areas, it eats apples and other commercially grown fruit. In winter it feeds on the twigs, buds, needles and inner bark of trees and shrubs, especially pole-sized trees (7 to 38 cm in diameter). In British Columbia, the Porcupine's major food trees are Douglas-fir, Lodgepole Pine, Ponderosa Pine and Western Hemlock. It selects trees on the basis of their availability, but also seems to show strong preferences for some species. In the northern coastal forests, it eats more Western Hemlock and Sitka Spruce than Western Redcedar and Amabilis Fir. Although the

Porcupine will girdle the base of trees, it usually climbs to feed in their upper branches and crown. It shaves off the dead, corky outer bark with its incisors, then eats the inner bark down to the woody cambium, macerating and chewing it before swallowing. Patches of girdled bark with distinctive horizontal or diagonal incisor marks (figure 26) on the exposed sapwood are a sure sign of Porcupine feeding activity.

The Porcupine can close its lips behind the incisors to avoid swallowing the outer bark as it gnaws. To help digest woody food, the Porcupine has an enormous caecum, nearly the size of its stomach, where bacteria ferment the cellulose in the bark. Unlike the Beaver, this rodent does not appear to re-ingest its feces. The Porcupine has a particular fondness for salt. In addition to natural mineral sources, such as waterlilies, it is attracted to salt accumulations from winter road maintenance; and it gnaws on human implements impregnated with salt from perspiration, such as wooden canoe paddles and tool handles. It also finds plywood glue tasty, gnawing large areas on

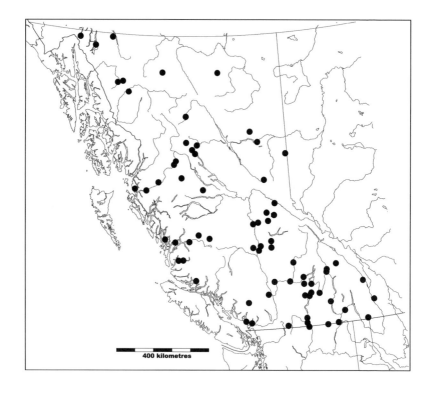

400 kilometres

buildings. The Porcupine will even gnaw on the rubber tires and lower parts of a motor vehicle.

No data are available on the Porcupine's breeding season in British Columbia, but in most other northern regions these mammals mate in October and November. The breeding strategy is unlike other B.C. rodents. Porcupines engage in an elaborate courtship, with both sexes producing an assortment of vocalizations, including screams, growls and chuckling sounds; and before mating, the male sprays the female with urine. A single young is born after a gestation period of 205 to 217 days. Weighing 340 to 640 grams, the newborn Porcupine is well developed and covered with soft quills. Its eyes are open and its teeth erupted. The nursing period lasts up to 120 days, but within a few weeks after birth, the young can survive on solid food. Some animals reach sexual maturity in their second year. Females can produce one young per year.

The Porcupine can survive up to 18 years in the wild. A major cause of death is highway accidents. Despite its protective body covering, this rodent is eaten by mammalian predators such as the Bobcat, Cougar, Coyote, Grey Wolf, Lynx, Red Fox, Fisher and Wolverine. Most predators have no special tactics for killing a Porcupine – they often have quills embedded in their facial areas. The most effective predator of Porcupines is the Fisher, an arboreal mammal well adapted for locating them in trees. A Porcupine-Fisher encounter is a "cat-and-mouse game": the Fisher tries to attack the Porcupine's face, the only area not protected by quills; the Porcupine protects itself by keeping its back towards the Fisher and hiding its face; over time, if the Fisher can inflict enough wounds on the Porcupine's face, it will weaken the rodent and eventually win the battle.

Range
The Porcupine is distributed across North America from northern Mexico to northern Alaska and Canada. It inhabits the entire mainland of British Columbia, but is not known to occur on any island in the province.

Taxonomy
Six subspecies are recognized; two occur in British Columbia:

Erethizon dorsatum myops Merriam – Alaska, the Yukon Territory, the western Northwestern Territories, northwestern Alberta, and extreme northern and northeastern B.C.

Erethizon dorsatum nigrescens Allen – Washington, northern Idaho, northwestern Montana, southwestern Alberta, and the entire north-central and southern mainland of B.C.

Conservation Status
This species is not of conservation concern.

Remarks
In northern British Columbia, the Porcupine was an important food source for aboriginal people. They also dyed Porcupine quills and wove them into elaborate patterns on clothing. Many coastal groups traded leather clothing decorated with Porcupine-quill embroidery. Unfortunately, many people today regard the Porcupine as a pest, because it damages buildings and automobiles. Forest managers also consider it a problem species in some parts of the province, because its winter bark feeding affects tree growth and survival. In the 1980s, Fishers were transplanted to the Khutzeymateen Inlet area of northwestern B.C. as a biological control for Porcupines.

Selected References: Krebs 1994, Roze 1989, Sullivan et al. 1986, Woods 1973, Zimmerling and Croft 2001.

APPENDIX 1

Scientific Names of Plants and Animals Mentioned in this Book

Trees

Amabilis Fir	*Abies amabilis*	Scrub Birch	*Betula glandulosa*
Beaked Hazelnut	*Corylus cornuta*	Sitka Spruce	*Picea sitchensis*
Bigleaf Maple	*Acer macrophyllum*	Subalpine Fir	*Abies lasiocarpa*
Black Spruce	*Picea mariana*	Subalpine Larch	*Larix lyallii*
Douglas-fir	*Pseudotsuga menziesii*	Trembling Aspen	*Populus tremuloides*
Engelmann Spruce	*Picea englemanni*	Western Hemlock	*Tsuga heterophylla*
Garry Oak	*Quercus garryana*	Western Larch	*Larix occidentalis*
Grand Fir	*Abies grandis*	Western Redcedar	*Thuja plicata*
Lodgepole Pine	*Pinus contorta*	Western White Pine	*Pinus monticola*
Mountain Hemlock		Whitebark Pine	*Pinus albicaulis*
	Tsuga mertensiana	White Oak	*Quercus alba*
Pacific Crab Apple	*Malus fusca*	White Spruce	*Picea glauca*
Paper Birch	*Betula papyrifera*	Yellow-cedar	
Ponderosa Pine	*Pinus ponderosa*		*Chamaecyparis nootkatensis*
Red Alder	*Alnus rubra*		

Shrubs

Antelope-Bush	*Purshia tridentata*	Gorse	*Ulex europaeus*
Kinnikinnick	*Arctostaphylos uva-ursi*	Himalayan Blackberry	*Rubus discolor*
Big Sagebrush	*Artemisiae tridentata*	Mallow Ninebark	
Black Gooseberry	*Ribes lacustre*		*Physocarpus malvaceus*
Black Hawthorn	*Crataegus douglasii*	Mock-orange	*Philadelphus lewisii*
Black Huckleberry		Oceanspray	*Holodiscus discolor*
	Vaccinium membranaceum	Partridgefoot	*Lutkea pectinata*
Choke Cherry	*Prunus virginiana*	Pink Mountain-heather	
Cloudberry	*Rubus chamaemorus*		*Phyllodoce empetriformis*
Common Juniper		Salal	*Gaulteria shallon*
	Juniperus occidentalis	Salmonberry	*Rubus spectabilis*
Common Rabbit-brush		Saskatoon	*Amelanchier alnifolius*
	Chrysothamnus nauseosus	Scotch Broom	*Cytisus scoparius*
Common Snowberry		Sheep Laurel	*Kalmia angustifolia*
	Symphoricarpos albus	Soapberry	*Shepherdia canadensis*
Crowberry	*Empetrum nigrum*	Squaw Currant	*Ribes cereum*

Shrubs (cont.)

Sweet Gale	*Myrica gale*	White-flowered Rhododendron	
Thimbleberry	*Rubus parviflorus*		*Rhododendron albiflorum*
Vine Maple	*Acer circinatum*	White Mountain-heather	
Western Bog-laurel			*Cassiope mertensiana*
	Kalmia microphylla		

Herbs & Forbs

Arctic Lupine	*Lupinus arcticus*	Prairie Sagewort	*Artemisia frigida*
Alfalfa	*Medicago sativa*	Russian Thistle	*Salsola kali*
Arrow-leaved Balsamroot		Silky Lupine	*Lupinus sericeus*
	Balsamorhriza sagittata	Silverleaf Phacelia	*Phacelia hastata*
Broad-leaved Willowherb		Sitka Burnet	*Sanguisorba canadensis*
	Epilobium latifolium	Sitka Valerian	*Valeriana sitchensis*
Bull Thistle	*Cirsium vulgare*	Small-flowered Blue-eyed Mary	
Common Dandelion			*Collinisia parviflora*
	Taraxacum officinale	Small-flowered Wood-rush	
Cow-parsnip	*Heracleum maximum*		*Luzula parviflora*
Fan-leaved Cinquefoil		Subalpine Daisy	*Erigeron peregrinus*
	Potentilla flabellifolia	Western Spring-beauty	
Fireweed	*Epilobium angustifolium*		*Claytonia lanceolata*
Great Mullein	*Verbascum thapsus*	Wild Strawberry	*Frageria virginiana*
Heart-leaved Arnica		Woolly Eriophyllum	
	Arnica cordifolia		*Eriophyllum lanatum*
Indian Hellebore	*Veratrum viride*	Yarrow	*Achillea millefolium*
Leafy Aster	*Aster foliaceus*	Yellow Glacier-lily	
Pearly Everlasting			*Erythronium grandiflorum*
	Anaphalis margaritacea		

Grasses, Ferns & Lichens

Alpine Timothy	*Phleum alpinum*	Lady Fern	*Athyrium filix-femina*
Alkali Saltgrass	*Distichlis spicata*	Methuselah's Beard	*Usnea longissima*
Bluebunch Wheatgrass		Needle-and-thread Grass	
	Agropyron spicatum		*Stipa comata*
Bracken Fern	*Pteridium aquilinum*	Old Man's Beard	*Usnea* sp.
Cheatgrass	*Bromus tectorum*	Sword Fern	*Polystichum munitum*
Creeping Bentgrass	*Agrostis stolonifera*	Timber Oat-grass	
Crested Wheatgrass			*Danthonia intermedia*
	Agropyron cristatum	Tufted Hairgrass	
Edible Horsehair	*Bryoria fremonti*		*Deschampsia cespitosa*

Birds

Ancient Murrelet	
	Synthliboramphus antiquus
Bald Eagle	*Haliaeetus leucocephalus*
Barn Owl	*Tyto alba*
Barred Owl	*Strix varia*
Burrowing Owl	*Athene cunicularia*
Cassin's Auklet	
	Ptychoramphus aleuticus
Cooper's Hawk	*Accipiter cooperi*
Golden Eagle	*Aquila chrysaetos*
Great Gray Owl	*Strix nebulosa*
Ferruginous Hawk	*Buteo regalis*
Great Blue Heron	*Ardea herodias*
Great Horned Owl	*Bubo virginianus*
Northern Hawk Owl	*Surnia ulula*

Long-eared Owl	*Asio otus*
Northern Goshawk	*Accipiter gentilis*
Northern Harrier	*Circus cyaneus*
Northern Saw-whet Owl	
	Aegolius acadicus
Red-tailed Hawk	*Buteo jamaicensis*
Rhinoceros Auklet	
	Cerorhinca monocerata
Rough-legged Hawk	*Buteo lagopus*
Short-eared Owl	*Asio flammeus*
Snowy Owl	*Nyctea scandiaca*
Spotted Owl	*Strix occidentalis*
Swainson's Hawk	*Buteo swainsoni*
Western Screech-Owl	*Otus kennicottii*

Mammals

Badger	*Taxidea taxus*
Black Bear	*Ursus americanus*
Bobcat	*Lynx rufus*
Coast Mole	*Scapanus orarius*
Cougar	*Puma concolor*
Coyote	*Canis latrans*
Domestic Cat	*Felis sylvestris*
Domestic Dog	*Canis familiaris*
Ermine	*Mustela erminea*
Fisher	*Martes pennanti*
Grey Wolf	*Canis lupus*
Grizzly Bear	*Ursus arctos*

Least Weasel	*Mustela nivalis*
Long-tailed Weasel	*Mustela frenata*
Lynx	*Lynx canadensis*
Marten	*Martes americana*
Mink	*Mustela vison*
Raccoon	*Procyon lotor*
Red Fox	*Vulpes vulpes*
Shrew-mole	*Neurotrichus gibbsii*
Striped Skunk	*Mephitis mephitis*
Western Spotted Skunk	
	Spilogale putorius
Wolverine	*Gulo gulo*

Other Vertebrates

Dolly Varden	*Salvelinus malma*
Tiger Salamander	
	Ambystoma tigrinum
Western Rattlesnake	
	Crotalus oreganus

APPENDIX 2

Distributions of Lagomorphs and Rodents in British Columbia's Biogeoclimatic Zones and Ecoprovinces.

ZONES	Collared Pika *Ochotona collaris*	American Pika *Ochotona princeps*	Snowshoe Hare *Lepus americanus*	White-tailed Jackrabbit *Lepus townsendii*	European Rabbit *Oryctolagus cuniculus*	Eastern Cottontail *Sylvilagus floridanus*	Nuttall's Cottontail *Sylvilagus nuttallii*	TOTAL SPECIES
Alpine Tundra	•	•	•					3
Bunchgrass		•	•				•	3
Boreal White and Black Spruce	•		•					2
Coastal Douglas-fir			•		•	•		3
Coastal Western Hemlock		•	•		•	•		4
Engelmann Spruce – Subalpine Fir	•	•	•					3
Interior Cedar–Hemlock		•	•					2
Interior Douglas-fir		•	•				•	3
Mountain Hemlock		•	•					2
Montane Spruce		•	•					2
Ponderosa Pine		•	•	•			•	4
Spruce–Willow–Birch	•		•					2
Sub-Boreal Pine–Spruce		•	•					2
Sub-Boreal Spruce		•	•					2

Table 1.

Distribution of lagomorphs in British Columbia's biogeoclimatic zones (see colour map, inside back cover).

ECOPROVINCES	Collared Pika *Ochotona collaris*	American Pika *Ochotona princeps*	Snowshoe Hare *Lepus americanus*	White-tailed Jackrabbit *Lepus townsendii*	European Rabbit *Oryctolagus cuniculus*	Eastern Cottontail *Sylvilagus floridanus*	Nuttall's Cottontail *Sylvilagus nuttallii*	TOTAL SPECIES
Coast and Mountains		•	•		•	•		4
Georgia Depression		•	•	•	•	•		5
Southern Interior		•	•	•			•	4
Central Interior		•	•					2
Southern Interior Mountains		•	•					2
Sub-Boreal Interior			•					1
Northern Boreal Mountains	•		•					2
Boreal Plains			•					1
Taiga Plains			•					1

Table 2.

Distribution of lagomorphs in British Columbia's terrestrial ecoprovinces (see map, figure 17).

Table 3.

Distribution of rodents in British Columbia's biogeoclimatic zones (see colour map, inside back cover).

ZONES	Mountain Beaver *Aplodontia rufa*	Northern Flying Squirrel *Glaucomys sabrinus*	Hoary Marmot *Marmota caligata*	Yellow-bellied Marmot *Marmota flaviventris*	Woodchuck *Marmota monax*	Vancouver Island Marmot *Marmota vancouverensis*	Eastern Grey Squirrel *Sciurus carolinensis*	Eastern Fox Squirrel *Sciurus niger*	Columbian Ground Squirrel *Spermophilus columbianus*	Golden-mantled Ground Squirrel *Spermophilus lateralis*	Arctic Ground Squirrel *Spermophilus parryii*	Cascade Mantled Ground Squirrel *Spermophilus saturatus*	Yellow-pine Chipmunk *Tamias amoenus*
Alpine Tundra	•	•	•			•			•	•	•	•	•
Bunchgrass		•		•				•	•			•	•
Bor. White & Black Spruce		•	•		•					•	•		
Coastal Douglas-fir		•					•						•
Coastal Western Hemlock	•	•	•	•	•		•					•	•
Eng. Spruce – Subalp. Fir	•	•	•	•	•				•	•	•	•	•
Interior Cedar–Hemlock		•	•	•	•				•	•			•
Interior Douglas-fir		•		•	•				•	•		•	•
Mountain Hemlock		•	•			•						•	•
Montane Spruce		•		•	•				•	•		•	•
Ponderosa Pine		•		•					•	•			•
Spruce–Willow–Birch		•	•								•		
Sub-Boreal Pine–Spruce		•	•	•	•				•				•
Sub-Boreal Spruce		•	•		•				•	•			•

Species														
Least Chipmunk *Tamias minimus*	•	•		•	•	•	•			•	•	•	•	
Red-tailed Chipmunk *Tamias ruficaudus*			•	•	•	•	•			•	•	•	•	
Townsend's Chipmunk *Tamias townsendii*								•	•					
Douglas' Squirrel *Tamiasciurus douglasii*		•		•	•			•						
Red Squirrel *Tamiasciurus hudsonicus*	•	•	•	•	•	•	•	•	•	•	•	•	•	•
Northern Pocket Gopher *Thomomys talpoides*		•	•	•	•	•	•	•	•	•	•	•	•	•
Great Basin Pocket Mouse *Perognathus parvus*		•						•						
Beaver *Castor canadensis*	•	•	•	•	•	•	•	•	•	•	•	•	•	•
Southern Red-backed Vole *Clethrionomys gapperi*	•	•	•	•	•	•	•	•	•	•	•	•	•	•
Northern Red-backed Vole *Clethrionomys rutilus*	•							•					•	•
Brown Lemming *Lemmus trimucronatus*	•							•					•	•
Long-tailed vole *Microtus longicaudus*	•	•	•	•	•	•	•	•	•	•	•	•	•	•
Montane Vole *Microtus montanus*			•					•						
Tundra Vole *Microtus oeconomus*	•							•	•	•	•	•	•	•
Creeping Vole *Microtus oregoni*			•	•				•						
Meadow Vole *Microtus pennsylvanicus*	•	•	•	•	•	•	•	•	•	•	•	•	•	•
Water Vole *Microtus richardsoni*	•		•	•				•					•	•
Townsend's Vole *Microtus townsendii*			•	•				•						
Muskrat *Ondatra zibethicus*	•	•	•	•	•	•	•	•	•	•	•	•	•	•
Heather Vole *Phenacomys intermedius*	•	•	•	•	•	•	•	•	•	•	•	•	•	•
Northern Bog Lemming *Synaptomys borealis*	•	•	•	•	•	•	•	•	•	•	•	•	•	•

Table 3 (cont.).

ZONES	House Mouse *Mus musculus*	Norway Rat *Rattus norvegicus*	Black Rat *Rattus rattus*	Bushy-tailed Woodrat *Neotoma cinerea*	Deer Mouse *Peromyscus maniculatus*	Keen's Mouse *Peromyscus keeni*	Western Harvest Mouse *Reithrodontomys megalotis*	Meadow Jumping Mouse *Zapus hudsonius*	Western Jumping Mouse *Zapus princeps*	Pacific Jumping Mouse *Zapus trinotatus*	Porcupine *Erethizon dorsatum*	TOTAL SPECIES
Alpine Tundra				•	•	•			•	•		30
Bunchgrass	•			•	•		•	•	•		•	21
Bor. White & Black Spruce	•			•	•	•		•			•	22
Coastal Douglas-fir	•	•	•		•	•				•		18
Coastal Western Hemlock	•	•	•	•	•	•			•	•	•	30
Eng. Spruce – Subalp. Fir	•			•	•			•	•	•	•	34
Interior Cedar–Hemlock	•			•	•			•	•		•	23
Interior Douglas-fir	•			•	•		•	•	•		•	27
Mountain Hemlock				?	•	•			•	•	•	22
Montane Spruce	•			•	•			•			•	22
Ponderosa Pine	•			•	•		•		•		•	20
Spruce–Willow–Birch				•	•			•	•		•	20
Sub-Boreal Pine–Spruce				?	•			•	•		•	19
Sub-Boreal Spruce	•			•	•	•		•	•		•	24

Table 4.

Distribution of rodents in British Columbia's terrestrial ecoprovinces (see map, figure 17).

ECOPROVINCES	Mountain Beaver *Aplodontia rufa*	Northern Flying Squirrel *Glaucomys sabrinus*	Hoary Marmot *Marmota caligata*	Yellow-bellied Marmot *Marmota flaviventris*	Woodchuck *Marmota monax*	Vancouver Island Marmot *Marmota vancouverensis*	Eastern Grey Squirrel *Sciurus carolinensis*	Eastern Fox Squirrel *Sciurus niger*	Columbian Ground Squirrel *Spermophilus columbianus*	Golden-mantled Ground Squirrel *Spermophilus lateralis*	Arctic Ground Squirrel *Spermophilus parryii*	Cascade Mantled Ground Squirrel *Spermophilus saturatus*	Yellow-pine Chipmunk *Tamias amoenus*
Coast and Mountains	•	•	•	•	•	•	•					•	•
Georgia Depression	•	•				•	•						•
Southern Interior	•	•	•	•					•	•	•		•
Central Interior		•	•	•	•					•		•	•
S. Interior Mountains		•	•		•				•	•			
Sub-Boreal Interior		•	•	•	•					•			•
N. Boreal Mountains		•	•		•						•		
Boreal Plains		•	•		•								
Taiga Plains		•			•								

Table 4 (cont.).

ECOPROVINCES	Least Chipmunk *Tamias minimus*	Red-tailed Chipmunk *Tamias ruficaudus*	Townsend's Chipmunk *Tamias townsendii*	Douglas' Squirrel *Tamiasciurus douglasii*	Red Squirrel *Tamiasciurus hudsonicus*	Northern Pocket Gopher *Thomomys talpoides*	Great Basin Pocket Mouse *Perognathus parvus*	Beaver *Castor canadensis*	Southern Red-backed Vole *Clethrionomys gapperi*	Northern Red-backed Vole *Clethrionomys rutilus*	Brown Lemming *Lemmus trimucronatus*	Long-tailed vole *Microtus longicaudus*	Montane Vole *Microtus montanus*
Coast and Mountains			•	•	•	•		•	•	•	•	•	
Georgia Depression			•	•	•			•	•			•	
Southern Interior			•	•	•	•	•	•	•			•	•
Central Interior					•			•	•	•		•	•
S. Interior Mountains	•	•			•	•		•	•			•	
Sub-Boreal Interior	•				•			•	•	•	•	•	
N. Boreal Mountains	•				•			•	•	•	•	•	
Boreal Plains	•				•			•	•		•	•	
Taiga Plains	•				•			•	•		•		

Species									
Tundra Vole *Microtus oeconomus*	•	•	•			•			•
Creeping Vole *Microtus oregoni*			•						
Meadow Vole *Microtus pennsylvanicus*	•	•	•	•	•	•	•	•	•
Water Vole *Microtus richardsoni*	•	•	•	•	•	•	•	•	•
Townsend's Vole *Microtus townsendii*			•	•		•			•
Muskrat *Ondatra zibethicus*	•	•	•	•	•	•	•	•	•
Heather Vole *Phenacomys intermedius*	•	•	•	•	•	•	•	•	•
Northern Bog Lemming *Synaptomys borealis*	•	•	•	•	•	•	•	•	•
House Mouse *Mus musculus*		•		•	•	•	•	•	•
Norway Rat *Rattus norvegicus*							•	•	•
Black Rat *Rattus rattus*							•	•	•
Bushy-tailed Woodrat *Neotoma cinerea*	•	•	•	•	•	•	•	•	•
Deer Mouse *Peromyscus maniculatus*	•	•	•	•	•	•	•	•	•
Keen's Mouse *Peromyscus keeni*			•			•	•		•
Western Harvest Mouse *Reithrodontomys megalotis*							•		
Meadow Jumping Mouse *Zapus hudsonius*	•	•		•	•	•	•	•	•
Western Jumping Mouse *Zapus princeps*		•	•	•	•	•	•		•
Pacific Jumping Mouse *Zapus trinotatus*	•		•	•	•	•	•	•	•
Porcupine *Erethizon dorsatum*	•	•	•	•	•	•	•	•	•
TOTAL SPECIES	16	19	22	23	24	25	31	22	35

GLOSSARY

Annulations Scaly rings on the tail.

Arboreal Capable of climbing trees.

Auditory bulla A bony capsule that covers the middle and inner ear.

Baculum The male genital bone (or penis bone); an important identification trait in chipmunks (figure 116).

Basilar length A measurement of skull length taken from the mid-ventral border of the foramen magnum to the anterior border of the alveoli of the median incisor (figure 51).

Basioccipital The ventral portion of the bone surrounding the foramen magnum.

Baubellum The female genital bone (or clitoris bone); an important identification trait in chipmunks (figures 117–19).

Biogeoclimatic zone An area with a relatively homogeneous climate and characteristic vegetation. British Columbia has 14 biogeoclimatic zones (see the colour fold-out map at the back of the book).

Body length The length of an animal excluding the tail; the total length minus the tail-vertebrae length (see figure 33).

Caecum A large blind pouch at the junction of the large and small intestines that acts as a fermentation tank.

Cheek teeth The premolars and molars.

Community An assemblage of organisms living together in a particular environment.

COSEWIC Committee on the Status of Endangered Wildlife in Canada, composed of national, provincial, territorial and scientific experts who designate species at risk in Canada.

Cusp A high peak or rounded area on the crown of a tooth.

Dentine A hard ivory-like substance beneath the tooth enamel. In some rodents, such as voles, the dentine is exposed on the crown or grinding surface of the molar teeth (figure 4).

DNA Deoxyribonucleic acid, the chemical molecule that carries the genetic code. Comparing the sequences of bases in a section of DNA is a powerful tool for revealing relationships among organisms.

Dorsal On the back (dorsum) or upper surface.

Ear length Length of the ear measured from the ear notch to the tip (figure 33).

Ecoprovince   A broad geographic area with consistent climate and terrain. British Columbia has nine terrestrial ecoprovinces (figure 17).

Enamel   The hard outer layer of the crown of a tooth (figure 4), usually white but may be yellow or brown in some rodents.

Fenestrations   Small openings or perforations (figure 55).

Feral   A wild animal that descended from a captive or domestic population.

Fossorial   Adapted to digging or burrowing and to living underground (e.g., moles and pocket gophers).

Foramen magnum   The large opening at the rear of the skull through which the spinal cord passes.

Forb   A broad-leaved herb, usually associated with meadows.

Gestation period   The length of pregnancy, from fertilization to birth.

Graminoid   A grass, sedge or other grass-like plant.

Hibernation   A state of lethargy characterized by a reduction in body temperature and metabolic rate.

Hind foot length   The length of the hind foot measured from the edge of the heel to the end of the longest claw (figure 33).

Home range   The area covered by an animal during its normal day-to-day activities.

Incisors   The front teeth in a mammals jaws (figures 1 and 2).

Incisive foramen   A pair of openings that pierce the palate behind the incisors (figures 1, 2 and 66).

Infraorbital opening   An opening that passes through the side of the rostrum to the front face of the orbit of the skull (figures 60–63).

Interorbital region   The dorsal area of the skull that lies between the orbits.

Interparietal bone   An unpaired bone at the back of the braincase (figure 57).

Irruption   A sudden, temporary increase in a local population.

IUCN   International Union for the Conservation of Nature, a group that publishes the Red Book of endangered species.

Mandibular   Of the lower jaw.

Mantle   The shoulders and back of the neck.

Maxilla   The bone in the upper jaw that supports the canine, premolar and molar teeth.

Maxillary toothrow length   The length of the upper toothrow (usually excluding incisors) taken at the alveoli (figures 51, 58 and 66).

Melanism (adjective: melanic)   A mutation where an animal's fur colour is black.

Molars   The rear teeth in a mammal's jaws (figures 1 and 2).

Monogamous   Having a single mate.

Pelage   The fur or hairs of a mammal.

Polygamous   Having more than one mate.

Postorbital process   Bony projection behind the orbit of the skull (figures 105 and 106).

Premolars   The teeth just forward of the molars (figures 1 and 2).

Re-entrant angle   Inward pointing angle along the margin of vole molars (figure 59).

Rostrum   The nasal area or snout of a skull.

Salient angle   Outward pointing angle along the margin of vole molars (figure 59).

Skull length   The length of the skull measured from posterior border of the occipital foramen to the anterior border of the rostrum (figure 58)

Subspecies   A population of a species that is geographically separated and taxonomically different from other populations; sometimes called geographic races.

Supraorbital process   A bony projection above the orbit of the skull; well developed in lagomorphs (figures 56 and 57).

Tail vertebrae length   The length of the tail measured from the base to last vertebra (figure 33).

Talus   A mass or sloping heap of rock debris at the base of a cliff.

Temporal ridge   A prominent ridge traversing the side or top of the braincase (figures 96 and 97).

Territory   A portion of an animal's home range that is actively defended to exclude other individuals.

Torpor   A short-term (daily) state of inactivity achieved by lowering the body temperature and reducing the metabolic rate in order to conserve energy.

Truffles   Fungi that produce fruiting bodies (sporocarps) under the ground.

Total length   The length of an animal from tip of the nose to last tail vertebra (figure 33).

Ventral   On the under or lower surface; underside.

Vibrissae   Long tactile hairs on the face; whiskers.

Voucher specimen   A whole animal or parts of it preserved to verify the identification of other specimens.

Witches' broom   Clumps of abnormal branch growth resulting from a fungal infection in the tree or shrub.

Wean   Accustom an infant mammal to finish nursing on mother's milk and to eat solid food.

Zygomatic arch   An arch of bone that extends across the orbit of the eye in the skull.

Zygomatic plate   A thin plate extending from the zygomatic arch to the rostrum (figure 70, 71).

Zygomatic process of maxilla   A projection of the maxilla bone that forms the anterior part of the zygomatic arch.

REFERENCES

General Books

Banfield, A.W.F. 1974. *The Mammals of Canada.* Toronto: University of Toronto Press.

Cowan, I.M., and C.J. Guiguet. 1965. *The Mammals of British Columbia.* Handbook 11. Victoria: British Columbia Provincial Museum.

Eder, T. and D. Pattie. 2001. *Mammals of British Columbia.* Edmonton: Lone Pine.

Maser, C. 1998. *Mammals of the Pacific Northwest.* Corvallis: Oregon Sate University Press.

Nowak, R.M. 1999. *Walker's Mammals of the World.* Sixth edition, 2 volumes. Baltimore: The John Hopkins University Press.

Verts, B.J., and L.N. Carraway. 1998. *Land Mammals of Oregon.* Berkeley: University of California Press.

Wilson, D.E., and S. Ruff. 1999. *The Smithsonian Book of North American Mammals.* Washington: Smithsonian Institution Press.

Selected References

Allard, M., and I. Greenbaum. 1988. Morphological variation and taxonomy of chromosomally differentiated *Peromyscus* from the Pacific Northwest. *Canadian Journal of Zoology* 66:2734-39.

Anderson, P.K., P.H. Whitney and J.P. Huang. 1976. *Arvicola richardsoni*: ecology and biochemical polymorphism in the front ranges of southern Alberta. *Acta Theriologica* 21:425-68.

Anderson, R.M. 1933. Five new mammals from British Columbia. In *Annual Report for 1931.* Ottawa: National Museum of Canada.

Anderson, S., and J.K.J. Jones. 1984. *Orders and Families of Recent Mammals of the World.* New York: John Wiley and Sons.

Arbogast, B.S. 1999. Mitochondrial DNA phylogeography of the New World flying squirrels (*Glaucomys*): implications for Pleistocene biogeography. *Journal of Mammalogy* 80:142-55.

Arbogast, B. S., R. A. Browne, and P. D. Weigl. 2001. Evolutionary genetics and Pleistocene biogeography of North American tree squirrels (*Tamiasciurus*). Journal of Mammalogy 82:302-319.

Armitage, K.B. 1991. Social and population dynamics of Yellow-bellied Marmots: results of long-term research. *Annual Review of Ecology and Systematics* 22:379-407.

Baker, R.J., L.C. Bradley, R.D. Bradley, R.D. Bradley, J.W. Dragoo, M.D. Engstrom, R.S. Hoffmann, C.A. Jones, F. Reid, D.W. Rice and others. 2003. *Revised Checklist of North American Mammals North of Mexico, 2003.* The Museum, Occasional Paper 229. Lubbock: Texas Tech University.

Banks, E.M., R.J. Brooks and J. Schnell. 1975. A radiotracking study of home range and activity of the Brown Lemming (*Lemmus trimucronatus*). *Journal of Mammalogy* 56:888-901.

Bartels, M.A., and D.P. Thompson. 1993. *Spermophilus lateralis. Mammalian Species* 440.

Batzli, G.O., and H. Henttonen. 1990. Demography and resource use by microtine rodents near Toolik Lake, Alaska. *Arctic and Alpine Research* 22:51-64.

Bee, J.W., and E.R. Hall. 1956. *Mammals of Northern Alaska.* Miscellaneous Publication 8. Lawrence: Museum of Natural History, University of Kansas.

Beg, M.A. 1969. Habitats, food habits, and population dynamics of the Red-tailed Chipmunk, *Eutamias ruficaudus.* PhD thesis, University of Montana.

———. 1971. Reproductive cycle and reproduction in Red-tailed Chipmunks, *Eutamias ruficaudus. Pakistan Journal of Zoology* 3:1-13.

Berry, R.J. 1981. Biology of the House Mouse. *Symposium of the Zoological Society of London* 47.

Bertram, D.F. 1995. The roles of introduced rats and commercial fishing in the decline of Ancient Murrelets on Langara Island, British Columbia. *Conservation Biology* 9:865-72.

Bertram, D.F., and D.W. Nagorsen. 1995. Introduced rats *Rattus* sp. on the Queen Charlotte Islands: implications for seabird conservation. *The Canadian Field-Naturalist* 109:6-10.

Best, T.L. 1993. *Tamias ruficaudus. Mammalian Species* 452.

Bihr, K.J., and R. J. Smith. 1998. Location, structure, and contents of burrows of *Spermophilus lateralis* and *Tamias minimus*, two ground dwelling sciurids. *The Southwestern Naturalist* 43:352-62.

Boonstra, R., and J.A. Hoyle. 1986. Rarity and coexistence of a small hibernator, *Zapus hudsonius*, with fluctuating populations of *Microtus pennsylvanicus* in the grasslands of southern Ontario. *Journal of Animal Ecology* 55:773-84.

Broadbrooks, H.E. 1958. *Life History and Ecology of the Chipmunk,* Eutamias amoenus, *in Eastern Washington.* Miscellaneous Publication 103. Ann Arbor: Museum of Zoology, University of Michigan.

———. 1965. Ecology and distribution of the pikas of Washington and Alaska. *The American Midland Naturalist* 73:299-335.

———. 1974. The nests of chipmunks with comments on associated behaviour and ecology. *Journal of Mammalogy* 55:630-39.

Bronson, M.T. 1979. Altitudinal variation in the life history of the Golden-mantled Ground Squirrel (*Spermophilus lateralis*). *Ecology* 60:272-79.

Brooks, A. 1947. The Brown Rat, *Rattus norvegicus*, in British Columbia. *The Canadian Field-Naturalist* 61:68.

Bryant, A.A. 1996. Reproduction and persistence of Vancouver Island Marmots (*Marmota vancouverensis*) in natural logged habitats. *Canadian Journal of Zoology* 74:678-87.

———. 1997. *Updated Status Report on Species at Risk in Canada: Vancouver Island Marmot,* Marmota vancouverensis. Ottawa: Committee on the Status of Endangered Wildlife in Canada.

Bryant, A.A., and D.W. Janz. 1996. Distribution and abundance of Vancouver Island Marmots (*Marmota vancouverensis*). *Canadian Journal of Zoology* 74:667-77.

Byrom, A., and C.J. Krebs. 1999. Natal dispersal of juvenile Arctic Ground Squirrels in the boreal forest. *Canadian Journal of Zoology* 77:1048-59.

Campbell, R.W. 1968. Alexandrian rat predation on Ancient Murrelet eggs. *The Murrelet* 49:38.

Campbell, R.W., D.A. Manuwal and A.S. Harestad. 1987. Food habits of the Common Barn Owl in British Columbia. *Canadian Journal of Zoology* 65:578-86.

Cannings, S.G., L.R. Ramsay, D.F. Fraser and M.A. Fraker. 1999. *Rare Amphibians, Reptiles and Mammals of British Columbia*. Victoria: Wildlife Branch, B.C. Ministry of Environment, Lands and Parks.

Carey, A.B., T.M. Wilson, C.C. Maguire and B.L. Biswell. 1997. Dens of Northern Flying Squirrels in the Pacific Northwest. *Journal of Wildlife Management* 61:684-99.

Carl, C.G., and C.J. Guiguet. 1972. *Alien Animals of British Columbia*. Handbook 14. Victoria: British Columbia Provincial Museum.

Carl, E.A. 1971. Population control in Arctic Ground Squirrels. *Ecology* 52:395-413.

Carraway, L.N., and B.J. Verts. 1985. *Microtus oregoni. Mammalian Species* 233.

———. 1993. *Aplodontia rufa. Mammalian Species* 431.

Carter, D., A. Harestad and F.L. Bunnell. 1993. Status of Nuttall's Cottontail in British Columbia. *B.C. Environment Wildlife Working Report* WR-56:26.

Chapman, J.A. 1975. *Sylvilagus nuttallii. Mammalian Species* 56.

Chapman, J.A., J.G. Hockman and M.M. Ojeda. 1980. *Sylvilagus floridanus. Mammalian Species* 136.

Chivers, D.J., and P. Langer. 1994. *The Digestive System in Mammals: Food, Form and Function*. New York: Cambridge University Press.

Conroy, C., and J. Cook. 2000. Phylogeography of a post-glacial colonizer: *Microtus longicaudus* (Rodentia: Muridae). *Molecular Ecology* 9:165-75.

Cornely, J.E., and B.J. Verts. 1988. *Microtus townsendii. Mammalian Species* 325.

Cotton, C.L., and K.L. Parker. 2000a. Winter activity patterns of Northern Flying Squirrels in sub-boreal forests. *Canadian Journal of Zoology* 78:1896-1901.

———. 2000b. Winter habitat and nest trees used by Northern Flying Squirrels in subboreal forest. *Journal of Mammalogy* 81:1071-86.

Couch, L. 1930. Notes on the pallid Yellow-bellied Marmot. *The Murrelet* 11:3-6.

Cowan, I.M. 1933. The British Columbia Woodchuck *Marmota monax petrensis. The Canadian Field-Naturalist* 47:57.

———. 1936. Nesting habits of the flying squirrel, *Glaucomys sabrinus. Journal of Mammalogy* 17:58-60.

———. 1946. Notes on the distribution of the chipmunks *Eutamias* in southern British Columbia and the Rocky Mountain region of southern Alberta with descriptions of two new races. *Proceedings of the Biological Society of Washington* 59:107-18.

———. 1954. The distribution of the pikas (*Ochotona*) in British Columbia and Alberta. *The Murrelet* 35:20-24.

Cowan, I.M., and M.G. Arsenault. 1954. Reproduction and growth in the Creeping Vole, *Microtus oregoni serpens* Merriam. *Canadian Journal of Zoology* 32:198-208.

Cowan, I.M., and J. Hatter. 1940. Two mammals new to the known fauna of British Columbia. *The Murrelet* 21:9.

Cowan, I.M., and J.A. Munro. 1946. Birds and Mammals of Revelstoke National Park. *Canadian Alpine Journal* 29:100-121.

Cranford, J.A. 1978. Hibernation in the Western Jumping Mouse (*Zapus princeps*). *Journal of Mammalogy* 59:496-509.

Cranford, J.A. 1983. Ecological strategies of a small hibernator, the Western Jumping Mouse *Zapus princeps*. *Canadian Journal of Zoology* 61:232-40.

Criddle, S. 1943. The little northern chipmunk in southern Manitoba. *The Canadian Field-Naturalist* 57:81-86.

Dearing, M.D. 1997. The function of haypiles of pikas (*Ochotona princeps*). *Journal of Mammalogy* 78:1156-63.

Demarchi, D.A., R.D. Marsh, A.P. Harcombe and E.C. Lea. 1990. The environment. In *The Birds of British Columbia* by R.W. Campbell, N.K. Dawe, I.M. Cowan, J.M. Cooper, G.W. Kaiser, M.C.E. McNall. Vancouver: University of British Columbia Press.

Demboski, B.J., and J. Sullivan. 2002. Extensive mtDNA variation within the Yellow-pine Chipmunk, *Tamias amoenus* (Rodentia: Sciuridae), and phylogeographic inferences for northwest North America. *Molecular Phylogenetics and Evolution* (in press).

Diersing, V.E. 1978. A systematic revision of several species of cottontails (*Sylvilagus*). PhD thesis, University of Illinois at Urbana-Champaign.

Drever, M.C. 1997. Ecology and eradication of Norway Rats on Langara Island, Queen Charlotte Islands. MScthesis, Simon Fraser, Burnaby.

Drever, M.C., L.K. Blight and D.F. Bertram. 2000. Predation on seabird eggs by Keen's Mice (*Peromyscus keeni*): using stable isotopes to decipher the diet of a terrestrial omnivore on a remote offshore island. *Canadian Journal of Zoology* 78:2010-18.

Drever, M.C., and A.S. Harestad. 1998. Diets of Norway Rats, *Rattus norvegicus*, on Langara Island, Queen Charlotte Islands, British Columbia: implications for conservation of breeding seabirds. *The Canadian Field-Naturalist* 112:676-83.

Driver, J.C. 1988. Late Pleistocene and Holocene vertebrates and paleoenvironments from Charlie Lake Cave, northeast British Columbia. *Canadian Journal of Earth Sciences* 25:1545-53.

———. 1998. Late Pleistocene Collared Lemmings (*Dicrostonyx torquatus*) from northeastern British Columbia. *Journal of Vertebrate Paleontology* 18:816-18.

Edwards, R.Y. 1955. The habitat preferences of the boreal *Phenacomys*. *The Murrelet* 36:35-38.

Eisenberg, J.F., and D.G. Kleiman. 1977. Communication in lagomorphs and rodents. In *How Animals Communicate*, edited by T.A. Sebeok. Bloomington: Indiana University Press.

———. 1983. *Advances in the Study of Mammalian Behavior*. Lawrence, Kansas: American Society of Mammalogists.

Elliott, C.L., and J.T. Flinders. 1991. *Spermophilus columbianus*. *Mammalian Species* 372.

Erb, J., N.C. Stenseth, and M.S. Boyce. 2000. Geographic variation in population cycles of Canadian Muskrats (*Ondatra zibethicus*). *Canadian Journal of Zoology* 78:1009-16.

Escherich, P.C. 1981. Social Biology of the Bushy-tailed Woodrat, *Neotoma cinerea*. *University of California Publications in Zoology* 110.

Ferrell, C.S. 1995. Systematics and biogeography of the Great Basin Pocket Mouse, *Perognathus parvus*. MSc thesis, University of Nevada, Las Vegas.

Ferron, J. 1985. Social behaviour of the Golden-mantled Ground Squirrel (*Spermophilus lateralis*). *Canadian Journal of Zoology* 63:2529-33.

———. 1997. How do Woodchucks (*Marmota monax*) cope with harsh winter conditions. *Journal of Mammalogy* 77:412-16.

Festa-Bianchet, M., and D.A. Boag. 1982. Territoriality in adult Columbian Ground Squirrels. *Canadian Journal of Zoology* 60:1060-66.

Finley, R.B.J. 1958. The wood rats of Colorado: distribution and ecology. University of Kansas Publications, Museum of Natural History 10:213-552.

Foster J.B. 1961. Life history of the *Phenacomys* vole. *Journal of Mammalogy* 42:181-98.

Fraker, M.A., B.A. Sinclair, D. Joly and M.V. Ketcheson. 1997. Distribution and Habitat Associations of Northern Pocket Gophers (*Thomomys talpoides*) in the Eastern Creston Valley, B.C. Unpublished report for Terramar Environmental Research, Sidney, B.C.

Frase, B.A., and R.S. Hoffman. 1980. *Marmota flaviventris. Mammalian Species* 135.

Fuller, W.A. 1969. Changes in numbers of three species of small rodent near Great Slave Lake, N.W.T. Canada, 1964-1967, and their significance for general population theory. *Annales Zoologici Fennici* 6:113-44.

Gannon, W.L. 1988. *Zapus trinotatus. Mammalian Species* 315.

Gilbert, B.S., and C.J. Krebs. 1991. Population dynamics of *Clethrionomys* and *Peromyscus* in southwestern Yukon 1973-1989. *Holarctic Ecology* 14:250-59.

Glass, B.P. 1951. *A Key to the Skulls of North American Mammals*. Stillwater, Oklahoma: Bryan P. Glass.

Gonzales, E.K. 2000. Distinguishing between modes of dispersal by introduced Eastern Grey Squirrels (*Sciurus carolinensis*). MSc thesis, University of Guelph.

Good, J.M., J. Demboski, D. Nagorsen and J. Sullivan. 2003. Phylogeography and introgressive hybridization: chipmunks (Genus *Tamias*) in the northern Rocky Mountains. *Evolution* 57:1900-1916.

Good, J.M., and J. Sullivan. 2001. Phylogeography of the Red-tailed Chipmunk (*Tamias ruficaudus*), a northern Rocky Mountain endemic. *Molecular Ecology* 10:2683-95.

Gray, D.R. 1967. The marmots of Spotted Nellie Ridge. B.Sc. thesis, University of Victoria.

Green, J.E. 1977. Population regulation and annual cycles of activity and dispersal in the Arctic Ground Squirrel. MSc thesis, University of British Columbia, Vancouver.

Grizzell, R.A. 1955. A study of the Southern Woodchuck, *Marmota monax monax. The American Midland Naturalist* 53:257-93.

Guiguet, C.J. 1952. Status of birds and mammals of the Osoyoos area in May, 1951. *Report of the British Columbia Provincial Museum 1951*, B25-B38.

———. 1953. *An Ecological Study of Goose Island, British Columbia, with Special Reference to Mammals and Birds*. Occasional Paper 10. Victoria: British Columbia Provincial Museum.

———. 1975. An introduction of the Grey Squirrel, *Sciurus carolinensis* (Gmelin), to Vancouver Island, British Columbia. *Syesis* 8:399.

Gyug, L.W. 2000. Status, distribution, and biology of the Mountain Beaver, *Aplodontia rufa*, in Canada. *The Canadian Field-Naturalist* 114:476-90.

Hafner, D.J., and R.M. Sullivan. 1995. Historical and ecological biogeography of Nearctic pikas (Lagomorpha: Ochotonidae). *Journal of Mammalogy* 76:302-21.

Hafner, D.J., E. Yensen and G.L.J. Kirkland, editors. 1998. *North American Rodents. Status Survey and Conservation Action Plan*. IUCN/SSC Rodent Specialist Group. Gland, Switzerland: IUCN.

Hall, E.R. 1927. An outbreak of house mice in Kern County, California. *University of California Publication in Zoology* 30:189-203.

Hanley, T.A., and J.C. Barnard. 1999. Spatial variation in population dynamics of Sitka mice in floodplain forests. *Journal of Mammalogy* 80:866-79.

Harestad, A.S. 1986. Food habits of Columbian Ground Squirrels: a comparison of stomach and fecal samples. *The Murrelet* 67:75-78.

Hatler, D.F. 2002. Beaver colony dynamics in the upper Nechako River watershed, British Columbia, 1989-2001. Unpublished report for Alcan Primary Metal, Kitimat, British Columbia.

Hawes, D.B. 1975. Experimental studies of competition among four species of voles. PhD thesis. University of British Columbia, Vancouver.

Heard, D.C. 1977. The behaviour of Vancouver Island Marmots, *Marmota vancouverensis*. MSc thesis, University of British Columbia, Vancouver.

Hebda, R.J., C.G. Warner and R.A. Cannings. 1990. Pollen, plant macrofossils, and insects from fossil woodrat (*Neotoma cinerea*) middens in British Columbia. *Geographie Physique et Quaternaire* 44:227-34.

Herman, T.B. 1979. Population ecology of insular *Peromyscus maniculatus*. PhD thesis, University of Alberta, Edmonton.

Hildebrand, M., D.M. Bramble, K.F. Liem and D.B. Wake. 1985. *Functional Vertebrate Morphology*. Cambridge, Mass.: Harvard University Press.

Hill, E.P. 1982. Beaver. In *Wild Mammals of North America: Biology, Management and Economics*, edited by J.A. Chapman and G.A. Feldhamer. Baltimore: John Hopkins University Press.

Hirakawa, H. 2001. Coprophagy in leporids and other mammalian herbivores. *Mammal Review* 31:61-80.

Hogan, K.M., M.C. Hedin, H.S. Koh, S.K. Davis and F. Greenbaum. 1993. Systematic and taxonomic implications of karyotypic, electrophoretic, and mitochondrial-DNA variation in *Peromyscus* from the Pacific Northwest. *Journal of Mammalogy* 74:819-31.

Holland, G.P. 1942. On the occurrence of *Reithrodontomys* in British Columbia. *The Murrelet* 23:60.

Holmes, W.G. 1979. Social behavior and foraging strategies of Hoary Marmots (*Marmota caligata*). PhD thesis, University of Washington, Seattle.

———. 1984. The ecological basis of monogamy in Alaskan Hoary Marmots. In *The Biology of the Ground-Dwelling Squirrels*, edited by J.O. Murie and G.R. Michener. Lincoln: University of Nebraska Press.

Innes D.G.L, and J.S. Millar. 1982. Life-history notes on the Heather Vole, *Phenacomys intermedius levis*, in the Canadian Rocky Mountains. *The Canadian Field-Naturalist* 96:307-11.

Iverson, S.L. 1967. Adaptations to arid environments in *Perognathus parvus* (Peale). PhD thesis, University of British Columbia, Vancouver.

Jannett, F.J.J. 1980. Social dynamics of the Montane Vole, *Microtus montanus*, as a paradigm. *The Biologist* 62(1-4):3-19.

Jenkins, S.H., and P.E. Busher. 1979. *Castor canadensis. Mammalian Species* 120.

Johns, D.W., and K.B. Armitage. 1979. Behavioural ecology of alpine Yellow-bellied Marmots. *Behavioural Ecology and Sociobiology* 5:133-57.

Johnstone, W.B. 1954. A revision of the pocket gopher *Thomomys talpoides* in British Columbia. *The Canadian Field-Naturalist* 68:155-64.

Jones, G.S. 1981. The systematics and biology of the genus *Zapus* (Mammalia, Rodentia, Zapodidae). PhD thesis, Indiana State University, Terre Haute.

Jones, G.S., J.O.J. Whitaker and C. Maser. 1978. Food habits of jumping mice (*Zapus trinotatus* and *Z. princeps*) in western North America. *Northwest Science* 52:57-60.

Keith, L.B. 1990. Dynamics of Snowshoe Hare populations. *Current Mammalogy* 2:119-95.

Kenagy, G.J., and B.M. Barnes. 1988. Seasonal reproductive patterns in four coexisting rodent species from the Cascade Mountains, Washington. *Journal of Mammalogy* 69:274-92.

King, J.A., editor. 1968. *Biology of* Peromyscus *(Rodentia)*. Special Publication 3. Lawrence, Kansas: American Society of Mammalogists.

Kirkland G.L., and J.M. Layne, editors. 1989. *Advances in the Study of* Peromyscus *(Rodentia)*. Lubbock: Texas Tech University Press.

Koprowski, J.L. 1994a. *Sciurus carolinensis*. *Mammalian Species* 480.

———. 1994b. *Sciurus niger*. *Mammalian Species*, 479.

Krebs, J.A. 1994. Porcupine populations and winter feeding damage in thinned and unthinned second-growth stands. MSc thesis, University of Alberta, Edmonton.

Krebs, C.J., S. Boutin, R. Boonstra, A.R.E. Sinclair, J.N.M. Smith., M.R.T. Dale, K. Martin and R. Turkington. 1995. Impact of food and predation on the Snowshoe Hare cycle. *Science* 269:111-1115.

Krebs, C.J., and I. Wingate. 1985. Population fluctuations in the small mammals of the Kluane Region, Yukon Territory. *The Canadian Field-Naturalist* 99:51-61.

Kwieckenski, G.G. 1998. *Marmota monax*. *Mammalian Species* 591.

Lambin, X., and C.J. Krebs. 1991. Spatial organization and mating system of *Microtus townsendii*. *Behavioral Ecology and Sociobiology* 28:353-64.

Lance, E.W., and J.A. Cook. 1998. Biogeography of Tundra Voles (*Microtus oeconomus*) of Beringia and the southern coast of Alaska. *Journal of Mammalogy* 79:53-65.

Leung, M.C. 1991. Status, range and habitat of the Cascade Mantled Ground Squirrel, *Spermophilus saturatus*, in British Columbia. MSc thesis, University of British Columbia, Vancouver.

Leung, M.C., and K.M. Cheng. 1997. The distribution of the Cascade Mantled Ground Squirrel, *Spermophilus saturatus*, in British Columbia. *The Canadian Field-Naturalist* 111:365-75.

Lim, B.K. 1987. *Lepus townsendii*. *Mammalian Species* 288.

Lindsay, S.L. 1982. Systematic relationship of parapatric tree squirrel species (*Tamiasciurus*) in the Pacific Northwest. *Canadian Journal of Zoology* 60:2149-56.

Ludwig, D.R. 1984a. *Microtus richardsoni*. *Mammalian Species* 223.

———. 1984b. *Microtus richardsoni* microhabitat and life history. In *Winter Ecology of Small Mammals*, edited by J.F. Merritt. Pittsburgh: Carnegie Museum of Natural History.

———. 1988. Reproduction and population dynamics of the Water Vole, *Microtus richardsoni*. *Journal of Mammalogy* 69:532-41.

Lyman, C.P., J.S. Willis, A. Malan and L.C.H. Wang. 1982. *Hibernation and Torpor in Mammals and Birds*. New York: Academic Press.

MacArthur, R.A., and M. Aleksiuk. 1979. Seasonal microenvironments of the Muskrat (*Ondatra zibethicus*) in a northern marsh. *Journal of Mammalogy* 60:146-54.

MacDonald, S.O., and C. Jones. 1987. *Ochotona collaris. Mammalian Species* 281.

Madison, D.M. 1980. An integrated view of the social biology of *Microtus pennsylvanicus. The Biologist* 80(1-4):20-33.

Martell, A.M., and R.J. Milko. 1986. Seasonal diets of Vancouver Island Marmots, *Marmota vancouverensis. The Canadian Field-Naturalist* 100:241-45.

Martin, P. 1971. Movements and activities of the Mountain Beaver (*Aplodontia rufa*). *Journal of Mammalogy* 52:717-23.

Maser, C., and R.M. Storm. 1970. *A Key to the Microtinae of the Pacific Northwest (Oregon, Washington, Idaho)*. Corvallis: Oregon Sate University Book Stores.

Maser, C., J.M. Trappe and R.A. Nussbaum. 1978. Fungal – small mammal inter-relationships with emphasis on Oregon coniferous forests. *Ecology* 59:799-809.

Maser, Z., C. Maser, and J.M. Trappe. 1985. Food habits of the Northern Flying Squirrel (*Glaucomys sabrinus*) in Oregon. *Canadian Journal of Zoology* 63:1084-88.

McAllister, J.A, and R.S. Hoffmann. 1988. *Phenacomys intermedius. Mammalian Species* 305.

McCabe, T.T. 1948. Beaver on the northern British Columbia islands. *The Canadian Field-Naturalist* 62:72-74.

McCabe, T.T., and I.M. Cowan. 1945. *Peromyscus maniculatus macrorhinus* and the problem of insularity. *Transactions of the Royal Canadian Institute* 25:117-216.

McKeever, S. 1964. The biology of the Golden-mantled Ground Squirrel, *Citellus lateralis. Ecological Monographs* 34:383-401.

McKenna, M.C., S.K. Bell, G.G. Simpson, R.H. Nichols, R.H. Tedford, K.F. Koopman, G.G. Musser, N.A. Neff, J. Shosani and D.M. McKenna. 1997. *Classification of Mammals Above the Species Level*. New York: Columbia University Press.

McLean, G., and A.J. Towns. 1981. Differences in weight changes and the annual cycle of male and female Arctic Ground Squirrels. *Arctic* 34:249-54.

McPhee, E.C. 1984. Ethological aspects of mutual exclusion in the parapatric species of *Clethrionomys gapperi* and *C. rutilus. Acta Zoologica Fennica* 172:71-73.

Meidinger, D., and J. Pojar. 1991. *Ecosystems of British Columbia*. Victoria: British Columbia Ministry of Forests.

Meier, P.T. 1992. Social organization of Woodchucks (*Marmota monax*). *Behavioral Ecology and Sociobiology* 31:393-400.

Melchior, H.R. 1971. Characteristics of Arctic Ground Squirrel alarm calls. *Oecologia*, 7:184-90.

Meredith, D.H. 1977. Interspecific agonism in two parapatric species of chipmunks, Eutamias. *Ecology* 58:423-30.

Merritt, J.F. 1981. *Clethrionomys gapperi. Mammalian Species* 146.

———. 1984. *Winter Ecology of Small Mammals*. Pittsburgh: Carnegie Museum of Natural History.

Merritt, J.F., and J.M. Merritt. 1978. Population ecology and energy relationships of *Clethrionomys gapperi* in a Colorado subalpine forest. *Journal of Mammalogy* 59:576-98.

Millar, J.S. 1970. The breeding season and reproductive cycle of the Western Red Squirrel. *Canadian Journal of Zoology* 48:471-73.

Moses, R.A., and J.S. Millar. 1992. Behavioural asymmetries and cohesive mother-offspring sociality in Bushy-tailed Woodrats. *Canadian Journal of Zoology* 70:597-604.

Munro, J.A. 1958. Distribution of the harvest mouse in British Columbia. *The Canadian Field-Naturalist* 72:146.

Murie, O.J. 1974. *A Field Guide to Animal Tracks*. Boston: Houghton Mifflin.

Murie, J.O., and M.A. Harris. 1978. Territoriality and dominance in male Columbian Ground Squirrels (*Spermophilus columbianus*). *Canadian Journal of Zoology* 56:2402-12.

Nagorsen, D.W. 1983. Winter pelage colour in Snowshoe Hares (*Lepus americanus*) from the Pacific Northwest. *Canadian Journal of Zoology* 61:2313-18.

———. 1987. Summer and winter food caches of the Heather Vole, *Phenacomys intermedius*, in Quetico Provincial Park, Ontario. *The Canadian Field-Naturalist* 101:82-85.

———. 1987. *Marmota vancouverensis*. *Mammalian Species* 270.

———. 1990. *The Mammals of British Columbia: A Taxonomic Catalogue*. Memoir 4. Victoria: Royal British Columbia Museum.

———. 1995. *Status of the Western Harvest Mouse in British Columbia*. Report WR-71. Victoria: British Columbia Ministry of Environment, Wildlife Branch.

———. 2002. *An Identification Manual to the Small Mammals of British Columbia*. Victoria: Ministry of Sustainable Resource Management; Ministry of Water, Land, and Air Protection; and Royal British Columbia Museum.

Nagorsen, D.W., N. Panter and M.A. Fraker. 2002. *Chipmunks of the Kootenay region, British Columbia: Distribution, Identification, Taxonomy, Conservation Status*. Nelson, B.C.: Columbia Basin Fish and Wildlife Compensation Program.

Negus, N.C., and P.J. Berger. 1998. Reproductive strategies of *Dicrostonyx groenlandicus* and *Lemmus sibiricus* in high-arctic tundra. *Canadian Journal of Zoology* 76:390-99.

Negus, N., P.J. Berger and L.G. Forslund. 1977. Reproductive strategy of *Microtus montanus*. *Journal of Mammalogy* 58:347-53.

Novak, M. 1987. Beaver. In *Wild Furbearer Management and Conservation in North America*, edited by M. Novak, J.A. Buke, M.E. Obbard and B. Mallard. Toronto: Ontario Ministry of Natural Resources.

O'Brien, J.P. 1988. Seasonal selection of coniferous trees by the Sewellel, *Aplodontia rufa*. *Mammalia* 52:325-30.

O'Farrell, T.P. 1965. Home range and ecology of Snowshoe Hares in interior Alaska. *Journal of Mammalogy* 46:406-18.

O'Farrell, T.P., R.J. Olson, R.O. Gilbert and J.D. Hedlund. 1975. A population of Great Basin Pocket Mice, *Perognathus parvus*, in the shrub-steppe of south central Washington. *Ecological Monographs* 45.

Patterson, B.D., and L.R. Heaney. 1987. Preliminary analysis of geographic variation in Red-tailed Chipmunks, *Eutamias ruficaudus*. *Journal of Mammalogy* 68:782-91.

Perry, H.R.J. 1982. Muskrats. In *Wild Mammals of North America: Biology, Management and Economics*, edited by J.A. Chapman and G.A. Feldhamer. Baltimore: John Hopkins University Press.

Pease, J.L., R.H. Vowles and L.B. Keith. 1979. Interaction of Snowshoe Hares and woody vegetation. *Journal of Wildlife Management* 43(1):43-60.

Powers, R.A., and B.J. Verts. 1971. Reproduction in the Mountain Cottontail rabbit in Oregon. *Journal of Wildlife Management* 35:605-13.

Proulx, G., and F.F. Gilbert. 1983. The ecology of the Muskrat, *Ondatra zibethicus*, in Luther Marsh, Ontario. *The Canadian Field-Naturalist* 97:377-90.

Quimby, D.C. 1951. The life history and ecology of the jumping mouse, *Zapus hudsonius*. *Ecological Monographs* 21:61-95.

Racey, K. 1928. Notes on *Phenacomys intermedius*. *The Murrelet* 9:54-56.

———. 1953. Cottontail rabbit in British Columbia. *The Murrelet* 34:9-10.

———. 1960. Notes relative to the fluctuation in numbers of *Microtus richardsoni richardsoni* about Alta Lake and Pemberton Valley, B.C. *The Murrelet* 41:13-14.

Racey, K., and I.M. Cowan. 1935. Mammals of the Alta Lake region of southwestern British Columbia. *Report of the British Columbia Provincial Museum 1935*: H15-H29.

Rausch, R.L. 1962. Notes on the Collared Pika, *Ochotona collaris* (Nelson), in Alaska. *The Murrelet* 42:22-24.

Redfield, J.A. 1976. Distribution, abundance, size and genetic variation of *Peromyscus maniculatus* on the Gulf islands of British Columbia. *Canadian Journal of Zoology* 54:463-74.

Reich, L.M. 1981. *Microtus pennsylvanicus*. *Mammalian Species* 159.

Reichman, O.J., and S.C. Smith. 1990. Burrows and burrowing behavior by mammals. *Current Mammalogy* 2:197-244.

Reimer, J.D., and M.L. Petras. 1968. Some aspects of commensal populations of *Mus musculus* in southwestern Ontario. *The Canadian Field-Naturalist* 83:32-42.

Robinson, D.J., and I.M. Cowan. 1954. An introduced population of the Grey Squirrel (*Sciurus carolinensis* Gmelin) in British Columbia. *Canadian Journal of Zoology* 32:261-82.

Rogowitz, G.L. 1997. Locomotor and foraging activity of the White-tailed Jackrabbit (*Lepus townsendii*). *Journal of Mammalogy* 78:1172-81.

Rogowitz, G.L., and M.L. Wolfe. 1991. Intraspecific variation in life-history traits of the White-tailed Jackrabbit (*Lepus townsendii*). *Journal of Mammalogy* 72:796-806.

Roze, U. 1989. *The North American Porcupine*. Washington: Smithsonian Institution Press.

Scheffer, V.B. 1995. Mammals of the Olympic National Park and vicinity (1949). *Northwest Fauna* 2:1-133.

Schreiber, R.K. 1978. Bioenergetics of the Great Basin Pocket Mouse. *Acta Theriologica* 23:469-78.

Schwartz, O.A., K.B. Armitage and D. van Vuren. 1998. A 32-year demography of Yellow-bellied Marmots (*Marmota flaviventris*). *Journal of the Zoological Society of London* 246:337-46.

Seidel, D.R., and E.S. Booth. 1960. Biology and breeding habits of the Meadow Mouse, *Microtus montanus*, in eastern Washington. *Walla Walla College Publications of the Department of Biological Sciences and the Biological Station* 29:1-14.

Shaw, W.T. 1944. Brood nests and young of two western chipmunks in the Olympic Mountains of Washington. *Journal of Mammalogy* 25: 274-84.

Sheppard, D.H. 1965. Ecology of the chipmunks *Eutamias amoenus luteiventris* (Allen) and *E. minimus oreocetes* Merriam, with particular reference to competition. PhD thesis, University of Saskatchewan, Saskatoon.

Sheppard, D.H. 1971. Competition between two chipmunk species *Eutamias*. *Ecology* 52:320-29.

Sheppe, W.A. 1959. Notes on the distribution and habitats of mammals in the Pacific Northwest. *The Murrelet* 40:1-4.

———. 1960. Systematic relations of *Clethrionomys* in the Pacific Northwest. *The Canadian Field-Naturalist* 74:171-73.

———. 1961. Systematic and ecological relations of *Peromyscus oreas* and *P. maniculatus*. *Proceedings of the American Philosophical Society* 105:421-46.

———. 1963. Population structure of the Deer Mouse, *Peromyscus*, in the Pacific Northwest. *Journal of Mammalogy* 44:180-85.

Slough, B.G. 1978. Beaver food cache structure and utilization. *Journal of Wildlife Management* 42:644-46.

Slough, B.G., and R.M.F. Sadleir. 1977. A land capability classification system for Beaver (*Castor canadensis* Kuhl). *Canadian Journal of Zoology* 55:1324-35.

Smith, A.T. 1987. Population structure of pikas: dispersal versus philopatry. In *Mammalian Dispersal Patterns: The Effects of Social Structure on Population Genetics*, edited by B.D. Chepko-Sade and Z.T. Halpin. Chicago: University of Chicago Press.

Smith, A.T., and M.L. Weston. 1990. *Ochotona princeps. Mammalian Species* 352.

Smith, C.C. 1968. The adaptive nature of social organization in the genus of tree squirrels *Tamiasciurus*. *Ecological Monographs* 38:31-63.

———. 1981. The indivisible niche of *Tamiasciurus*: An example of non-partitioning of resources. *Ecological Monographs* 51:343-63.

Smith, F.A. 1997. *Neotoma cinerea. Mammalian Species* 564.

Smith, H., and E.J. Edmonds. 1985. The Brown Lemming, *Lemmus sibiricus*, in Alberta. *The Canadian Field-Naturalist* 99:99-100.

Smolen, M.J., and B.L. Keller. 1987. *Microtus longicaudus. Mammalian Species* 271.

Steele, M.A. 1998. *Tamiasciurus hudsonicus. Mammalian Species* 586.

———. 1999. *Tamiasciurus douglasii. Mammalian Species* 630.

Steppan, S.J., M.R. Akhverdyan, E.A. Lyapunova, D.G. Fraser, N.N. Voronstov, R.S. Hoffmann and M.J. Braun. 1999. Molecular phylogeny of the marmots (Rodentia: Sciuridae): tests of evolutionary and biogeographic hypotheses. *Systematic Biology* 48:715-34.

Stevens, W.F., and A.R. Weisbrod. 1981. The biology of the European Rabbit on San Juan Island, Washington, U.S.A. In *Proceedings of the World Lagomorph Conference Held in Guelph, Ontario, August 1979*, edited by K. Myers and C.D. MacInnes. Guelph, Ontario: University of Guelph.

Storer, T.I., and D.E. Davis. 1953. Studies on rat reproduction in San Francisco. *Journal of Mammalogy* 34:365-73.

Sullivan, T.P. 1977. Demography and dispersal in island and mainland populations of the Deer Mouse, *Peromyscus maniculatus*. *Ecology* 58:964-78.

———. 1979. Demography of populations of Deer Mice in coastal forest and clear-cut (logged) habitats. *Canadian Journal of Zoology* 57:1636-48.

———. 1990. Responses of Red Squirrel (*Tamiasciurus hudsonicus*) populations to supplemental food. *Journal of Mammalogy* 71:579-90.

Sullivan, T.P., A.S. Harestad and B.M. Wikeem. 1990. Control of mammal damage. In *Regenerating British Columbia's Forests*, edited by D.P. Lavender, R.P. Parish, C.M. Johnson, G. Montgomery, A. Vyse, R.A. Willis and R.A. Winston. Vancouver: University of British Columbia Press.

Sullivan, T.P., and E.J. Hogue. 1987. Influence of orchard floor management on vole and pocket gopher populations and damage in apple orchards. Journal of the American Society for Horticultural Science 112:972-977.

Sullivan, T.P., W.T. Jackson, J. Pojar and A. Banner. 1986. Impact of feeding damage by the Porcupine on Western Hemlock - Sitka Spruce forests of north-coastal British Columbia. *Canadian Journal of Forest Research* 16:642-47.

Sullivan, T.P., B. Jones and D.S. Sullivan. 1989. Population ecology and conservation of the Mountain Cottontail, *Sylvilagus nuttallii nuttallii*, in southern British Columbia. *The Canadian Field-Naturalist* 103:335-340.

Sullivan, T.P., and W. Klenner 2000. Response of Northwestern Chipmunks (*Tamias amoenus*) to variable habitat structure in young Lodgepole Pine forest. *Canadian Journal of Zoology* 78: 283-93.

Sullivan, T.P., and C. Krebs. 1981. Microtus population biology: demography of *M. oregoni* in southwestern British Columbia. *Canadian Journal of Zoology* 59:2092-2102.

Sullivan, T.P., R.A. Lautenschlager and R.J. Wagner. 1999. Clearcutting and burning of northern spruce-fir forests: implications for small mammal communities. *Journal of Applied Ecology* 36:327-44.

Sullivan, T.P., and W.L. Martin. 1991. Influence of site factors on incidence of vole and lemming feeding damage to forest plantations. *Western Journal of Applied Forestry* 6:64-67.

Sullivan, T.P., and R.A Moses. 1986. Red Squirrel populations in natural and managed stands of Lodgepole Pine. *Journal of Wildlife Management* 50:595-601.

Sullivan, T.P., and D.S. Sullivan. 1982a. Barking damage by Snowshoe Hares and Red Squirrels in Lodgepole Pine stands in central British Columbia. *Canadian Journal of Forest Research* 12:443-48.

———. 1982b. Population dynamics and regulation of the Douglas' Squirrel (*Tamiasciurus douglasii*) with supplemental food. *Oecologia* 53:264-70.

———. 1988. Influence of alternative foods on vole populations and damage in apple orchards. *Wildlife Society Bulletin* 16:170-75.

Sullivan, T.P., D.S. Sullivan and E.J. Hogue. 2001. Reinvasion dynamics of Northern Pocket Gopher (*Thomomys talpoides*) populations in removal areas. *Crop Protection* 20:189-98.

Sullivan, T.P., D.S. Sullivan and C.J. Krebs.1983. Demographic responses of a chipmunk, (*Eutamias townsendii*), population with supplemental food. *Animal Ecology* 52: 743-56.

Sullivan, T.P., D.S. Sullivan and P. Lindgren. 2000. Small mammals and stand structure in young pine, seed-tree, and old-growth forest, southwest British Columbia. *Ecological Applications* 10:1367-83.

Sustainable Resource Management, Ministry of. 2002. The Vertebrates of British Columbia: Scientific and English Names. Unpublished report: Standards for Components of British Columbia's Biodiversity 2. Resources Inventory Committee, Victoria.

Sutton, D.A. 1982. The female genital bone of chipmunks genus *Eutamias*. *The Southwestern Naturalist* 27:393-402.

———. 1992. *Tamias amoenus*. *Mammalian Species* 390.

———. 1993. *Tamias townsendii*. *Mammalian Species* 435.

Taitt, M.J. 1981. The effect of extra food on small rodent populations: I. Deermice (*Peromyscus maniculatus*). *Journal of Animal Ecology* 50:111-24.

Taitt, M., and C.J. Krebs. 1985. Population dynamics and cycles. In *Biology of the New World* Microtus, edited by R.T. Tamarin. Lawrence, Kansas: American Society of Mammalogists.

Taulman, J.T. 1977. Vocalizations of the Hoary Marmot, *Marmota caligata*. *Journal of Mammalogy*, 58:681-83.

Taylor, E.W. 1999. Abundances of small mammals in different successional stages of Western Hemlock forest on the Olympic Peninsula, Washington. *Northwestern Naturalist* 80:39-43.

Teipner, C.L., E.O. Garton and L.J. Nelson. 1983. *Pocket Gophers in Forest Ecosystems*. U.S. Department of Agriculture, Forest Service, Intermountain Forest and Range Experiment Station.

Thomas, B. 1971. Evolutionary relationships among *Peromyscus* from the Georgia Strait, Gordon, Goletas and Scott islands of British Columbia, Canada. PhD thesis, University of British Columbia, Vancouver.

Thomas, W.K., and A.T. Beckenbach. 1986. Mitochondrial DNA restriction site variation in the Townsend's Vole, *Microtus townsendii*. *Canadian Journal of Zoology* 64:2750-56.

Thompson, D.C. 1978. The social system of the Grey Squirrel. *Behaviour* 64:305-28.

Thompson, S.D. 1985. Subspecific differences in metabolism, thermoregulation, and torpor in the Western Harvest Mouse *Reithrodontomys megalotis*. *Physiological Zoology* 58:430-44.

Topping, M.G., and J.S. Millar. 1996. Home range size of Bushy-tailed Woodrats, *Neotoma cinerea*, in southwestern Alberta. *The Canadian Field-Naturalist* 110:351-53.

Trethewey, D.E., and B.J. Verts. 1971. Reproduction in Eastern Cottontail rabbits in western Oregon. *The American Midland Naturalist* 86:463-76.

Trombulak, S.C. 1987. Life history of the Cascade Golden-mantled Ground Squirrel (*Spermophilus saturatus*). *Journal of Mammalogy* 68:544-54.

———. 1988. *Spermophilus saturatus*. *Mammalian Species* 322.

Valle, R. 1962. Swimming ability and behavior on rafts as factors in the distribution of *Peromyscus maniculatus austerus*. MSc thesis, Wall Walla College, Wall Walla, Washington.

Vander Wall, S.B. 1990. *Food Hoarding in Animals*. Chicago: University of Chicago Press.

Van Horne, B. 1982. Demography of the Longtail Vole *Microtus longicaudus* in seral stages of coastal coniferous forest, southeast Alaska. *Canadian Journal of Zoology* 60:1690-1709.

Vaughan, T.A. 1974. Resource allocation in some sympatric, subalpine rodents. *Journal of Mammalogy* 55:764-95.

Verts, B.J., and L.N. Carraway. 1999. *Thomomys talpoides*. *Mammalian Species* 618.

———. 2001. *Tamias minimus*. *Mammalian Species* 653.

Verts, B.J., and G.L. Kirkland. 1988. *Perognathus parvus*. *Mammalian Species* 318.

Ward, B.C., M.C. Wilson, D.W. Nagorsen, D.E. Nelson, J.C. Driver and R.J. Wigen. 2003. Port Eliza cave: North American west coast interstadial environment and implications for human migrations. *Quaternary Science Reviews* 22:1383-88.

Webster, A.B., and R.J. Brooks. 1981. Daily movements and short activity of free-ranging Meadow Voles. *Oikos* 37:80-87.

Webster, W.D., and J.K.J. Jones. 1982. *Reithrodontomys megalotis. Mammalian Species* 167.

Wells-Gosling, N., and L.R. Heaney. 1984. *Glaucomys sabrinus. Mammalian Species* 229.

West, S.D. 1977. Midwinter aggregation in the Northern Red-backed Vole, *Clethrionomys rutilus. Canadian Journal of Zoology* 55:1404-09.

———. 1982. Dynamics of colonization and abundance in central Alaskan populations of the Northern Red-backed Vole, *Clethrionomys rutilus. Journal of Mammalogy* 63:128-43.

West, S.D., R.G. Ford and J.C. Zasada. 1980. Population response of the Northern Red-backed Vole (*Clethrionomys rutilus*) to differentially cut White Spruce forest. Report PNW-RN-362. U.S. Forest Service.

Weston, M.L. 1981. The *Ochotona alpina* complex: a statistical re-evaluation. In *Proceedings of the World Lagomorph Conference Held in Guelph, Ontario, August 1979*, edited by K. Myers and C.D. MacInnes. Guelph, Ontario: University of Guelph.

Whitaker, J.O., Jr. 1963. A study of the Meadow Jumping Mouse, *Zapus hudsonius* (Zimmermann), in central New York. *Ecological Monographs* 33:215-54.

———. 1972. *Zapus hudsonius. Mammalian Species* 11.

White, J.A. 1951. A practical method for mounting the bacula of small mammals. *Journal of Mammalogy* 32:125.

———. 1953. The baculum in the chipmunks of western North America. Museum of Natural History, University of Kansas Publications 5:611-31.

Whitney, P.H. 1976. Population ecology of two sympatric species of subarctic microtine rodents. *Ecological Monographs* 46:85-104.

Willner, G.R., G.A. Feldhamer E.E. Zucker and J.A. Chapman. 1980. *Ondatra zibethicus. Mammalian Species* 141.

Wilson, D.E., F.R. Cole, J.D. Nichols, R. Rudran and M.S. Foster. 1996. *Measuring and Monitoring Biological Diversity: Standard Methods for Mammals.* Washington: Smithsonian Institution Press.

Wilson, D.E., and D.M. Reeder. 1993. *Mammal Species of the World: A Taxonomic and Geographic Reference.* Washington: Smithsonian Institution Press.

Wolff, J.O. 1978. Food habits of Snowshoe Hares in interior Alaska. *Journal of Wildlife Management* 42:148-53.

Woods, C.A. 1973. *Erethizon dorsatum. Mammalian Species* 29.

Wright, G.M., and J.W. Weber. 1979. Range extension of the Fox Squirrel in southeastern Washington and into adjacent Idaho. *The Murrelet* 60:73-75.

Yensen, E., J.A. Cook and D.W. Nagorsen. 1998. Rodents of northwestern North America. In *North American Rodents: Status Survey and Conservation Action Plan*, edited by D.J. Hafner, E. Yensen and G.L.J. Kirkland. Gland, Switzerland: IUCN Publications.

Zimmerling, T.N., and C.D. Croft. 2001. Resource selection by Porcupines: winter den site location and forage tree choices. *Western Journal of Applied Forestry* 16:53-57.

ACKNOWLEDGEMENTS

I am indebted to many people who helped make this book possible. I especially thank Andrew Harcombe, Laura Friis and Gerry Truscott for their encouragement and support. Doug Burles, Mark Fraker, Carlos Galindo-Leal, Emily Gonzales, Les Gyug, John Krebs, Walt Klenner, Denis Knopp, Dave Lowe, Anna Roberts, Mike Sarell, Dale Seip, Thomas Sullivan, Doug Steventon, Todd Zimmerling and Gustavo Zuleta generously shared information, some unpublished, from their field research on small mammals in the province. Mike Badry provided historical fur-harvest data. Dave Hatler deserves special mention for his collections of small mammals made from various remote areas of the province over the years and providing unpublished body weights for some of the larger rodent species. Grant Keddie provided information on aboriginal use of these animals. Andrew Bryant contributed the striking cover photo. Yorke Edwards calculated coordinates for the many locality records, diligently tracked down many obscure historical museum records and extracted useful information from historical field notes. The Ministry of Sustainable Resource Management conducted a GIS analysis of species occurrences in the province's biogeoclimatic zones and ecoprovinces. The book benefited from a number of discussions I had with Nick Panter about rodent and lagomorph distributions and habitats in Alberta and British Columbia.

The illustrations were drawn by Bill Adams, Donald Gunn and Michael Hames, whose patience and helpful suggestions were much appreciated.

This book would not have been possible without the assistance of co-op students from the University of Victoria and Camosun College: Candice Ford did a bibliographic search on the 52 species; Katia Gauvin, Andrea Fayrashal, Danielle Flath, Robyne Lumley, Shannon Tong and Tammy Welyk entered records into a computer database and verified locality coordinates. Nancy Nagorsen assisted with data entry during my work at the American Museum of Natural History, Canadian Museum of Nature, Cowan Vertebrate Museum, Museum of Vertebrate Zoology at Berkeley, Royal Ontario Museum and the National Museum of Natural History.

The development of the identification keys, species descriptions, measurements and range maps were based largely on museum specimens. The following museums allowed me to examine in their collections or provided data for British Columbian specimens housed in their collections: American Museum of Natural History, New York; Burke Museum, University of

Washington, Seattle; California Academy of Sciences, San Francisco; Canadian Museum of Nature, Ottawa; Carnegie Museum of Natural History, Pittsburgh; Cowan Vertebrate Museum, University of British Columbia, Vancouver; Field Museum of Natural History, Chicago; James R. Slater Museum of Natural History, University of Puget Sound, Tacoma; Museum of Comparative Zoology, Harvard University; Museum of Natural History, University of Kansas, Lawrence; Museum of Vertebrate Zoology, Manitoba Museum of Man and Nature, Winnipeg; Museum of Vertebrate Zoology, University of California, Berkeley; Museum of Zoology, University of Michigan, Ann Arbor; National Museum of Natural History, Washington, D.C.; Philadelphia Academy of Sciences, Philadelphia; Royal Ontario Museum, Toronto; Texas Cooperative Wildlife Collection, Texas A&M University, College Station; University of Alberta, Edmonton; University of California, Los Angeles; University of Montana, Missoula.

I thank Michelle Goslin and David Campbell (Canadian Museum of Nature), Christine Adkins (Cowan Vertebrate Museum), Robert Fisher (National Museum of Natural History) and Susan Woodward (Royal Ontario Museum) for their help and hospitality during my visits to their museums. Photocopies of historical field notes were provided by the American Museum of Natural History, Canadian Museum of Nature, Museum of Vertebrate Zoology and the Smithsonian Archives.

About the Author

David Nagorsen is a biological consultant based in Victoria and an associate in the Centre for Biodiversity and Conservation at the Royal Ontario Museum. He has studied small mammals in British Columbia for many years and has published several books, including two other Royal BC Museum Handbooks.

Rodents and Lagomorphs of British Columbia

Drawings:
Bill Adams: figures 33-50, 52-57, 59-63, 67-78, 81-86, 88-100, 102-115 (© Province of British Columbia).
Donald Gunn: figures 1, 2, 4, 5, 7, 31, 32, 51, 64-66, 79, 80, 87, 101, 116, 117 and all skulls in Species Accounts (© Province of B.C.).
Michael Hames ©: figure 6 and all whole animals in Species Accounts.
David Nagorsen: all range maps in Species Accounts (© Royal BC Museum).
Royal BC Museum ©: figures 16, 23, 24 and 25.
David Nagorsen ©: figures 118 and 119.
B.C. Ministry of Forests, Forest Science Program ©: colour insert.
B.C. Ministry of Sustainable Resource Development ©: figure 17.
Photographs:
Andrew Bryant ©: front cover and figure 9.
Robert Cannings ©: back cover and figure 21.
David Nagorsen ©: figures 3, 10, 13, 14, 22 and 28.
Thomas P. Sullivan ©: 26, 27, 29 and 30.
Other figures: 8 – © R. Yorke Edwards; 11 – © Alison Beal; 12, 15 – © Karl Larsen; 18 – © Brent Ward; 19 – © Jonathan Driver; 20 – © B.C. Ministry of Sustainable Resource Development, Surveys and Resource Mapping Branch (photo 495:101).

Other books in the Mammals of British Columbia series:

Volume 1: *Bats of British Columbia*
    by David W. Nagorsen and R. Mark Brigham (1993)
Volume 2: *Opossums, Shrews and Moles of British Columbia*
    by David W. Nagorsen (1996)
Volume 3: *Hoofed Mammals of British Columbia*
    by David Shackleton (1999)

For a complete list of books in print by the Royal BC Museum, go to our website: www.royalbcmuseum.bc.ca
or write to: Publishing, Royal BC Museum, 675 Belleville Street, Victoria, British Columbia, Canada, V8W 9W2.

INDEX

Cottontail, Mountain 106
Nuttall's 9, 21, 30, 31, 39, 45, 55,
85-86, 92, **106-9**, 376, 377
Cougar 30, 94, 103, 108, 114, 121,
127, 134, 146, 236, 371, 375
Coyote 81, 89, 94, 103, 108, 114,
121, 127, 134, 139, 157, 163,
168, 178, 222, 255, 260, 271,
275, 286, 291, 297, 331, 337,
361, 371, 375
dandelion 139, 167, 178, 183, 197,
221, 374
Dicrostonyx sp. 19
Dipodidae Family 4, 37, 41, 352
Dipodomys ordii 22
Dog, Domestic 157, 375
Eagle, Bald 103, 146, 297, 375
Golden 81, 89, 103, 114, 127,
134, 146, 163, 168, 375
ecoprovinces 17, 18, 20-22, 377,
381-83, 385, 401
Erethizon dorsatum 41, **367-72**
Erethizontidae Family 4, 41, 366
Ermine 35, 75, 81, 89, 103, 113,
115, 121, 168, 174, 191, 208,
215, 243, 248, 255, 260, 266,
271, 275, 286, 291, 303, 331,
337, 350, 361, 364, 375
ferns 75, 98, 113, 143, 144, 374
Fir, Amabilis 112, 177, 207, 208,
301, 334, 369, 373
fir, Douglas- 5, 112, 113, 118, 150,
177, 182, 201, 206, 207, 208,
213, 214, 241, 274, 289, 328,
329, 334, 341, 342, 369, 373
Fir, Grand 118, 150, 207, 213, 373
Subalpine 80, 112, 118, 125, 143,
166, 177, 182, 188, 195, 201,
207, 213, 241, 283, 301, 308, 373
fish 255, 318, 336, 375
Fisher 87, 371, 372, 375
forbs and other herbs 6, 13, 75, 80,
87, 94, 98, 102-3, 108, 113, 120,
127, 133, 143, 144, 160, 162,
167, 173, 178, 183, 190, 197,
202, 220, 221, 226, 235, 242,
259, 264, 270, 274, 283, 285,
290, 302, 309, 348, 349, 355,
359, 360, 369, 374, 385

fossils 2, 3, 17, 19, 147, 262, 287, 293
fossorial 8, 112, 218, 222, 274, 385
Fox, Red 75, 81, 89, 139, 174, 248,
255, 260, 271, 297, 355, 371, 375
Fox Squirrel – *see* Squirrel, Fox
fruit, fruit trees 4, 5, 6, 28, 31, 34,
95, 133, 151, 156, 189, 196, 197,
208, 214, 221, 222, 265, 267,
281, 318, 335, 360, 364, 369; *see
also* berries
fungi 4, 6, 28, 120, 123, 144, 151,
167, 173, 178, 183, 189, 202,
206, 214, 243, 274, 302, 318,
330, 343, 349, 355, 360, 364,
386; *see also* truffles
fur harvest/trade 34, 217, 238, 299
Geomyidae Family 4, 40, 218
Glaucomys sabrinus 40, **117-23**
Gopher, Northern Pocket 6, 7, 8, 9,
12, 28, 30, 33, 34, 40, 50, 64,
110, **218-24**, 265, 379, 382
gophers, pocket 40, 218-24
Goshawk, Northern 81, 89, 103,
127, 152, 157, 163, 174, 178,
184, 191, 208, 215, 375
graminoids (gen.) 80, 127, 253, 259,
270, 285, 385
grasses 93, 98, 108, 144, 162, 190,
226, 264, 265, 348, 374
Groundhog 130, 136
ground squirrels – *see* squirrels, ground
Hare, Arctic 20
Snowshoe 7, 9, 20, 23, 27, 30, 31,
32, 33, 37, 38, 39, 44-45, 45, 55,
71, 74, 84, **85-91**, 92, 96, 97,
101, 106, 113
Varying 37, 85
hares and rabbits 1, 2, 7, 8, 14, 15,
31, 39, 44, 54, 72, 84-109
Harrier, Northern 174, 275, 291,
297, 355, 375
Hawk, Cooper's 191, 203, 208, 215,
375
Ferruginous 157, 222, 375
Red-tailed 81, 157, 163, 174, 178,
191, 291, 375
Rough-legged 275, 291, 303, 375
Swainson's 127, 375
hawks 99, 168, 173

moss 118, 127, 144, 206, 212, 242, 248, 253, 302, 308, 309, 329, 354, 364
Mountain Beaver 9, 21, 30, 31, 38, 39, 50, 64, 110, **111-16**, 137, 364, 378, 381
mountain beavers 4, 39, 110-16
Mouse, Cascade Deer 333
  Columbian 37
  Deer 10, 12, 25, 26, 27, 28, 33, 35, 37, 41, 52, 64, 226, 227, 290, 291, 333-34, 335, 337, 339, **340-46**, 347, 380, 383
  Great Basin Pocket 6, 9, 11, 13, 21, 22, 30, 31, 32, 40, 50, 57, **225-31**, 348, 379, 382
  House 14, 26, 32, 33, 41, 51, 63, **311-15**, 325, 347, 380, 383
  Keen's 10, 21, 23, 25, 26, 27, 28, 32, 33, 37, 41, 52, 64, 320, **333-39**, 340, 342, 344, 345, 346, 380, 383
  Lemming 307
  Meadow Jumping 30, 41, 51, 58, **353-57**, 358, 380, 383
  Pacific Jumping 41, 51, 57, 358, 361, **363-65**, 380, 383
  Sitka 37, 333
  Western Harvest 12, 21, 22, 30, 31, 32, 33, 41, 52, 64, 227, 312, 340-41, **347-51**, 380, 383
  Western Jumping 11, 41, 51, 57n, 58, 353-54, **358-62**, 363, 365, 380, 383
  White-footed 340
Muridae Family 4, 37, 40, 239
Murinae Subfamily 37, 41, 239, 311
Murrelet, Ancient 318, 320, 322, 375
Mus musculus 41, **311-15**
Muskrat 6, 8, 10, 13, 26, 29, 34, 41, 46, 59, 239, 287, **294-99**, 379, 383
Neotoma cinerea 41, **327-32**
nuts – *see* fruit *and* seeds
oaks 150, 151, 152, 154, 156, 373
Ochotona alpina 77
  collaris 39, **73-77**
  princeps 39, 76, **78-83**

Ochotonidae Family 2, 39, 73
Ondatra zibethicus 41, **294-99**
Oryctolagus cuniculus 39, **96-100**
Owl, Barn 121, 275, 291, 314, 318, 324, 364, 375
  Barred 152, 331, 375
  Burrowing 229, 266, 375
  Great Gray 331, 375
  Great Horned 89, 103, 108, 121, 152, 157, 174, 178, 203, 208, 222, 291, 297, 303, 331, 355, 364, 375
  Long-eared 178, 222, 229, 266, 355, 364, 375
  Northern Hawk 303, 375
  Northern Saw-whet 266, 303, 350, 375
  Short-eared 229, 266, 275, 291, 303, 375
  Snowy 291, 375
  Spotted 121, 122, 208, 331, 364, 375
  Western Screech- 152, 229, 266, 355, 375
owls 99, 114, 156, 173, 222, 309, 329, 331, 350, 361, 364
Packrat 327
Perognathus parvus 40, **225-31**
Peromyscus 37, 52n, 334, 335, 337, 345
Peromyscus keeni 41, **333-39**
  maniculatus 41, **340-46**
  oreas 37, 337
  sitkensis 37, 337, 338
Phenacomys intermedius 41, **300-306**
Pika, American 13, 15, 22, 23, 30, 35, 38, 39, 45, 54, 73-74, 77, **78-83**, 169, 376, 377
  Collared 22, 23, 39, 45, 54, 73-77, 78, 272, 376, 377
  Common 78
  Rocky Mountain 78
pikas 1, 2, 7, 10, 13, 14, 15, 26, 33, 35, 39, 44, 54, 72, 73-83
Pine, Lodgepole 27, 118, 166, 182, 190, 207, 208, 212, 213, 236, 241, 246, 301, 334, 369, 373
  Ponderosa 166, 182, 183, 197, 207, 213, 226, 369, 373